# Unless Recalled Earlier

### Date Due

| JAN 2 0 1991 | | | |
|---|---|---|---|
| | | | |
| | | | |
| | | | |
| | | | |
| | | | |
| | | | |
| | | | |
| | | | |
| | | | |
| | | | |
| | | | |
| | | | |
| | | | |
| | | | |
| | | | |

# Structure and Bonding in Crystals

## VOLUME I

# Contributors

Aaron N. Bloch

Leo Brewer

Jeremy K. Burdett

Marvin L. Cohen

G. V. Gibbs

Walter A. Harrison

B. G. Hyde

E. P. Meagher

Marshall D. Newton

M. O'Keeffe

Linus Pauling

J. C. Phillips

Gina C. Schatteman

Mark A. Spackman

Robert F. Stewart

D. K. Swanson

Alex Zunger

# Structure and Bonding in Crystals

## VOLUME I

### Edited by

## MICHAEL O'KEEFFE
## ALEXANDRA NAVROTSKY

*Department of Chemistry*
*Arizona State University*
*Tempe, Arizona*

1981

**ACADEMIC PRESS**
*A Subsidiary of Harcourt Brace Jovanovich, Publishers*

New York   London   Toronto   Sydney   San Francisco

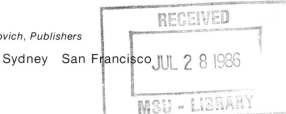

ACADEMIC PRESS, INC.
111 Fifth Avenue, New York, New York 10003

*United Kingdom Edition published by*
ACADEMIC PRESS, INC. (LONDON) LTD.
24/28 Oval Road, London NW1   7DX

Library of Congress Cataloging in Publication Data
Main entry under title:

Structure and bonding in crystals.

Includes bibliographies and index.
1. Solid state chemistry. 2. Crystallography.
3. Chemical bonds.   I. O'Keeffe, Michael.
II. Navrotsky, Alexandra.
QD478.S77          541'.0421              81-7924
ISBN  0-12-525101-7 (v. 1)               AACR2

# Contents

*List of Contributors*     *xi*

*Preface*     *xiii*

*Contents of Volume II*     *xvii*


**1  Historical Perspective**

LINUS PAULING

Text     1


**2  Quantum Theory and Crystal Chemistry**

J. C. PHILLIPS

Text     13
References     24


**3  Pseudopotentials and Crystal Structure**

MARVIN L. COHEN

I. Introduction     25
II. Pseudopotentials     26

III. Conclusions and the Future    46
     References    48

## 4 Quantum-Defect Orbital Radii and the Structural Chemistry of Simple Solids

AARON N. BLOCH AND GINA C. SCHATTEMAN

I. Introduction    49
II. Review of Some Fundamental Concepts    50
III. Relationship of the Orbital Radii to Hartree-Fock
     Wavefunctions    55
IV. Orbital Electronegativities and Renormalized Orbital
     Radii    57
V. The Problem of the Octet Binary Compounds    63
VI. Conclusions    70
     References    71

## 5 A Pseudopotential Viewpoint of the Electronic and Structural Properties of Crystals

ALEX ZUNGER

I. Introduction    73
II. Pseudopotentials and Structural Scales    77
III. First-Principles Density-Functional
     Pseudopotentials    83
IV. Trends in Orbital Radii    102
V. Separation of Crystal Structural of 565 Binary AB
     Compounds    117
VI. Summary    131
     References    132

## 6 Elementary Quantitative Theory of Chemical Bonding

WALTER A. HARRISON

I. Introduction    137
II. The Formulation    138

III. The Bonding Energy     141
IV. Fourth-Order Terms     143
V. The Chemical Grip     143
VI. AB Compounds     145
VII. The Role of Noble Metal d States, Ion
    Distortion     146
VIII. The s–p Hybridization Energy     146
IX. The Oxygen Bridge     151
X. Tetrahedral Complexes     152
XI. Perovskites; The s–d Hybridization Energy     153
XII. Central-Atom Hybrids and Bond Orbitals     153
    References     154

# 7  The Role and Significance of Empirical and Semiempirical Correlations

LEO BREWER

I. Introduction     155
II. Bonding Models     156
III. Correlation of Electronic Configuration and Crystal
    Structure     170
IV. Application of a Semiempirical Correlation     171
    References     174

# 8  Theoretical Probes of Bonding in the Disiloxy Group

MARSHALL D. NEWTON

I. Introduction     175
II. Theoretical Considerations     176
III. Computational Details     184
IV. Results and Discussion     185
V. Conclusions     191
    References     192

# 9 A Comparison of Experimental and Theoretical Bond Length and Angle Variations for Minerals, Inorganic Solids, and Molecules

G. V. GIBBS, E. P. MEAGHER, M. D. NEWTON, AND D. K. SWANSON

I. Introduction     195
II. Molecular Orbital Method     197
III. The Molecular Structure of Orthosilicic Acid, $Si(OH)_4$: A Comparison with the Shapes of $SiO_3(OH)^{3-}$ and $SiO_2(OH)_2^{2-}$ Anions in Hydrated Silicates     199
IV. Force Constants and Optimized Geometry for the Disiloxy Unit of the Pyrosilicic Acid Molecule, $H_6Si_2O_7$: A Comparison with Experimental Geometries and the Bulk Modulus of the Silica Polymorphs     202
V. Si—O Bridging Bond Length–Bond Strength Sum and Angle Variations     211
VI. Geometries of Molecules and Related Groups in Solids     216
VII. Conclusions     221
     References     222

# 10 The Role of Nonbonded Forces in Crystals

M. O'KEEFFE AND B. G. HYDE

I. Introduction     227
II. Structures Derived from That of Cristobalite     229
III. Nonbonded or "One-Angle" Atomic Radii     237
IV. Applications of Nonbonded Radii in Crystal Chemistry     239
V. Nonbonded Interaction Potentials     244
VI. What Is the Size of an Atom?     251
     References     253

# 11 Molecules within Infinite Solids

JEREMY K. BURDETT

I. Introduction    255
II. Perturbation Theory    256
III. Geometries of Inorganic Molecules    258
IV. Simple versus Exact Theories    263
V. Molecular Orbitals in Solids    264
VI. Reactions in the Solid State    274
   References    276

# 12 Charge Density Distributions

ROBERT F. STEWART AND MARK A. SPACKMAN

I. Introduction    279
II. Electrostatic Properties from Diffraction Structure
   Factors    281
III. Pseudoatoms in Diatomic Molecules    283
IV. Valence Densities from Pseudoatoms    286
V. Electrical Field Gradients    296
VI. Conclusions    297
   References    298

# 13 Some Aspects of the Ionic Model of Crystals

M. O'KEEFFE

I. Introduction    299
II. The Energy    300
III. Bond Lengths    301
IV. Ionic Radii    303
V. Structural Predictions    304
VI. Close Packing (Eutaxy)    307
VII. van Der Waals Energy and Structure    308
VIII. The Bulk Modulus and Elastic Constants    308
IX. The Volume (Density)    310
X. Polarizability and Polarization    311

XI.  Madelung Potentials and Energy Levels     316
XII.  Some Conclusions       319
      References       320

*Index*     323

# List of Contributors

*Numbers in parentheses indicate the pages on which the authors' contributions begin.*

*Aaron N. Bloch* (49), Department of Chemistry, The Johns Hopkins University, Baltimore, Maryland 21218, and Corporate Research-Science Laboratories, Exxon Research and Engineering Company, Linden, New Jersey 07036

*Leo Brewer* (155), Materials and Molecular Research Division, Lawrence Berkeley Laboratory and Department of Chemistry, University of California, Berkeley, California 94720

*Jeremy K. Burdett* (255), Department of Chemistry, The University of Chicago, Chicago, Illinois 60637

*Marvin L. Cohen* (25), Department of Physics, University of California and Materials Molecular Research Division, Lawrence Berkeley Laboratory, Berkeley, California 94720

*G. V. Gibbs* (195), Department of Geological Sciences, Virginia Polytechnic Institute and State University, Blacksburg, Virginia 24061

*Walter A. Harrison* (137), Department of Applied Physics, Stanford University, Stanford, California 94305

*B. G. Hyde* (227), Research School of Chemistry, The Australian National University, Canberra ACT 2600, Australia

*E. P. Meagher* (195), Department of Geological Sciences, University of British Columbia, Vancouver, British Columbia V6T 2B4, Canada

*Marshall D. Newton* (175, 195), Chemistry Department, Brookhaven National Laboratory, Upton, New York 11973

*M. O'Keeffe* (227, 299), Department of Chemistry, Arizona State University, Tempe, Arizona 85281

*Linus Pauling* (1), Linus Pauling Institute of Science and Medicine, Menlo Park, California 94025

*J. C. Phillips* (13), Bell Laboratories, Murray Hill, New Jersey 07974

*Gina C. Schatteman* (49), Department of Chemistry, The Johns Hopkins University, Baltimore, Maryland 21218, and Corporate Research-Science Laboratories, Exxon Research and Engineering Company, Linden, New Jersey 07036

*Mark A. Spackman* (279), Department of Chemistry, Carnegie-Mellon University, Pittsburgh, Pennsylvania

*Robert F. Stewart* (279), Department of Chemistry, Carnegie-Mellon University, Pittsburgh, Pennsylvania

*D. K. Swanson* * (195), Department of Geological Sciences, Virginia Polytechnic Institute and State University, Blacksburg, Virginia 24061

*Alex Zunger* (73), Solar Energy Research Institute, Golden, Colorado 80401

* Present address: Department of Earth and Space Science, State University of New York, Stony Brook, New York 11294

# Preface

The last few years have witnessed a remarkable convergence of interest in the structure and bonding in crystals with complex structures. Physicists have developed methods of dealing with the electronic and other properties of the increasingly complex solids that have been found to have unusual properties and that are being used in device technology. Solid state chemists have shown a renewed interest in synthesis — sparked, in part, by the development of new techniques for structure investigation. Geoscientists at the same time have found that the new insights into structure and bonding in silicates are tremendously useful to mineralogy and earth physics.

Before this renaissance, virtually the only guide to the factors determining crystal structure was the principles set out in the early days of solid state chemistry and put in definitive form in the chemists "bible"— Pauling's "Nature of the Chemical Bond." Indeed in dealing with complex silicates, it is little exaggeration to state that there had been little progress (until recently) since the formulation of the celebrated "Pauling's Rules" 50 years ago.

In January 1980, we took the opportunity of this golden anniversary to convene a diverse group of scientists with a common interest in principles of structure and bonding in complex solids, to discuss and assess recent development. The meeting (generously supported by the Solid State Chemistry program of the National Science Foundation [grant DMR-

7827019] and by the Center for Solid State Science at Arizona State University) brought together solid state physicists and chemists, metallurgists, ceramists, and geologists, who found that not only did they have common interests, but often spoke similar languages.

These volumes are a direct outgrowth of that meeting; virtually all the contributors were participants, although the books are by no means "conference proceedings." In the light of their experience at the meeting, the contributors were encouraged to present overviews of their fields of research and its bearing on related fields.

We benefited greatly from the presence of Linus Pauling—he was always ready with penetrating questions and comments. In addition, he presented a delightful *ad lib* account of the early days of crystal chemistry, particularly at Pasadena, and this is recorded, essentially *verbatim,* here.

Chemists have long had the intuition that atoms retain, in some sense, their own special identity in compounds. To the solid state chemist at least, one of the more interesting advances in solid state physics has been the development of atomic pseudopotentials and the discovery that they could be applied directly to atoms in crystals. The first few chapters (1–7) reflect this approach in which the electronic properties of the isolated atoms are used to make useful and often rather exact calculations and predictions of the structure, stability, and properties of solids of many different types.

A related and parallel development has been the discovery that the methods of the "molecular" chemists (MO theory, etc.) lend themselves rather readily and revealingly to the discussion of solids; a related group of chapters (8–12) discuss recent developments in these fields.

There is new understanding too of the older topics such as bond length, bond strength, ionic radii, etc., and some of these developments are described in chapters 13–19. These concepts have been used by geochemists and geophysicists to systematize and predict phase transitions at high pressure, as the last three papers in this section discuss.

A final group of chapters (19–26) deals with the problems of classifying complex solids and with systematic descriptions of the relationships between their structures. This is a particularly important problem when structures and sometimes stoichiometries are at first sight apparently very complex, but when "correctly" described are much easier to appreciate and understand.

We are indebted to our colleagues both for making very helpful suggestions and for graciously accepting editorial input to their contributions.

Pamela O'Keeffe and a number of A.S.U. students and research associates helped enormously in transporting the meeting participants from Phoenix to the desert oasis of Castle Hot Springs, and in many other ways.

*Michael O'Keeffe*
*Alexandra Navrotsky*

# Contents of Volume II

14 The Bond-Valence Method: An Empirical Approach
to Chemical Structure and Bonding
I. D. BROWN

15 Interatomic Distance Predictions for Computer
Simulation of Crystal Structures
WERNER H. BAUR

16 Bond Distance in Sulfides and a Preliminary
Table of Sulfide Crystal Radii
R. D. SHANNON

17 Energetics of Phase Transitions in AX,
$ABO_3$, and $AB_2O_4$ Compounds
ALEXANDRA NAVROTSKY

18 Crystal Chemical Effects on Geophysical
Equilibria
JOHN C. JAMIESON, MURLI H. MANGHNANI AND L. C. MING

19  Module Structure Variation with Temperature,
    Pressure, and Composition: A Key to the
    Stability of Modular Structures
    R. M. HAZEN AND L. W. FINGER

20  Theoretical Prediction of Order
    Superstructures in Metallic Alloys
    J. M. SANCHEZ AND D. de FONTAINE

21  Graph Theoretic Enumeration of Structure
    Types: A Review
    T. J. McLARNAN AND P. B. MOORE

22  Polytypism in Complex Crystals: Contrasts
    between Mica and Classical Polytypes
    JAMES F. THOMPSON, JR.

23  The Influence of Cation Properties on the
    Conformation of Silicate and Phosphate Anions
    FREDRICH LIEBAU

24  The Description of Complex Alloy Structures
    STEN ANDERSSON

25  Structural Features of Rare-Earth-Rich–
    Transition-Metal Alloys
    E. PARTHÉ

26  On Poly Compounds: Polycationic and
    Polyanionic Tetrelides, Pnictides, and
    Chalcogenides
    F. HULLIGER

Index

# 1

# Historical Perspective

## LINUS PAULING

Let's go back before my birth. I remember 50 years ago reading the papers that the English amateur scientist Barlow published in the period 1883 to 1887 after he had worked out the theory of space groups. He got interested in individual structures and described a number of them. Lord Kelvin had asked the question of what is the densest packing of equivalent spheres. By equivalent spheres he meant spheres that are related to one another by a translational operation. He said that there is only one, what we call cubic closest packing. But Barlow, with his knowledge of space groups and of the nature of symmetry operations, said there is another: hexagonal closest packing. So back in 1883, 30 years before x-ray diffraction was discovered, Barlow assigned copper, silver, and gold to cubic closest packing, and magnesium, zinc, and cadmium to hexagonal closest packing. He described the sodium chloride structure and said that sodium chloride and potassium chloride had this structure. He described the cesium chloride structure, which he allocated to cesium chloride (I no longer remember how he could have done this—it seems to me to be virtually impossible). Then he went on to describe the sphalerite structure, which he also ascribed to sphalerite, and the wurtzite structure, which he ascribed to wurtzite. He described the diamond structure but he didn't assign it to diamond. I made a literature search and talked with P. P. Ewald and others about whether anyone had assigned the diamond structure to diamond before the Braggs had determined the structure. It seems to me that it should have been obvious after van't Hoff and LeBel in 1873 had said that the carbon atom has the tetrahedral arrangement and that there are four single bonds in diamond. If any of you finds a statement to this effect in the literature, please let me know. It would increase my faith in the intelligence of the human race. Probably this failure resulted from a lack of interest in solid-state chemistry and physics before 1913.

We know about the work of von Laue and the Braggs, so I shan't say anything about that, but since we are here in the United States I'd like to discuss the beginnings of x-ray crystallography in the United States. Arthur

Structure and Bonding in Crystals, Vol. I

Amos Noyes is the man who started to develop physical chemistry at MIT along about 1900 and supported it himself. The total support was $10,000 a year, and he gave $5,000 of it from his own pocket to keep the laboratory going. He then gradually moved to Pasadena in the period from 1913 to 1916, and from 1916 on he was in Pasadena permanently. About that time one of his former students, named C. Lalor Burdick, was in Europe getting a Ph.D. in organic chemistry. The war was on and he decided to leave Germany, and Noyes recommended that he go to England and learn about x-ray crystallography from the Braggs. He did, and with E. A. Owen reported on the structure of silicon carbide. Shortly after his return in 1916 he moved to Pasadena and built an x-ray spectrometer. With this he and James H. Ellis, who had gotten his Ph.D. at MIT, made an investigation of chalcopyrite and published it in 1917. This was the first crystal structure determination to be carried out in the United States, except for the essentially simultaneous work done by the powder method on iron and other metals by A. W. Hull of the General Electric Company. Ellis died long ago, but Burdick is still alive. He went to Chile and was there copper mining for a couple of years, came back to the United States, pretty soon signed up with DuPont and became an important man, and after a while retired. He is still alive and running the Lalor Foundation, which his uncle had set up (his uncle after whom he was named). So the original American x-ray crystallographer is still among us.

In Pasadena the Burdick x-ray spectrometer was still around when I arrived in 1922. R. G. Dickinson was a graduate student from about 1917 to 1920. He used the spectrometer to get some data from several crystals and got his Ph.D. in 1920. He was the first person to get a Ph.D. degree from the California Institute of Technology, and it was with an x-ray diffraction thesis, and in chemistry. I think he knew a lot of physics and mathematics too. I think that this fact illustrates the difference between x-ray crystallography in the United States and that in England and other European countries. In the United States x-ray crystallography was done from the beginning in chemistry departments, and in Europe it was done in physics departments.

When I got to Pasadena in 1922, as a beginning graduate student, I had already read the book by W. H. and W. L. Bragg, *X-Rays and Crystal Structure*, which I got by mail from the state library in Salem—I was working in the summer on the Oregon coast, near Seaside. Noyes had written to me saying he thought that it would be good if I were to work with Dickinson on x-ray diffraction. In fact, he suggested to me that it would be a good idea to study the structure of lithium hydride, to see whether or not there is a negative hydrogen ion, $H^-$, in this crystal. I spent my first month as a graduate student in the shop making a piece of apparatus by means of which I could cause lithium to combine with hydrogen in the absence of oxygen. But then a paper appeared by Bijvoet in Holland on the structure of lithium hydride, so that I didn't continue with that problem.

On Dickinson's recommendation I started reading the literature to look for cubic crystals. Dickinson had determined the structure of potassium chlorostannate, a one-parameter structure. You could determine one parameter, or even two, in those days, but it was hard to determine more than two. Cubic crystals have enough symmetry elements so that there is a reasonable chance that the atoms will lie on them and be in an invariant position, or perhaps have just one parameter to be determined. I looked through the five volumes of Groth's *Chemische Kristallographie* and picked out all the cubic compounds and set about making those that I thought might be worth investigating. I made $CaHgBr_4$—that's easy enough, calcium bromide and mercuric bromide—and took some x-ray photographs under Dickinson's direction and found that there were 32 molecules (192 atoms) in the unit cube and that to determine the structure was clearly impossible!

I was interested in intermetallic compounds—I had worked in metallography when I was an undergraduate. I needed to melt metals, so I built a furnace. I got an alundum crucible and wound some nichrome wire around it and packed it with asbestos in a can and got a rheostat and a little crucible. I dissolved some cadmium in an excess of molten sodium and let it cool slowly, then put it in absolute ethanol. Sodium reacts with absolute ethanol but sodium dicadmide is stable enough to resist attack by this reagent, so that crystals about 6 or 7 mm on edge were left. It was then thought that you really needed large crystals to do an x-ray diffraction job, and these were big enough. I found, however, that the cubic unit cell in this crystal contained about 1100 atoms, another impossible job!

The next thing that I tried was to melt up some potassium sulfate and some anhydrous nickel sulfate in this crucible and let it cool. I made a whole lot of compounds of this sort with formulas such as $K_2Ni_2(SO_4)_3$, and I determined the size of the units and the space group symmetry and found that there were 19 parameters required—far too many. The magnesium compound $K_2Mg_2(SO_4)_3$ is a mineral (langbeinite) whose structure was determined many years later.

I went on making more and more cubic crystals, and each time they turned out to be too complicated to do anything with. After about a couple of months' effort along this line Dickinson said, "Let's try something else." He went to the stock room and got from the shelves a cardboard box that contained nodules of molybdenite, looking like graphite. With a razor blade he cleaved a nodule, getting a flat cleavage face, and then reflected a beam of x-rays from this face. This was before the theory of layer-line diagrams had been worked out, but a beam of x rays reflected from the face gave the value of some multiple of the length of the hexagonal axis—the edge of the unit in the hexagonal direction. We knew the axial ratio of the crystal from crystallographic measurements of developed faces, so we had possible values for the $a$-axis also—just this one reflection was all we needed to get started. Then we took some Laue photographs. These photographs showed a lot of reflections. Of course they showed symmetry if the x-ray beam came along the hexagonal axis. We would take one of these photographs and made a gnomonic projection, giving the pole corresponding to each x-ray reflection as it appeared on the photograph. In this projection zones are represented by straight lines, so that it is easy to assign indices to the Laue reflections and for each one to calculate the value of $n\lambda$, knowing the angle of reflection and the possible sizes of the unit cell. We looked for the smallest hexagonal unit that would explain the reflections on the photograph. There's a lot of work involved in analyzing these photographs, but fortunately I had a helper, a very smart one, who would plot out the graphs, assign indices, and calculate the $n\lambda$ values. Later on in Munich when I was writing a paper on sizes of ions and she was drawing the figures, she would look up and say, "Are you sure that that ion is the right size the way you've drawn it in your sketch?"

This x-ray method was very powerful. For example, using this method, Dickinson investigated tin tetraiodide, $SnI_4$, and found that the cubic unit contained 32 molecules. But Hermann Mark in Germany also was studying $SnI_4$ at the same time, not using the Laue technique, and he said that the unit contained only four molecules, and published a paper about it. In fact, his structure was only an approximation to the actual structure. The actual structure, with 32 molecules, depends upon five parameters. So here was Dickinson faced with the problem of a cubic crystal determined by five parameters. I thought he was very clever, because he not only calculated the structure factors for the simple structure but also calculated the derivatives of the structure factors with respect to each of the five parameters and in this way was able to evaluate the changes from the ideal structure needed to get the complete structure. But that five-parameter structure was about the limit with the methods of calculation that were available then, when you had to look up every cosine or sine in tables. There was quite an improvement later on when the Beevers–Lipson strips came along—they gave sines and cosines to two decimal places, so that you could move along fast but with rather diminished accuracy. I had a student in the 1930s who was interested in trying to do something about this, and for a year he worked with me building a harmonic synthesizer that we could program and then, by turning a crank, have the little wires run back and forth and turn out the values of the structure factors. That job never got completed because, when I applied for some money from some granting agency in order to finish the machine, a man at IBM asked "Why don't you use IBM cards?" We then got some IBM machines and punched out a lot of cards that gave values of sines and cosines, and this speeded up things immensely. We published a couple of papers about making calculations with IBM cards. That was really a great contribution to the technique of making x-ray crystallographic calculations.

But now to go back to 1922. Things were really wonderful. Nine years had gone by ('13 to '22) since the discovery of x-ray diffraction and there had been many structure determinations made, but really a pretty small number when you consider how many crystals there are in the world. In each crystal structure that was determined there was almost certainly some information of interest. In the case of molybdenite, much to our surprise, the molybdenum atom turned out to have six sulfur atoms around it at the corners of a trigonal prism instead of the octahedron that we had expected. The year before, Richard Bozorth, working with Dickinson, had determined the cadmium iodide structure, a layer structure involving octahedra. Dickinson had determined the potassium chlorostannate structure, and Wyckoff had determined the potassium chloroplatinate structure, also involving octahedra. Dickinson had also determined the structure of potassium zinc tetracyanide—it had a tetrahedral arrangement—and also potassium tetrachloroplatinate and palladate, with square planar coordination.

Dickinson and an undergraduate, Albert Raymond, then made the first crystal structure determination of an organic compound. This compound was hexamethylene tetramine, which has a cubic structure with two parameters, one for carbon and another for nitrogen. This was a very good structure determination, verifying the tetrahedral arrangement of the four bonds formed by the carbon atom.

I don't think that there was a single mistake made in all of the x-ray investigations that were carried out in these early years in Pasadena. The reason for this is that the technique that was used, developed largely by R. W. G. Wyckoff and S. Nishikawa, was a remarkable one. During the war Nishikawa, who had been studying the theory of space groups, was in Cornell, where Wyckoff was a student in chemistry. Wyckoff, who was a brilliant and precocious student (he got his Ph.D. at age 22), came under Nishikawa's influence and began developing the technique. Wyckoff spent one year, 1921 to '22, in Pasadena, but he was gone by the time that I arrived in the fall of 1922.

When it came to calculating the intensities of the x-ray reflections, one problem was that the x-ray scattering powers of the various atoms weren't known. The one thing we were sure about was that the form factor fell off monotonically for all atoms with increasing values of $\sin \theta / \lambda$. Only qualitative comparisons of intensities were used. We hunted to find two reflections which happened to be at such an angle that they had the same value of $n\lambda$—their spacing and angles were such as to give the same wavelength of the x-rays being reflected. Then we could compare their intensities, and if the reflection corresponding to the plane with the smaller spacing had a larger intensity than the one with the larger spacing we knew that its structure factor had to be larger. All the other factors would tend to make it smaller— the Lorentz polarization factor, the Debye factor, the temperature factor, and the form factor all fall off with decreasing interplanar distance. Also we could see very weak reflections on the Laue photographs so that I think we never were fooled into accepting a unit cell smaller than the actual one.

The second structure that I determined was that of an intermetallic compound, $Mg_2Sn$—this turned out very nicely. I had the little furnace, and I melted some magnesium and tin together in stoichiometric ratios and let the furnace cool slowly over many hours. Then I took the piece of solid alloy out of the crucible and hit it with a hammer—it is brittle—and I got cleaved plates out of it. I took x-ray photographs and worked out the structure (the fluorite structure), which was a very simple and straightforward task. I was involved in determining several other intermetallic-compound structures over the years. People working with me on alloys have included David Shoemaker and Sten Samson. Year after year I kept trying to find the structure of sodium dicadmide. It was both a pleasure and a disappointment for me when Sten Samson finally came up with that structure, 30 years after I had begun to work on it.

I worked with Dickinson for 2 years, and then he went to Europe on a Rockefeller Fellowship and I was alone for awhile. When he came back he decided to go into a different field in chemistry and left x-ray diffraction to me. I became impatient with the limitations of the technique. My impatience took a form different from what it often took in other people. Other people tried to find a way of solving the phase problem. I did something else. In 1922, when I first got to Pasadena, I read W. L. Bragg's paper on atomic radii and was very interested in it. I immediately began collecting values of atomic radii from all of the published crystal structure papers and thinking about the problem of interatomic distances. The empirical information made it clear that there was some possible systematization to the interatomic distances and also to other aspects of the crystal structures, and I thought perhaps we had come to the time when we could predict what the structures are without x-ray diffraction patterns. For example, as I began to study the mineral topaz I made x-ray photographs to determine cell and space-group symmetry. I then used my knowledge about how the atoms of aluminum, silicon, oxygen, and fluorine might interact with one another to predict a structure, which when tested by comparison with the observed x-ray intensities turned out to be correct.

I called this the stochastic method, from the Greek stochastikos, meaning apt to divine the truth by conjecture. If you make a stochastic hypothesis you are supposed to go immediately and check it by comparison of some prediction it makes with experiment or observation (the x-ray diffraction patterns, in our case). The rule is that you have only one guess, because if you guess long enough you might by accident run across a wrong structure that fits. Applying this method was a lot of fun.

This meeting ought to be a memorial meeting to Willy Zachariasen. I shall talk a little about him because I enjoyed my friendship with him over many years. Around 1930 I thought that we ought to have Zachariasen in Pasadena as a member of the CIT faculty and I found (I wasn't chairman of the chemistry department then) that Dr. Noyes was willing to pay half of his salary, so I went to John Buwalda, the chairman of the Division of the Geological Sciences, and asked if Geology would be willing to pay half his salary. After all, he had gotten his Ph.D. with V. M. Goldschmidt in geology back in Oslo. But Buwalds said, "That isn't geology." You know geologists in those days were in some ways a pretty backward lot. So Willy didn't come to Pasadena. I'm not sure what life would have been like if he had—I think he had a wonderful time as it was, but he might well have liked Pasadena better than Chicago, and perhaps we could have got him to quit smoking if he had come to Pasadena.

Quantum mechanics came along in the early years and I was really fortunate. I don't think that there was any place that would have been better for me to be in than Pasadena—this new school where I was the seventh person to get a Ph.D. degree and where there were such extraordinarily able people. Noyes himself was a rather old-fashioned chemist but he wrote very good textbooks in analytical chemistry, qualitative analysis, and physical chemistry. Richard Tolman was a remarkably able man who had been a student of Noyes at MIT. In mathematical physics there was Harry Bateman—I had several courses with him on Newtonian potential theory, vector analysis, and integral equations. The first year that I was there, 1922–23, Tolman ran a course on atomic structure with a textbook with a nice title, *The Origin of Spectra*, by P. D. Foote and F. L. Mohler about the old quantum theory, the electronic structure of atoms, and the old quantum theory in general—a very nice course. The second year, '23–'24, we had a seminar using a book in German, *Atombau and Spektrallinien*, by Arnold Sommerfeld. The next year the English edition became available, and we repeated the seminar with it.

Lots of interesting things were going on. Quantum mechanics was discovered in 1925. My wife and I went to Europe on a Guggenheim Fellowship in April 1926, just as Schrödinger's papers were being published. I had decided to go to Sommerfeld's institute in Munich. The program that I had put down for my Guggenheim Fellowship was to investigate the topology of the interior of the atom. I wanted to find out how to handle many-electron atoms in such a way as to be able to make statements about the chemical and physical properties of substances and the structures of substances. My wife and I arrived in Munich, and Sommerfeld immediately began giving lectures on wave mechanics. This was really remarkable—the same year that Schrö-dinger was publishing his papers on wave mechanics. In 1926 I published my first paper on quantum mechanics. I had seen a paper, just published, by Gregor Wentzel in which he had used a method of treating many-electron atoms that Schrödinger himself had described in 1922. This was to replace the two K electrons by a surface distribution of negative charge, so that you had a Coulomb potential inside a certain distance from the nucleus and a different Coulomb potential outside, then you replace the L-shell electrons by a surface distribution and so on for the other shells. Wentzel had calculated the screening constants for the relativistic or magnetic doublets in the various atoms and they didn't agree with experiment, so he said that there was a difficulty with the theory of the spinning electron. I thought that it was wonderful that there was a way of setting up pseudopotentials and calculating the properties of atoms—this was just what I was looking for. So I sat down and started working through Wentzel's paper in order to understand it, and pretty soon I found that he had neglected some terms in his expansion of the azimuthal quantum number—without justification he had assumed that the azimuthal quantum number had the same value in all of these segments. When I carried out the theory correctly I found that the calculated values agreed very well with experiment. I was then able to use this theory to calculate screening constants for all sorts of atoms and all sorts of properties. I not only published this paper on the magnetic-doublet screening constants in 1926, but also a long paper on other properties—polarizability, diamag-netic susceptibility, sizes of ions, and other properties—in the Proceedings of the Royal Society in 1927, and a paper on ionic radii, pretty much theoretical, in the Journal of the American Chemical Society in 1927, and then a long paper on form factors for atoms and various other properties based upon this reasonably simple method of handling many-electron atoms and ions. It was a great period and it was a lot of fun. Quantum mechanics brought in a lot of new ideas and helped to solve many problems that had existed with respect to valence, chemical bonding, molecular structure, and crystal structure. Just the quantum-mechanical ideas themselves were enough to get one started on looking at the empirical information and interpreting it.

In 1930 we were again in Europe. I worked for a month or 6 weeks in Manchester in W. L. Bragg's laboratory. I was interested in the structure of epidote and also that of tourmaline, but I didn't succeed in solving these structures. Bragg invited me to see how the x-ray ionization-chamber appa-ratus worked and to see how they determined crystal structures in his laboratory. So I went up every day with my crystal of epidote, which we set up on the apparatus. There was an x-ray tube in a lead-covered box and an ionization chamber. We set the ionization chamber at a certain angle with wide slits and then the crystal face at a little less than half that angle and moved the crystal through the diffraction angle, to get the integrated intensity of the reflection. But in order to make this quantitative the crystal needed to be moved at a constant speed. It was moved by rotating a little rod that had attached at its end a little capstan, about 3 centimeters in diameter and with four spokes. There was a metronome sitting on top of the lead box. So I sat there with two fingers on the spokes of the capstan, moving it at each tick of the metronome, for one revolution of this capstan. This was the experimental technique for x-ray crystallography that was being used in the Bragg Institute in 1930!

In 1930 we went on to Zurich to attend a small meeting on x-ray crystallography with Ewald, Hermann (Hermann was Ewald's assistant in getting out the Strukturbericht), W. L. Bragg, J. D. Bernal, Niggli (he was a geologist in Zurich), and Wyckoff. Also in 1930 I talked to Hermann Mark. Mark during the preceding months had had the idea about electron diffraction by gas molecules, and his assistant had built an apparatus with which they had made some electron diffraction photographs of carbon tetrachloride and a few other gas molecules. Mark was not planning to continue this work; he had other things to do. He kindly gave me plans for the apparatus, which I took back to Pasadena, in order to initiate the extensive series of studies in this field that we then carried out.

James Holmes Sturdivant was, I suppose, my first real graduate student, the first person who got his doctorate for work done completely with me. In 1928 he and I published a paper on the structure of brookite. This orthorhombic titanium dioxide structure was a pretty difficult one for those days. It was one that we predicted, and it was predicted in an interesting way. In rutile there are $TiO_6$ octahedra sharing two edges and in anatase, the other tetragonal form of $TiO_2$, octahedra with four shared edges. I thought brookite ought to have octahedra with three shared edges, so we put octahedra together in the simplest way with three shared edges, and the space group and the dimensions came out right for brookite. Also in rutile the shared edge lengths drop to 2.50 Å from the average 2.70 Å, and in anatase too the shared edges drop to 2.50 Å. I published a theoretical paper in which these values were explained in terms of the Madelung constant and the repulsive forces between the ions. I thought that the three shared edges in brookite must also be shortened to 2.50 Å. How could we find what values of the parameters would do that? With no computers at hand we didn't see how to do it analytically, so we made models in which the properly distorted octahedra were made out of a moderately thick sort of drawing paper, and my wife sewed them together with needle and thread. Then we sewed these octahedra together into the structure and measured the dimensions to get the coordinates of the atoms. I think that our 1928 paper was the first one in the literature to show a drawing (and a photograph) in which polyhedra represent a structure.

Another early student of mine was M. D. Shappell—he got a Ph.D. in geology. He worked out the structure of bixbyite, $(Fe, Mn)_2O_3$, which was pretty interesting. Willy Zachariasen had published a structure of the C-modification of the rare-earth sesquioxides, which have the bixbyite structure. While working on the x-ray diffraction pattern of bixbyite, I noticed that the structure factor had the same value for two different values of the metal atom (for example, plus or minus one-fourth), but the structures were really different for minus one-fourth and plus one-fourth. While I was contemplating that fact I got a copy of Physical Review Notices, containing abstracts of papers to be given at a meeting of Physical Society. Lindo Patterson was going to give a paper on a proof that no two nonidentical structures can have the same Patterson diagrams—that is, can have the same set of interatomic-distance vectors. But here in bixbyite we had a pair of structures that are different from one another but that have exactly the same set of interatomic distances and hence the same x-ray pattern and Patterson diagram. I wrote to Patterson about this fact, and when the meeting came he gave a paper on "The Conditions Under Which Two Different Structures Have the Same Patterson Diagram"—Zachariasen had selected the wrong value of the parameter, so that his structure was wrong.

I have determined 67 crystal structures, I had assistance with some of them, but the early ones I did all by myself, or rather, with the help of my wife only. Zachariasen, if he were here, would probably say that he had determined 670 crystal structures, and I wouldn't be surprised if that were approximately correct—he was really remarkable in many ways, and one of them is the amount of work that he got done.

I was interested in the silicate minerals, starting with topaz and then going on to various other silicate minerals, and also in the sulfide minerals. After the molybdenite structure I didn't do anything about sulfides for a long time. About 1930 I determined the structure of enargite $Cu_4AsS_4$. It's very simple; you just have the wurtzite framework and replace one of the coppers by arsenic. I thought that there must be a number of similar sulfide minerals and that we could determine their structures and get some new information about interatomic distances. So I got some sulvanite, $Cu_3VS_4$, and, not expecting to be surprised, determined the structure. It is cubic, and the sulfur atoms are in the same places as in sphalerite. The three copper atoms also are in the same places as three of the zinc atoms in sphalerite, but the fourth position is empty. The vanadium atom instead is found in the middle of an octahedron of copper atoms, so that the sulfur atom forms three bonds to copper and one bond to vanadium, all in the same general direction, with no bonds on the other side (but of course that's where a shared electron pair of the sulfur atom is). Every 10 years I write a paper of the structure of sulvanite—the nature of the bonds in sulvanite—because my ideas change from time to time! We determined the structure of tetrahedrite also. That involved a lot of sulfur atoms tetrahedrally coordinated, but also one sulfur atom that is in the center of an octahedron of six copper atoms—that was a surprise too.

So, about 1932, I was thinking that this is a pretty interesting field. I thought that I knew all about the silicates by that time, but I didn't understand sulfides at all and I thought that we ought to be determining a lot of sulfide structures. I read that a good sum of money had been given to the American Geological Society—the Penrose fund—to support research, so I made out an application to the American Geological Society for $3000, which would pay for an assistant for me (perhaps only $2000—you know it didn't cost much in 1930 to pay for an assistant) to help in the study of the crystal structure of the sulfide minerals. I got back a letter of rejection and the suggestion that perhaps this wasn't geology. I wrote again to the old geologist at MIT, explaining why it was important to do work of this sort, and submitted my application again, but I didn't get any reply from that letter. I did, however, get a grant from the Rockefeller Foundation, and we were able to do a couple of sulfide structures. This grant from the Rockefeller Foundation didn't carry any strings with it, but it also became clear to me in a year or two that the Rockefeller Foundation wasn't really interested in the structure of sulfides. What they were interested in was biology—biochemistry. In fact, the man who was in charge of these things with the Rockefeller Foundation, Warren Weaver (who had been an instructor of mathematics in Pasadena up to 1921 or '22), was the first person to make use of the term "molecular biology," along in the 1930s. Applying for a grant to Rockefeller Foundation wasn't like applying to the National Science Foundation or National Institutes of Health, where you have to write down just how you'll build your apparatus, and just how you'll carry out your experiments, and just what discoveries you'll make. So I wrote and said, "I want to work on the magnetic properties of hemoglobin" and I got money, much more than I was getting before to work on sulfide minerals. [At this point Professor Pauling gave an account of how this led on to his very well known work on proteins, protein structure, etc.]

It was wonderful to be at Pasadena at that time, to just be beginning scientific work at that time when new techniques had come along, first x-ray and then a little later the wonderful technique of electron diffraction, when it was possible to go into a field and to turn out an interesting result very quickly, every few weeks to have something more to think about, and when quantum mechanics was just developing, too, to help formulate the theoretical ideas needed to explain and correlate the experimental results. I don't know whether there will ever be another time similar to it. But at least I know that there are still things to be discovered about the nature of the world, and I think that there always will be things to be discovered.

I used to tell my students, "It's all right for you to understand things, but remember, it's the facts that you keep in your head that count. It isn't knowing where you can find something in the book, it's what you have in your head that counts. So don't be satisfied just to learn that there's some place in the book where you can look something up."

# 2

# Quantum Theory and Crystal Chemistry

## J. C. PHILLIPS

The beauty, regularity, and variety of crystal structures have fascinated scientists for more than 150 years. With modern diffraction techniques, scientists have determined the structures of hundreds of thousands of crystals, some containing as many as hundreds of atoms per unit cell. Probably more information about the behavior of nature is available in this subject than in any other field of human endeavor. Thus natural scientists from many different disciplines have been attracted to this field, each hoping to deepen and to extend his understanding.

From the outset it is obvious that the very great scope of the subject makes it most suitable for theories which are predominantly empirical. Any approach which attempts to incorporate the ideas of quantum mechanics runs the risk of describing far too few structures at the expense of far too large an investment in time and computational complexity, most of which will remain forever inaccessible to all but, at best, a few specialists. The central themes of quantum mechanics—the distinctions between ionic, metallic, and covalent bonding—have long since been incorporated into empirical atomic radii by Goldschmidt, Pauling, and Zachariasen. What does a fundamental approach have to offer beyond the great systematizations already achieved by classical crystal chemistry?

To answer this question we can choose to follow Newton's advice and try to stand on the shoulders of giants. In reality, the statistical successes of classical crystal chemistry have been achieved primarily because chemical bonding in most solids is predominantly simple, i.e., almost purely ionic, metallic, or covalent in most cases. There are relatively few cases of complex mixed bonding, but when we do look at these cases we find that then the empirical theories are less satisfactory. Most classical minerals exemplify simple bonding (usually ionic) because they are stable with respect to pressure, temperature, and disproportionation. Some of the most interesting synthetic materials are, however, nearly unstable and therefore lie in borderline regions of complex mixed bonding. Some obvious examples of such

**13**

Structure and Bonding in Crystals, Vol. I
Copyright © 1981 by Academic Press, Inc.
All rights of reproduction in any form reserved.
ISBN 0-12-525101-7

materials are metallic superconductors and solid electrolytes, such as the many double salts derived from AgI. I suspect that many borderline materials exist with novel and unexpected properties, and that discovery of these materials can be expedited by clarifying the origin of structural boundaries in terms of differences in chemical bonding, especially incipient covalent instabilities. However, this is a very large task and if progress is to be made in this direction the collaborative efforts of many materials scientists will be required.

If we wish to refine our understanding of crystal chemistry beyond the level of models of ionic, metallic, and covalent radii, then we must augment these concepts with other quantitative measures of interatomic interactions. The most general and most successful extension of this kind has been Pauling's table of elemental electronegativities, which can be correlated with heats of formation. This extension is very general and is at least qualitatively successful for both molecules and crystals. The extent to which it is successful is, in fact, quite a good measure of the extent to which we are justified in regarding electrons as localized on atoms or in chemical bonds (apart from resonance between alternative configurations).

When we make use of atomic radii or electronegativity tables we are aware, of course, that drastic simplifications are being made for the sake of convenience. In principle full quantum-mechanical calculations of structural energies and other properties should be carried out. We must then ask ourselves, how great is the price that we have paid for simplicity? The answer very much depends on the circumstances.

Suppose, for example, that we are considering a series of solids all of which are (in some sense) chemically homologous and safely confined to one kind of material—ionic, covalent, or metallic. In this case our empirical radii and electronegativities should explain trends in structure and heats of formation quite well. On the other hand, if the series includes two kinds of solids—say ionic and covalent, or covalent and metallic—then we may well expect to encounter difficulties that the empirical approach was not designed to handle.

The simple empirical approach has proved its utility in thousands of structural determinations. It continues to be a mainstay in the analysis of complex ionic crystals, if only because it excludes many structural models as unreasonable, reducing to a manageable number the possibilities to be considered. However, by extending this approach to include new properties we may hope to obtain fresh perspectives on both the successes and the limitations of empirical crystal chemical methods.

When physicists think of connecting crystal chemistry to quantum mechanics, they generally tend to construct models of the electronic energy levels of the crystal. They often do not appreciate that isolated energy levels

may have little significance in determining structure and other bulk properties. The greatest advances in broadening the views of physicists of crystalline
properties in the last 25 years have come about through the introduction of
pseudoatoms, pseudopotentials, and pseudocharge densities, as discussed
elsewhere in this volume by M. L. Cohen. The pseudopotentials are weak
and in the case of simple metals (like Al) they can be used to calculate isolated
energy levels by hand, a point emphasized especially by W. A. Harrison.
More generally, the pseudopotential form factors are transferable characteristics of valence electrons which can be used to calculate many properties
(such as lattice vibration frequencies) in an elementary way. Cohen has
shown that almost all electronic properties of crystalline semiconductors
(and their surfaces and interfaces) can be discussed realistically using elemental pseudoatom form factors that are the quantum analog of Pauling's
electronegativity table.

Although pseudoatoms represent a great conceptual advance, especially
for physicists (who are always in need of being reminded of the importance
not only of accuracy but also of generality), they still do not make a direct
connection with chemical concepts such as covalency and ionicity. To make
such connections one must take advantage of data which have recently
become abundant largely because of the successes of pseudopotential calculations of electronic structures. I refer primarily to optical data which probe
the electronic polarizabilities of simple solids (such as Si and NaCl) over an
energy range comparable to the widths ( $\sim 10$ eV) of occupied valence states—
i.e., data which have been obtained far into the vacuum ultraviolet.

As these data became plentiful it became apparent that there were chemical
trends in absorption energies which surely reflected the competing effects of
covalent and ionic interactions. Cohen and co-workers fitted these optical
spectra with pseudoatom form factors containing three parameters per atom.
At this point I asked myself what the connection was between the chemical
concepts of covalency and ionicity and these form factors. Clearly covalency
between atoms A and B involved a kind of average of the A and B form
factors, while ionicity involved a difference. But how were the average and
the difference to be calculated from the form factors?

The answer to this question is that any procedure that one can define is
to some extent arbitrary. There is no arbitrariness in bond lengths or heats
of formation, which are observables, and what I needed was an observable
related to the optical spectra. There is one, and only one, such electronic
property—the low-frequency limit of the electronic dielectric constant (ex-
cluding the lattice polarizability in the far infrared), denoted by $\varepsilon_0$.

Now in molecules the connection between $\varepsilon_0$ and the electronic energy
levels is greatly complicated by the difference between the external electro-
magnetic field and the internal field which varies from atom to atom (so-called

local field effects). These differences are quite small in all crystals except those with very large energy gaps and small dielectric constants $\varepsilon_0 \lesssim 5$ (such as the silicates), and even there their effect is probably still small (in general of order $1/\varepsilon_0$ or less). Thus in crystals $\varepsilon_0$ is a very valuable experimental observable, and it can be used as the basis of an empirical theory, providing—providing what?

The sophisticated reader will guess the answer immediately, but the answer came as a great shock to me. The catch is that $\varepsilon_0$ (the square of the optical index of refraction $n$) must actually have been measured. In general this has not been the case. For the simplest family of binary solids, the octet family $A^N B^{8-N}$, which includes Si, Ge, GaAs, ZnS, NaCl, KBr, and so on, and which contains about 80 members, nearly all with cubic or hexagonal structures, $\varepsilon_0$ has been measured for only about 40 (or half) of the compounds. The measurement is not difficult, but it is not glamorous either, and it has often not been made. For less common solids the situation is, of course, much worse.

This is not the place to discuss every aspect of dielectric theory, but we can look at the main ideas. The basic equation is

$$\varepsilon_0 = 1 + (\hbar\omega_p)^2/(E_h^2 + C^2) \tag{1}$$

where $\omega_p^2 = 4\pi N e^2/m$ and $N$ is the number of valence electrons/unit volume. The *average* (not smallest) energy gap between occupied and unoccupied valence states is

$$E_g^2 = E_h^2 + C^2 \tag{2}$$

which contains a covalent (or homopolar) part $E_h$ and an ionic (or charge transfer) part $C$.

What are the conditions for the validity of Eq. (1)? First, the plasma frequency $\omega_p$ should be well defined. This is always the case for ionic and covalent solids, and it shows up as a narrow peak in the fast electron energy loss $\Delta E$ spectrum at $\Delta E = \hbar\omega_p$ (which is about 16 eV in Ge). It is not the case for molecules, where as a result local field effects are not completely negligible.

The second condition for the validity of (1) is that there should be only one kind of bond in the solid, describable by an average energy gap $E_g$. If there are lone-pair electrons as well as bonding electrons, then Eq. (1) is no longer valid; the treatment of lone-pair systems (such as $SiO_2$, $GeO_2$, $GeSe_2$ $As_2Se_3$) is an interesting problem which at present has not been treated satisfactorily. The proper generalization of (1) to ternary compounds also remains to be established.

By now some of the limitations of quantum-mechanical theories—even grossly simplified quantum-mechanical theories—are becoming apparent.

Even such a simple relation in Eq. (1) is much more involved in practice than atomic radii or heats of formation. Does the theory offer any rewards to compensate for its inconvenience?

It does. To begin with, $E_h$ is found to play the same role in solids as the resonance integral $\beta$ does in Hückel theory, while $C$ is similar to the Pauling electronegativity difference $\Delta X$ except that it is about 10 times more accurate for the $A^N B^{8-N}$ binary compounds (and for $A^N B^{10-N}$ compounds such as PbS, which are also semiconductors). Thus $E_h$ is a function of bond length $d$, and $C$ is found to fit the simple expression

$$C(A, B) = 1.5(Z_A/r_A - Z_B/r_B)\exp[-k_s(r_A + r_B)/2] \tag{3}$$

where $Z_A$ and $Z_B$ are the core charges of atoms A and B and $r_A$ and $r_B$ the atomic radii. The exponential screening factor is the one given by Thomas–Fermi screening theory in terms of the valence electron density $N$.

Now if we are given $E_h$ for diamond, Si, Ge and grey Sn, together with the factor 1.5 in Eq. (3), we can *predict* the dielectric constants of all 80 $A^N B^{8-N}$ compounds. The predicted values agree with experiment to better than 5%, which in many cases represents the experimental accuracy. Since the theory was created, several new values of $\varepsilon_0$ have been obtained which of course agree well with the predicted values.

There are by now many applications of dielectric theory to predict and to correlate bulk properties of all $A^N B^{8-N}$ compounds, as discussed in my review article (Phillips, 1970) and book (Phillips, 1973). To the surprise of nearly all physicists and almost no chemists, by far the most important variable is the ionicity $f_i$ of the chemical bond defined by

$$\begin{aligned} f_i &= 1 - E_h^2/(E_h^2 + C^2) \\ &= C^2/(E_h^2 + C^2) \end{aligned} \tag{4}$$

In spite of the limitations of the dielectric theory—many of them, by the way, are extrinsic and are due to the paucity of experimental data on $\varepsilon_0$—it still constitutes by far the most accurate and well-defined model of chemical bonding available for any family of homologous materials, molecular or crystalline, organic or inorganic. One of the great successes of the theory is shown in Fig. 1, where a population plot of the $A^N B^{8-N}$ compounds is given on a Cartesian $(E_h, C)$ plot. If $\tan \phi = C/E_h$, than $f_i = \sin^2 \phi$. Note that all compounds with $f_i \leqslant 0.785 \pm 0.01$ have four-fold coordination, while all compounds with $f_i \geqslant 0.785$ have 6-fold coordination.

In models of interatomic forces in covalent systems, the best results are obtained with VFF (valence force fields) consisting of bond-stretching ($\alpha$) and bond-bending forces ($\beta$). A plot of $\beta/\alpha$ vs $f_i$ is shown in Fig. 2, with $\beta/\alpha \to 0$ as $f_i \to 0.785$. This covalent softening explains many of the remarkable properties of AgI in double-salt solid electrolytes.

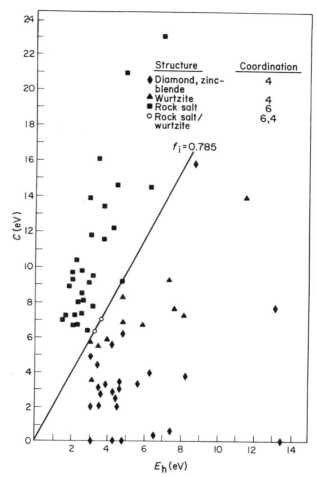

**Fig. 1.** The structures of some 80-odd $A^N B^{8-N}$ compounds are placed on an $(E_h, C) =$ (covalent, ionic) average energy gap plot. Here $E_h$ and $C$ have been determined from Eq. (1) for $\varepsilon_0$ and from the bond length $d$. The values of $C$ obtained in this way are also described by Eq. (3). [As an example, in GaAs, $(E_n, C)$ are $(4.32, 2.90)$ eV.] The line $f_i = C^2/(E_n^2 + C^2) = 0.785$ separates covalent structures with CN = 3 or 4 from ionic structures with CN = 6 or 8.

It may be said that there are two irreproachable sciences: crystallography and thermodynamics. The remarkable accuracy of dielectric theory is reflected in the fact that it has been used successfully to predict (Phillips, 1973) not only the heats of formation of $A^N B^{8-N}$ compounds (which are of order 1 eV) but also the heats of mixing of two such compounds (which are of order 0.1 eV), which are very important thermodynamically but are generally

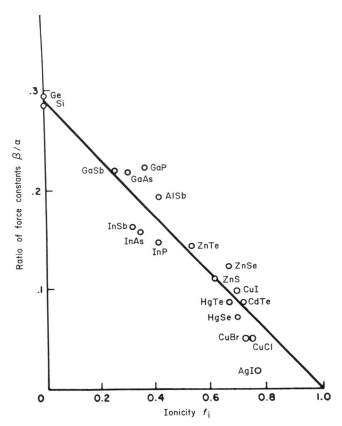

**Fig. 2.** The dimensionless ratio of bond-bending ($\beta$) and bond-stretching ($\alpha$) force constants as defined by P. Keating and R. M. Martin from the elastic constants of covalent $A^N B^{8-N}$ compounds with CN = 4. The covalent nature of the bond-bending forces is illustrated dramatically.

regarded as too small and too complex to be treated quantum-mechanically. The theory also produces a linear correlation between the coordination (6 vs. 4) preference energy, as measured in high-pressure and alloy experiments, and the dielectric ionicity, as discussed by A. Navrotsky and myself (Navrotsky and Phillips, 1975).

I would like to mention some of the more salient developments in dielectric theory since publication of the book. The theory has predicted (Shay *et al.*, 1976) the value of the interfacial dipole at CdS–InP interfaces within 10%— that is to say, 0.05 eV out of 0.5 eV—which is quite remarkable accuracy of 1% considering that the work functions of the two crystals are of order 5 eV.

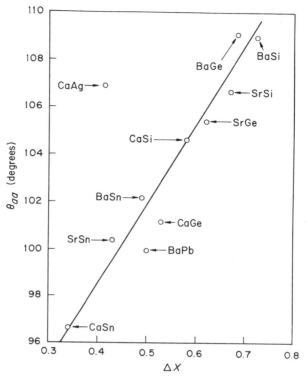

**Fig. 3.** Helical chain angle for CrB-type compounds formed from simple metals and metalloids, plotted against the Bloch–Simons orbital electronegativity difference. On several structural plots CaAg appears to be anomalous. It is also one of the Raney metallic compounds used in catalysis.

And it has recently been used by a graduate student at Cambridge University, P. Littlewood, to explain structural trends in the $A^N B^{10-N}$ family of small-gap semiconductors.

The impact of Fig. 1 on our thinking about the structural properties of complex solids has been substantial. In the hands of A. N. Bloch and co-workers it has generated tables of orbital radii ($r_s, r_p, r_d$) for each element A. From these orbital radii one can construct linear combinations which measure covalency and orbital electronegativity. One would expect the appropriate linear combination to vary from one solid type to another, which holds out the possibility of very accurate analyses of complex structures which were hitherto inaccessible to analysis using classical atomic radii, the Pauling electronegativity, or electron/atom ratios. A successful correlation

(Chelikowsky and Phillips, 1978) of the chain angle in CrB-type compounds formed from normal metals and metalloids (e.g., CaSi) is shown in Fig. 3. The chain angle is linear in $X(A) - X(B)$, where $X = (r_s^{-1} + r_p^{-1} + r_d^{-1})$.

Each year sees the appearance of new empirical formulations and reformulations of crystal chemical regularities. Do these new approaches describe genuine electronic interactions which are truly characteristic of chemical observables or are they merely numerology? Without some basis for independent judgment the scientific community has inclined to the latter view. As a result several major new contributions have not been properly appreciated. Orbital radii can be used to test empirical theories of electronic interactions and atomic structure of metallic compounds (Chelikowsky and Phillips, 1978). Because the orbital radii are based on atomic term values they form an extended and largely unambiguous reference frame for carrying out such tests. It is true, of course, that the functional relations between empirical parameters and orbital radii are themselves empirical [compare Eqs. (3)]. Nevertheless, these relations may be algebraically simple, and may contain only a few adjustable parameters. Then if they connect two very large (and previously unrelated) sets of crystalline and atomic parameters, we can expect not only to increase our confidence in the empirical model but also to gain additional insight into its physical content.

An empirical theory of the heats of formation of intermetallic compounds and alloys has recently been developed by A. R. Miedema and co-workers. His theory is global in the sense that it embraces more that 25 simple metals and more than 25 transition metals and considers more than 500 binary alloy systems. This is a very large data base. The theory is based on the regular solution approximation, i.e., it discusses only the contributions to the heat of formation of an AB compound which are symmetric in A and B and omits A–A and B–B interactions which are important in the A-rich or B-rich alloys. In fact, only the central nearest neighbor A–B energies are considered, while the structural energies associated with changes in coordination number are omitted.

Miedema's model represents a refinement of Pauling's model of heats of formation in one very essential respect. While Pauling focuses on the charge transfer term that makes a negative (potential energy) contribution to $\Delta H_f$, Miedema adds a positive (kinetic energy) contribution which can be used to explain the immiscibility ($\Delta H_f \gtrsim RT$) of certain (A, B) pairs (about 15% in all). The form of this kinetic energy term varies greatly between molecules, insulators, metals, and covalent solids. Thus Miedema's model, unlike Pauling's, is restricted to intermetallic compounds. Moreover, Miedema places his main emphasis on the sign of $\Delta H_f$ and does not try to fit the magnitude (which is usually small in the metallic case) accurately. Miedema's theory is a truly linearized theory which focuses on the regime where the positive and negative contributions to $\Delta H_f$ almost cancel one another, whereas Pauling's approach is better suited to the very wide range of stable systems which are predominantly (but not necessarily completely) ionic.

In order to include the kinetic energy terms (which in metals arise from the mismatch in atomic charge densities averaged over the surfaces of the atomic spheres), Miedema introduces two parameters for each element, one a measure of the electronegativity of atoms in metals (denoted by $\phi^*$ to suggest an elemental work function) and another representing the atomic charge density (or something related to it) at the atomic radius, $n(r_a)$. Thus Miedema's theory contains over 100 adjustable parameters, leading many people to doubt its significance.

J. R. Chelikowsky was able to show (Chelikowsky and Phillips, 1978) that nearly all of Miedema's parameters for normal metals could be fitted by simple algebraic functions of $r_s$, $r_p$, and $r_d$, and that the algebraic form of $\phi^*$ was merely a refinement of the expression $r_s^{-1} + r_p^{-1} + r_d^{-1}$ previously derived for the Pauling electronegativity by Bloch. [As expected, the expression for $n(r_a)$ is much more complicated.] Thus Chelikowsky has demonstrated that Miedema is not merely doing numerology, as asserted by some rather superficial so-called "first principles" analysis by energy-band theorists (whom I prefer not to cite).

One of the interesting by-products of Chelikowsky's demonstration of the otherwise hidden atomic orbital content of Miedema's parameters is the implication that these quantities are superior coordinates for discussing other properties of metallic alloys. Specifically they should be superior to the Hume–Rothery and Darken–Gurry coordinates (size and Pauling electronegativity) for discussing atomic solubilities in metals. Chelikowsky (1979) has demonstrated that this is indeed the case, and he has shown explicitly in the case of close-packed divalent metals how the host structure influences the solubility patterns, an entirely new aspect of the problem not included in Miedema's regular solution formulation of the alloy problem. Thus quantum theory has greatly extended and amplified the content of what appeared (and, unfortunately, still does appear) to many scientists to be merely empirical numerology. Incidentally, one of the most satisfying aspects of this work is its ability to treat scores or even hundreds of compounds without even constructing Schrödinger's equation in the solid.

I should like to mention a very great accomplishment in the area of metallic coordination chemistry in which the Miedema coordinates play a crucial role. E. N. Kaufmann and co-workers have implanted some 25 elements into close-packed Be and have identified (Kaufmann et al., 1977) the metastable equilibrium sites of each element as either substitutional or (tetrahedral or octahedral) interstitial. Classically one would have expected almost all the elements (which are larger than Be) to prefer the larger interstitial site

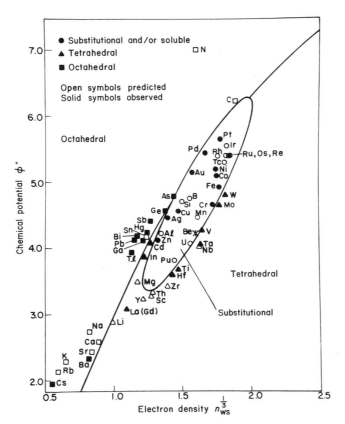

**Fig. 4.** Population map of site preferences of impurities implanted in Be. The boundaries are elliptical or hyperbolic curves. See Chelikowsky (1979) and Kaufmann *et al*, (1977) for further solubility plots and discussion. [From Kaufmann *et al*. (1977).]

(octahedral), but the data shows that the elements distribute themselves about equally among the three possibilities. A Darken–Gurry plot using Hume–Rothery coordinates (size and Pauling electronegativity) fails to separate the three groups. However, a plot using Miedema's coordinates is completely successful, as shown in Fig. 4.

In conclusion, I believe that with modern experimental techniques there are many new opportunities for crystal chemistry. One of the newly emerging areas in which structural arguments are likely to be of very great importance is thin films, both crystalline and noncrystalline (Poate *et al.*, 1978). In order to transform this subject from black art to science one must be able to analyze the structure on several scales, including a molecular scale of order 10 Å. Here the knowledge of which structures are reasonable (Phillips, 1979) will once again play a central role, as it has so often in the past.

## REFERENCES

Chelikowsky, J. R., and Phillips, J. C. (1978). *Phys. Rev. B* **17**, 2453.
Chelikowsky, J. R. (1979). *Phys. Rev. B* **19**, 686.
Kaufmann, E. N., Vianden, R., Chelikowsky, J. R., and Phillips, J. C. (1977). *Phys. Rev. Lett.* **39**, 1671.
Navrotsky, A., and Phillips, J. C. (1975). *Phys. Rev. B* **11**, 1583.
Phillips, J. C. (1970). *Rev. Mod. Phys.* **42**, 317.
Phillips, J. C. (1973). "Bonds and Bands in Semiconductors." Academic Press, New York.
Phillips, J. C. (1979). *J. Non-Cryst. Solids* **34**, 153.
Poate, J. M., Tu, K. N., and Mayer, J. W. (1978). "Thin Films." Wiley, New York.
Shay, J. L., Wagner, S., and Phillips, J. C. (1976). *Appl. Phys. Lett.* **28**, 31.

# 3

# Pseudopotentials and Crystal Structure

## MARVIN L. COHEN

|       |                                                          |     |
| ----- | -------------------------------------------------------- | --- |
| I.    | Introduction                                             | 25  |
| II.   | Pseudopotentials                                         | 26  |
|       | A. Phillips–Van Vechten Dielectric Theory                | 30  |
|       | B. Electronic Charge Densities                           | 32  |
|       | C. Hard-Core and Density-Functional Pseudopotentials     | 35  |
|       | D. The "Direct Approach"                                 | 40  |
| III.  | Conclusions and the Future                               | 46  |
|       | References                                               | 48  |

## I. INTRODUCTION

I will consider the contribution to this volume by J. C. Phillips to be the true introduction to my discussion, which will attempt to present more details on how the pseudopotential is being used to explore problems in crystal structure and bonding. Although I would like to supply you with simple ideas and rules for describing complex structures, some of our work is better described as complex ideas for simple structures. In defense of the approach, I should emphasize that it is based on a stable and logical conceptual foundation (quantum mechanics) and it yields a great deal beyond structural information. Some of this information forms the base for our understanding of the electronic and optical properties of solids. I will touch on some of the developments in these areas, but I will focus primarily on questions related to structure.

Our goal is to analyze and predict structurally related properties of solids using microscopic theory and a knowledge of the quantum physics of the constituent atoms. Only simple structures containing one, two, or three different types of atoms have been used to test the ideas and techniques. The approach can be extended to more complex structures, and at present, the only limitation appears to be computational effort (i.e., computer time). The

**25**

Structure and Bonding in Crystals, Vol. I

method is based on a pseudoatom or pseudopotential model, and most of the underlying concepts are physically motivated; that is, the physics is not "lost in the computer."

The pseudopotential approach was devised to provide a simpler and more accurate way to compute the electronic structure of crystals. The extensions to problems involving crystal structure face a standard obstacle. The total energy differences between different crystal structures are a small fraction of the energy associated with the sum of single electron energies. Therefore, a computation which is considered accurate for energy levels may not be sufficiently accurate for choosing the minimum energy structure. Hence, attempts were made to circumvent the "direct approach" of calculating total energies to determine crystal structure. Some of these have led to new physical insights into the physics of these problems.

In this paper, I will briefly describe the pseudopotential approach and its use in structural energy problems. Before giving the recent results using the "direct approach" of calculating total energies, I will describe a few avenues which have been very successful in clarifying the structural problems. These include the dielectric formalism for structure of Phillips and Van Vechten, charge density contour maps and chemical bonds, and hard-core pseudopotentials. I will supply references to the original work in place of giving the details and underlying formalism.

## II. PSEUDOPOTENTIALS

A pseudopotential represents the potential which an outer or valence electron is subject to in an atom or a solid. The inner or core electrons are assumed to be "frozen" and unresponsive to the perturbations associated with bonding effects. Using this approximation, properties associated with valence electrons, e.g., optical spectra, chemical bonding, etc., may be analyzed directly without the necessity of calculating properties associated with the core electrons. A disadvantage of this approach is that no knowledge of the properties of the core electrons is obtained.

Because the valence electrons in a solid are relatively free—that is, not bound tightly to the nuclei—the pseudopotential which describes their interactions with the core is relatively weak when compared to the Coulomb potential for a charged nucleus. The origin of the cancellation of the very attractive ionic potential is rooted in the Pauli principle. Because the valence electrons cannot occupy the same states as the core electrons, there is a "Pauli force" which keeps the valence electrons out of the core region. This repulsive potential cancels much of the attractive Coulomb core potential, leaving a net weak potential or pseudopotential. This cancellation was first shown by

Phillips and Kleinman (1959), and it can be calculated directly. The usual approach, however, is not to calculate the repulsive Pauli term but rather to obtain the potential from experimental data or from a constraint forcing the potential to reproduce atomic wavefunctions for the valence electrons.

Because the pseudopotential is constructed for valence electrons which are relatively free, the potential itself is weak. In Fig. 1, a comparison is made between a pseudopotential for valence electrons and an all-electron ionic Coulomb potential. The major difference between the pseudopotential and the all-electron potential is that the pseudopotential does not contain the strong attractive portion of the ion potential close to the core. Nevertheless, the pseudopotential is capable of accurately reproducing the energy levels and wavefunctions for the valence electrons. In Fig. 2, a comparison is made between the soft-core and hard-core varieties of pseudopotentials. The differences and advantages of each will be discussed later.

In principle, electrons having different angular momenta, $l$, will feel different potentials. Hence, the pseudopotential depends on $l$, and in general, there will be an s, p, and d pseudopotential for each angular momentum. The $l$-dependent potentials are called nonlocal pseudopotentials. Often a local approximation is made, and one potential is used for s, p, and d electrons.

Because the pseudopotentials are relatively weak, a basis set composed of plane wave components is usually sufficient to represent the valence wavefunctions. For cases where the wavefunctions are more localized (e.g., d-states), Gaussian or other localized orbitals can be used to augment the plane wave basis set.

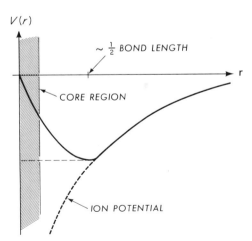

**Fig. 1.** Schematic plot of a pseudopotential for valence electrons compared with an all-electron ionic potential. Away from the core region, the two potentials merge.

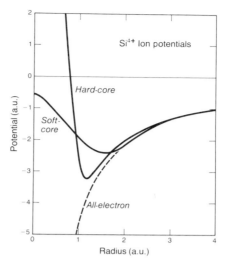

**Fig. 2.** Soft-core and hard-core pseudopotentials for valence electrons and an all-electron ionic potential for silicon.

Usually the Fourier transform of the potential is most convenient to use. Because of the lattice periodicity, the potential can be represented as a Fourier sum

$$V(\mathbf{r}) = \sum_{\mathbf{G}} V(\mathbf{G})S(\mathbf{G})e^{i\mathbf{G} \cdot \mathbf{r}} \qquad (1)$$

where $\mathbf{G}$ is a reciprocal lattice vector, and $S(\mathbf{G})$ is the structure factor which locates the atoms in the crystal. The coefficients $V(\mathbf{G})$ are the pseudopotential form factors. These are obtained from the Fourier transform of $V(\mathbf{r})$. A schematic curve is shown in Fig. 3. The weakness of $V(\mathbf{r})$ in the core region leads to a small $V(q)$ for large $q$. Hence, the sum in Eq. (1) can be truncated after only a few $\mathbf{G}$ terms. In fact, three form factors are usually sufficient to describe a potential accurately.

In the Empirical Pseudopotential Method (EPM), the potentials are obtained from fitting the form factors to data based on atomic or solid properties (Cohen and Bergstresser, 1966). Many potentials have been determined in this way, and the optical properties of a variety of solids have been interpreted in terms of transitions between electronic energy levels which in turn were determined by the potential. One of the important properties of the EPM form factors was their transferable nature (Cohen and Heine, 1970). Within reasonable limits, the potentials obtained from analyzing one crystal could

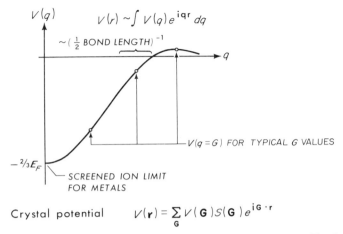

Fig. 3. Schematic plot for a pseudopotential in reciprocal or $q$-space. The form factors $V(\mathbf{G})$ determine the electronic structure for bulk crystalline solids.

be used for another. In some cases, potentials could be extracted for elements based on analyses of compounds. For example, an analysis of the optical data of InSb yields the form factors for In and Sb separately. When these are used to compute properties of In and Sb, a metal and a semimetal, the results are surprisingly good. Hence, it is possible to shine light on InSb, analyze the reflectivity, and then compute the superconducting transition temperature and Fermi surface topology of In.

At present, the EPM form factors generally give the most accurate determination of the electronic band structure (Chelikowsky and Cohen, 1976), whereas potentials obtained from atomic wavefunctions—that is, density functional pseudopotentials (DFP)—are best suited to calculate ground-state properties. Because the optical properties of a solid involve electronic excitations, the DFP are not expected to describe these non-ground-state properties accurately. The consequences of this feature and other aspects of the DFP approach will be discussed later.

Once the pseudopotential has been determined, the most "direct approach" to study structure is to calculate the total energy corresponding to a given structural configuration and compare this result with other structural configurations. In principle, the configuration with the minimum energy should correspond to the observed structure. Unfortunately, the energies involved are large and their differences are small, so accuracy in determining the total energies is essential. The current state of the art using all-electron potentials for metals is described in Moruzzi et al. (1978). The problem is compounded when an all-electron approach is used because the energies of the core electrons are included giving even larger total energies than in the pseudopotential approach, which involves only the valence electrons. Recently, it has become possible to do accurate pseudopotential calculations using the "direct approach" of calculating the total energy. These results will be discussed later. First, a few nondirect approaches will be described.

## A.  Phillips–Van Vechten Dielectric Theory

A nondirect approach to structure was developed in a series of publications by Phillips and Van Vechten (Phillips, 1970, 1973). Although total energies are not calculated in this model, the theory connects the dielectric response to the bonding nature of crystals, and hence, it is predictive and yields some new insight into structural questions. Harrison has introduced a somewhat similar approach, the Bond Orbital Method, which he describes in this volume and in his recent book (Harrison, 1980). I will summarize some of the results of Phillips and Van Vechten (PVV) here since these bear directly on a few central aspects of pseudopotential and charge density calculations discussed in the next section.

The PVV theory takes advantage of a connection between the dielectric function for a semiconductor and its average energy gap (Penn, 1962)

$$\varepsilon_0 = 1 + (h\omega_p)^2/E_g^2 \qquad (2)$$

where $\omega_p^2$ is the plasma frequency, $4\pi n e^2/m$, $n$ is the number of valence electrons per unit volume, and $E_g$, the Phillips gap, is the average optical gap. The minimum gap in a typical semiconductor is $\sim 1$ eV, but $E_g \sim 4$ eV. The Phillips gap represents the average separation between the occupied valence and empty conduction states in a semiconductor. It can also be associated with the main peak in the reflectivity spectrum.

The contributions to $E_g$ arise from two sources—a homopolar contribution and an ionic contribution. These can best be illustrated using pseudopotentials. Three related compounds from the Ge row of the periodic table, Ge, GaAs, and ZnSe, are useful to illustrate the PVV approach. These compounds form in the diamond and zinc blende structures with the same number of valence electrons and approximately the same lattice constants. The core electronic structure is the same for the atoms of all five elements in these materials, and the differences arise from the nuclear charge and the number of valence electrons associated with each atom. The pseudopotential for the two atoms in the unit cell of these crystals can be separated into an average or symmetric potential and a difference or antisymmetric potential. For Ge, since the two atoms in the unit cell are the same, the antisymmetric potential is zero. For GaAs and ZnSe, the average or symmetric potential is well approximated by the Ge potential (Cohen and Bergstresser, 1966), and the antisymmetric potential which measures the differences in the potentials is approximately twice as large in II–VI compounds like ZnSe than in III–V compounds such as GaAs. The antisymmetric potential causes a charge shift from the less attractive Ga or Zn to the more attractive As or Se cores. This gives an ionic component to the bonding, and the antisymmetric potential is a measure of the ionic nature of the crystal.

Returning to the PVV approach, if one pictures a free-electron system which is perturbed by the symmetric and antisymmetric potentials, these potentials will produce a gap between the occupied and empty electronic energy levels. This gap will have two components which are related to the symmetric (homopolar) potential and the antisymmetric (charge transfer) potential. The Phillips gap of Eq. (2), $E_g$, then has two components which PVV write as

$$E_g^2 = E_h^2 + C^2 \tag{3}$$

where $E_h$ and $C$ are the homopolar and charge transfer components. For Ge, $C = 0$ and $E_g = E_h = 4.3$ eV. For GaAs and ZnSe, $E_h = 4.3$ eV and $C = 2.9$ and $5.6$ eV respectively. PVV make the anzatz that the ionicity, $f_i$, of the bond is

$$f_i = \frac{C^2}{E_g^2} \tag{4}$$

For Ge, $f_i = 0$; for GaAs, $f_i = 0.31$; and for ZnSe, $f_i = 0.63$. When $f_i$ becomes large, the bond is very ionic, and it is expected that the charge transfer is too large for a fourfold coordination appropriate for covalent structures to survive. A sixfold coordination (i.e., like NaCl) is expected, and PVV analyze this fourfold/sixfold transition using structural data for a series of compounds to obtain the critical ionicity for the transformation, $f_c = 0.785$ (Phillips, 1973).

The important input to the PVV theory is the dielectric constant which is a measure of the polarizability of the chemical bond. The relation between the electronic density in the bond, its ionic character, the band gaps in the optical spectra, and so on all give a calculational bridge to explore the relationships between structural properties and electronic properties. This approach also circumvents the problems and effort associated with doing full scale electronic calculations. However, to visualize the charge distribution and the properties of the bonds associated with the fourfold/sixfold $f_c$ transition, it is useful to compute the electronic charge density for various crystals having different ionicities.

## B.  Electronic Charge Densities

The description of the relation between the symmetric and antisymmetric components of the crystalline pseudopotential and the PVV homopolar and ionic gaps exploits the concept of charge transfer, but this aspect of the physical properties of the system is not computed directly in the PVV model. However, such a calculation can be done once the crystal pseudopotential is known. The approach is straightforward. First, the potential is used to compute the energy eigenvalues and wavefunctions for the various electronic states. The energy levels in the solid are grouped into bands, and the wavefunctions corresponding to these bands can be computed for each momentum state, $\mathbf{k}$, in the band $n$. The wavefunction, $\psi_{n,\mathbf{k}}(\mathbf{r})$, can be squared and summed over $\mathbf{k}$ to produce the charge density, $\rho_n(\mathbf{r})$, for band n

$$\rho_n(\mathbf{r}) = e \sum_k \left| \psi_{n,\mathbf{k}}^*(\mathbf{r})\psi_{n,\mathbf{k}}(\mathbf{r}) \right| \tag{5}$$

The total electronic charge density for a solid would require another sum over all the occupied levels. For a semiconductor or insulator, this sum would be over the occupied bands, and the total valence charge density, $\rho(\mathbf{r})$, becomes

$$\rho(\mathbf{r}) = \sum_{\substack{n \\ \text{occupied} \\ \text{bands}}} \rho_n(\mathbf{r}). \tag{6}$$

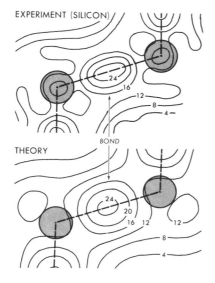

EXPERIMENT (SILICON)

THEORY

BOND

**Fig. 4.**  Valence charge densities for Si. The theoretical results were based on a pseudopotential calculation. The experimental values (Yang and Coppens, 1974) were determined using x-ray measurements. The shaded regions around the atoms represent the core regions where the pseudopotential results are not expected to be accurate. The contours of constant charge density are measured in units of $e/\Omega_c$ where $\Omega_c$ is the unit cell volume.

The first calculations of this kind were done using wavefunctions based on the EPM, and a variety of systems were explored (Cohen, 1973). More recent calculations give very similar results with refinements (Chelikowsky and Cohen, 1976).

A plot of the total charge density for Si in a (110) plane containing the two silicon atoms is shown in Fig. 4. An experimental determination of $\rho(\mathbf{r})$ is also given. The experimental results were obtained (Yang and Coppens,

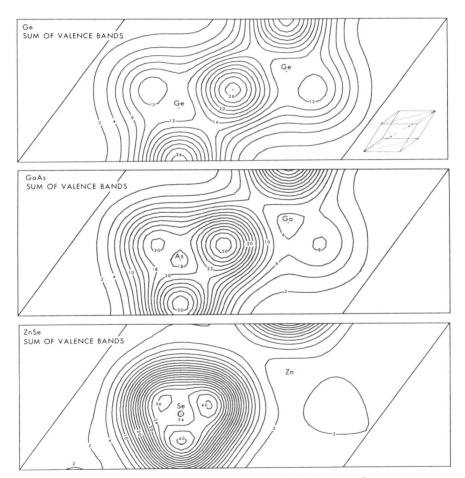

**Fig. 5.**  The valence charge density for Ge, GaAs, and ZnSe.

1974) using x-ray measurements. The covalent bond is clearly shown in this plot as a pile-up of charge halfway between the two Si atoms. The same covalent topology of $\rho(\mathbf{r})$ is seen for C, Si, Ge, and $\alpha$-Sn. However, as the antisymmetric potential becomes finite, there is some charge transfer, and the bonding becomes partially ionic. This is clearly illustrated by an early EPM plot (Cohen, 1973) displaying the charge density for Ge, GaAs, and ZnSe (Fig. 5).

Figure 5 displays the charge density in the (110) plane containing the two atoms in the unit cell. The charge density for Ge is similar to Si and is covalent. GaAs is partially ionic, and the charge contour lines indicate that more charge surrounds the As than the Ga. The peak in the charge distribution between the atoms is the remnant of the covalent bond. The position of the peak moves away from the midpoint between the atoms in the direction of the As atom. This effect is magnified in ZnSe where the covalent contribution has decreased substantially from Ge.

To obtain structural information and make the connections with PVV theory which were indicated earlier, the charge in the "covalent region" was integrated and compared with the ionicity

$$Z_b = \int_{\Omega_0} (\rho - \rho_0) d^3r \tag{7}$$

where $Z_b$ is the charge in the bond. The integration is over a volume, $\Omega_0$, containing the bond charge, and the background charge, $\rho_0$, is subtracted out. The separation between covalent bond charge and background charge density is ambiguous, but the results for $Z_b$ are not very sensitive to the choice of $\rho_0$. The choice made here is that the last closed contour lines in Fig. 5 are chosen to be the boundaries between the bond charge and background charge regions. It is clear from Fig. 5 that $Z_b$ for Ge is largest, and it decreases for GaAs and is smallest for ZnSe. If the PVV ionicity scale is used, the dependence of $Z_b$ on ionicity, $f_i$, can be demonstrated. This is shown in Fig. 6 for the Ge, GaAs, ZnSe series and for the Sn, InSb, CdTe row of the periodic table. In both cases, the extrapolations to vanishing bond charge give a critical ionicity of $f_c \sim 0.79$.

Although the above calculation is not sufficiently accurate to give a precise $Z_b$ for $f_c$ and it does not determine the minimum value of $Z_b$ at which the fourfold/sixfold transition takes place, it is clear from the results that $Z_b$ is approaching zero as $f_i$ approaches $f_c$. This is consistent with the view that it is the bonding charge which stabilizes the fourfold covalent structure and that the vanishing of $Z_b$ results in a transition to electrostatic bonding and sixfold coordination.

So, a connection between the PVV approach and the charge density approach can be made. The same pseudopotentials which can be used to

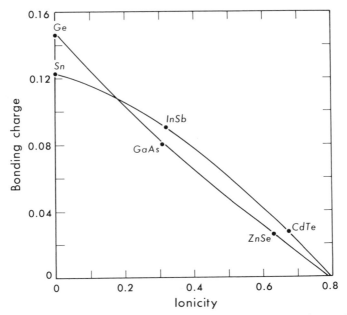

**Fig. 6.**   Dependence of the bond charge $Z_b$ on ionicity (Phillips–Van Vechten scale) for two isoelectronic series. The bond charge becomes small or vanishes as the Phillips–Van Vechten critical value of the ionicity ($f_c = 0.785$) is approached.

compute the optical properties can also be used to obtain wavefunctions and charge densities. The optical properties can be compared with experiment, and these can also be used to compute the average gap in the PVV theory. On the other hand, the charge densities can be used to describe the structural phase transition explored using the PVV theory. The final results are not sensitive to the ionicity scale used. If the Pauling scale is used, then $f_c$ changes slightly, but the major effects are the same.

Although PVV theory has been applied primarily to compounds with eight valence electrons, recent extensions to 10-valence-electron compounds have been made (Littlewood, 1979).

## C.   Hard-Core and Density-Functional Pseudopotentials

An example of a hard-core pseudopotential (HCP) is shown in Fig. 2. The most distinguishing property of the HCP is the zero of the potential at small **r**. The hard-core nature arises from the Pauli force discussed earlier. Because of the orthogonality of the valence electron states to the core states, the valence electrons are pushed out of the core region, and this is simulated by the strong small **r** component of the HCP. The soft-core pseudopotential (SCP) achieves this to a lesser extent and often does not give accurate wavefunctions for the region near the core, but since we are normally more interested in the electronic structure at larger **r**—that is, the bonding region—than for many properties of solids, the SCP suffices.

I will digress briefly to describe some technical aspects of the HCP and SCP. For regions closer to the core and inside the core, the HCP is generally more accurate than the SCP. Then, why bother with the SCP at all? The advantage of the SCP is computational convenience. If a plane-wave basis set is used for the electronic wavefunctions, the number of plane waves needed to reproduce the eigenstates of a SCP is far less than those required for a HCP. In fact, in some cases, the HCP problem cannot be solved within standard 1980 computer memory capacity. However, mixed basis sets (Louie *et al.*, 1979), which use localized functions together with plane waves, can accomodate many problems. The limitations imposed by computational memory and speed may be only a temporary problem. As computers evolve, perhaps the problems described above will not be serious, and the computational advantages of the SCP over HCP will not be relevant. However, at present, these advantages are important, and even when the HCP is used, it is "cut off" or "smoothed" at small **r** to eliminate the strongest part of the core region. For example, the potential shown in Fig. 2 continues to rise unbounded for small **r**. In practice, this potential would be smoothly extrapolated to a constant value at **r** = 0.

Returning to physical properties, a major advantage of the HCP is its ability to supply structural indicies. Just as in the PVV scheme in which crystal structure types were associated with the homopolar and charge transfer gaps, $E_h$ and $C$, the HCP determines a core radius for each atom of a compound. In fact, if the HCP is nonlocal (*l*-dependent), then radii can be associated with each of the *l*-dependent potentials yielding $r_s$, $r_p$, and $r_d$, that is, core radii for each atomic potential. Linear combinations of these radii can, in turn, be formed to provide structural indicies like $E_h$ and $C$.

A. Bloch and co-workers (Simons and Bloch, 1973; St. John and Bloch, 1974) were the first to develop an HCP scheme of the type described above for studying structural properties. The Simons–Bloch potential, $V_{SB}$, is nonlocal, and it can be written in analytic form

$$V_{SB}(r) = -\frac{Z_v}{r} + \frac{B_l}{r^2} + \frac{l(l+1)}{2r^2} \tag{8}$$

where $Z_v$ is the ionic charge felt by the valence electrons, $B_l$ is a parameter, and $l$ is the angular momentum. The first term is the attractive Coulomb potential associated with the charged ion. The second term represents the repulsive Pauli term, and the third gives the usual quantum centrifugal potential for different angular momenta.

The orbital radii determined from Eq. (8) can be used as structural indicies. The procedure is simply to find $r_l$ such that $V_{SB}(r_l) = 0$. For binary compounds composed of atoms A and B with s and p valence electrons, the relevant radii are $r_s^A$, $r_p^A$, $r_s^B$, and $r_p^B$. The most commonly used combinations of these radii for structure studies are

$$
\begin{aligned}
R_\sigma^{AB} &= (r_p^A + r_s^A) - (r_p^B + r_s^B) \\
R_\pi^{AB} &= (r_p^A - r_s^A) + (r_p^B - r_s^B)
\end{aligned}
\tag{9}
$$

St. John and Bloch (1974) have used a structural map of the Mooser–Pearson or Phillips–Van Vechten type to separate the valence 8 (or octet) binary compounds into diamond, zinc blende, wurtzite, and graphite structures. The $R_\sigma^{AB}$ and $R_\pi^{AB}$ are the coordinates for the map. Machlin *et al.* (1977) successfully extended this study to suboctet nontransition-metal compounds, that is, $A^N B^{P-N}$ ($3 \leqslant P \leqslant 6$), using the same radii. Extensions of this approach to represent the empirical Miedema (1976) coordinates for heats of formation were made by Chelikowsky and Phillips (1978). So, the use of the Simons–Bloch radii for structural information became evident, and theoretical efforts were made to improve the accuracy of the radii, justify the empirical approach, extend the applicability of the scheme, and apply the approach to wider classes of compounds.

The first systematic refinement of the Simons–Bloch potential was made by Andreoni *et al.* (1978). These authors modified the repulsive $r^{-2}$ term using wavefunction matching criteria and improved the accuracy of the potentials using the first row of the periodic table as a prototype series for the calculation. Many other refinements and extensions came from the development of the so-called "first-principles or density-functional pseudopotentials" (Zunger and Cohen, 1978, 1979). This avenue of research provided a more first-principles derivation of structural radii, $r_l$, and extended the domain of elements, compounds, and crystal structures studied.

The density-functional pseudopotential (DFP), of which the Zunger–Cohen variety is just one specific example, is developed from an ionic pseudopotential screened by the electronic response to this potential. The electronic component of the potential contains the usual Hartree screening and an exchange-correlation potential. One useful approximation for these electronic potentials is to express them as functionals of the electronic charge density. Local approximations have been studied extensively, and these appear to give good results for ground state properties. The density-functional formalism is described and developed in the papers of Hohenberg and Kohn (1964) and Kohn and Sham (1965). The scheme for developing the DFP and references to other authors who developed this approach are given in Zunger (this volume, Chapter 5).

The Zunger-Cohen DFP is a HCP with a zero at relatively small $r$ which determines the structural radii, $r_l$, as in the Simons-Bloch scheme. The potentials and radii were evaluated for atoms of the first five rows of the periodic table. The potentials were shown to give good electronic band structures and wavefunctions for a series of solids in addition to accurate determinations of atomic energy levels and wavefunctions. The structural radii were used in a similar manner to the approach of St. John and Bloch (1974), and this application provided a structural separation of both the octet $A^N B^{8-N}$ and the suboctet $A^N B^{P-N}$ ($3 \leqslant P \leqslant 6$) crystal structures. A structural

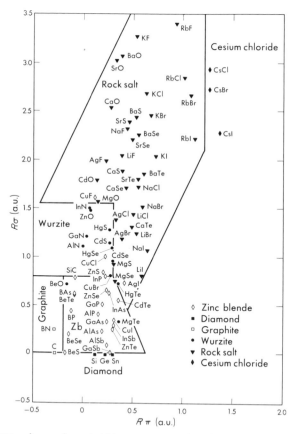

**Fig. 7.** Structural map for octet binary compounds using $R_\sigma$ and $R_\pi$ derived from the Zunger–Cohen density-functional pseudopotential.

map for the octet series appears in Fig. 7. A more extensive compilation and other structural maps are given by Zunger in this volume.

The potentials described above can be used for structural studies through the $r_l$ and directly by calculating the electronic structure of the material of interest. The potentials are determined numerically, and they have been displayed graphically (Zunger and Cohen, 1978). Recently, an analytic fit to the potentials has been constructed (Lam *et al.*, 1980). The analytic form is relatively simple, and the fitting parameters used reflect the chemical trends of the elements. Using only three parameters, the regularities of the periodic table are reproduced. The availability of these analytic potentials should increase the usefulness and accessibility of this approach using the HCP model.

The density-functional approach can also be used to generate soft-core pseudopotentials. For the reasons stated before, the SCP is usually easier to use for electronic calculations. First, the SCP was developed to study surfaces and local configurations. Because of the complexity of these problems, it was originally felt that it was necessary to limit the calculations to local potentials. The density-functional approach was used, and the calculations—for example, surface electronic structure calculations—were done self-consistently. This meant that the electronic charge density generated for a specific surface geometry was used to produce the screening and exchange-correlation potentials, and a feedback loop was established (Fig. 8). For silicon, the two most popular potentials were the Appelbaum–Hamann potential (1976) and the "Berkeley" potential (Schlüter *et al.*, 1975). These

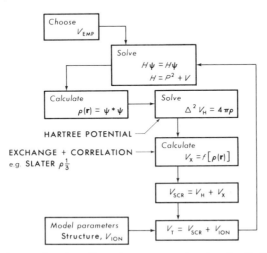

**Fig. 8.**  Flow diagram for the self-consistent approach using density-functional pseudo-potentials for electronic calculations on surfaces and other localized configurations.

potentials were widely used for surface calculations. They are quite "soft" and were sufficiently accurate to answer many major questions in surface physics. However, for accurate calculations of the type needed for total energy comparisons (of different crystal types), a more precise pseudopotential is needed.

The Zunger–Cohen potentials could suffice, but many felt that a nonlocal SCP would be more efficient. Schemes to compute the SCP were recently developed by Hamann *et al.* (1979) and by Kerker (1980). These nonlocal potentials accurately reproduce the energy levels and wavefunctions (outside the core) relative to all-electron calculations. Therefore, the current choices for pseudopotentials include the following: empirical pseudopotentials (soft-core and hard-core) which are derived with some experimental input, soft-core pseudopotentials (local and nonlocal) with variable softness, and hard-core pseudopotentials. The latter SCP and HCP are derived from a density-functional scheme.

Only the empirical HCP (Simons–Bloch) and the density-functional HCP (Zunger–Cohen) have turning points or radii, $r_l$, which can be used for structural studies, but all the pseudopotentials can be used in the "direct approach" of calculating total energies.

## D. The "Direct Approach"

As described previously, the most straightforward scheme for studying structures using pseudopotentials is to compute the total energy of hypothetical structures and compare them. The lowest energy structure should be the observed one. In addition, a great deal of structural information like cohesive energies, bulk and shear moduli, lattice constants, vibrational spectra, and phase diagrams can be computed theoretically. Because of the accuracy needed for the electronic structure, a self-consistent calculation using a density-functional pseudopotential is the most appropriate. Some of the recent results have been spectacular, and a great deal has been learned about calculations of this type.

As a first attempt, the Appelbaum and Hamann potential and the "Berkeley" potential derived primarily for surface calculations were used to compute structural properties. In a non-self-consistent calculation, Wendel and Martin (1979) showed that the Appelbaum and Hamann potential was too attractive and that the lattice constant, for example, was too small. Nevertheless, these authors could explore some structural properties and examine the behaviour of the charge density when the silicon lattice is distorted by a lattice wave.

Self-consistent calculations using the Berkeley potential were done (Ihm and Cohen, 1979, 1980) based on a new method for calculating total energies

(Ihm *et al.*, 1979). One advantage of this new approach is that the total energy calculation is done in momentum space. Electronic energy band structures are also calculated in momentum space, and Fourier components of the electronic charge density and potentials are outputs of the band calculations. Since these are the inputs for the total energy calculation when computed in momentum space, both calculations can be done together, and a higher accuracy can be achieved.

The results using the Berkeley potential for the lattice constant and bulk modulus is 5.32 Å and $1.8 \times 10^{12}$ dyn/cm$^2$; the experimental values are 5.43 Å and $0.99 \times 10^{12}$ dyn/cm$^2$. These results are not as accurate as those obtained using the Zunger–Cohen potential (Zunger, 1980) or the Hamann–Schlüter–Chiang potential (Yin and Cohen, 1980) because nonlocality is not explicitly taken into account and the potential is not accurate in the region near the core. These results were, nevertheless, very satisfying since they showed that a calculation of this type could be done. Another aspect of this calculation which yielded physical insight concerning the role of bonding in these crystals was a comparison for different crystal structures of the total energies and electronic charge densities. The diamond structure was compared with the body-centered cubic (bcc), face-centered cubic (fcc), hexagonal close-packed (hcp), and white tin structures. The diamond structure configuration yielded the lowest total energy, and the charge density plots revealed the strongest covalent bonds for this structure (Figs. 9 and 10). The structure having the next highest energy is the white tin structure. As can be seen from Fig. 10, this structure also has reasonably strong covalent bonding.

The above calculational approach was also applied to the silicon (111) surface. Using the Berkeley potential, it is possible to compute the total valence electronic charge density for a surface and illustrate the redistribution of the charge at the surface (Schlüter *et al.*, 1975). The results which are shown in Fig. 11 were calculated for an ideal (111) surface. They illustrate the rearrangement of charge at a surface and the diffusing of charge from the bond charge region along the surface. The calculational procedure is illustrated in Fig. 8. In addition to the pseudopotential itself, it is necessary to input the position of the atoms at the surface. If the ideal arrangement of Figure 11 is assumed, this surface has metallic properties. However, the surface is, in reality, a semiconducting surface, and this behaviour originates from the reconstruction of the surface.

It is believed that the geometry of the Si (111) surface reconstruction can essentially be described by a buckled surface (Fig. 12) with alternate layers of atoms being raised and lowered respectively. Electronic calculations based on this model (Schlüter *et al.*, 1975) yield a semiconducting surface, and other calculated properties are in agreement with measurements. Low-energy electron diffraction data (LEED) is the most common tool used to determine

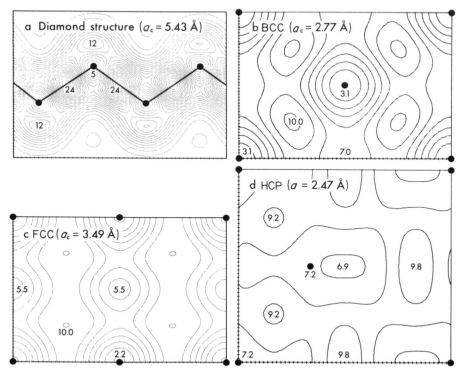

**Fig. 9.** The total valence electronic charge density contours for bulk silicon in the diamond, bcc, fcc, and hcp structures near their respective equilibrium lattice constants. The contour maps of the electronic charge density are in the (110) plane for the diamond, bcc, and fcc structures and in the (1$\bar{1}$20) plane for hcp. Heavy dots represent atomic sites. Normalization corresponds to eight electrons per cell.

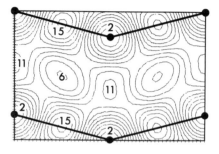

**Fig. 10.** The total valence electronic charge density contours for bulk silicon in the white tin structure. The plot is presented in the (010) plane. Heavy dots represent atomic sites, and solid lines connect nearest neighbors. Normalization corresponds to eight electrons per cell.

**Fig. 11.** Total valence electronic charge distribution for an unrelaxed Si (111) surface. The charge is plotted as contours in a (110) plane intersecting the (111) surface at right angles. The plotting area starts in the vacuum and extends about $4\frac{1}{2}$ atomic layers into the crystal. Atomic positions and bond directions are indicated by dots and heavy lines. Contours are normalized to electrons per Si bulk unit cell volume.

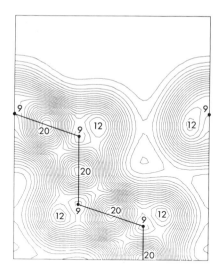

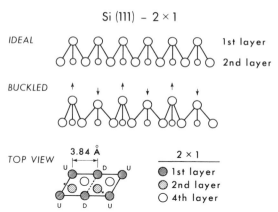

Si (111) – 2 × 1

**Fig. 12.** Schematic representation of the ideal and (2 × 1) reconstructed Si (111) surface. The buckled configuration is indicated by the arrows. There are slight lateral shifts of second-layer atoms.

surface structure, but often the results are not definitive, and the calculations associated with the measurements are difficult. It would, therefore, be desirable to be able to compute surface structures using total energy calculations similar to those described above for the bulk.

At present, an accurate pseudopotential calculation of the Si (111) buckled geometry has not been done, but a calculation of the effect of relaxation of the outermost atoms on a Si (111) surface has been completed (Ihm and Cohen,

1979; Ihm and Cohen, 1980). This calculation yields an optimal relaxation distance of 0.16 Å in reasonable agreement with experiment. The calculated energy gain is 0.13 eV per atom.

For a reconstructed surface, a total energy calculation has been done for the Si (001) surface (Ihm *et al.*, 1980). This self-consistent pseudopotential calculation uses the results of a tight-binding calculation (Chadi, 1979) to provide some likely geometries to test for this surface. There are several choices for the reconstruction. It is thought that the atoms on a Si (001) surface pair or dimerize. Models using symmetric dimer geometries where the two atoms in the pair move in an equivalent manner have dominated the proposed models for this surface. Chadi (1979), using a tight-binding approximation for the electronic structure, estimated that allowing the atoms in the dimer to move asymmetrically could lead to a lowering of the total energy. The calculation by Ihm *et al.* (1980) verifies the tight-binding results and estimates that ∼0.12 eV per dimer is gained in going from a symmetric to asymmetric geometry. The charge density for the asymmetric dimer model is shown in Fig. 13; this plot clearly displays the strong covalent bond charge which exists between the members of a pair on this surface.

To date, the most extensive total energy structural calculations done using the "direct approach" are those of Yin and Cohen (1980). The SCP determined by Hamann *et al.* (1979) was used, and in principle, the only input for the calculation is the atomic number for Si and the fact that Si crystallized in the

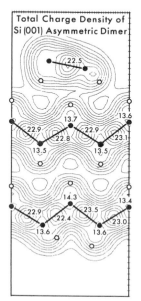

**Fig. 13.** Contour plot of the total valence charge density for the asymmetric dimer model of Si (001) plotted in a (110) plane cutting the surface at right angles. The solid circles represent the Si atoms lying on the plane, and the solid lines represent the hypothetical covalent bonds. Si atoms not on the plane are denoted by open circles. Normalization is determined by the number of electrons per bulk Si unit cell ($\omega = 270$ a.u.), and the contour spacing is 2.

**TABLE I**

**Comparison of Calculated and Measured Structural Properties of Silicon**

|  | Lattice constant (Å) | Crystal energy (Ry) | Cohesive energy (eV) | Bulk modulus ($10^{12}$ erg/cm$^3$) | Shear modulus $(C_{11}-C_{12})$ ($10^{12}$ erg/cm$^3$) |
|---|---|---|---|---|---|
| Calculated | 5.46 | −7.909 | 4.67 | 0.97 | 1.09 |
| Experiment | 5.43 | −7.919 | 4.63 | 0.99 | 1.11 |
| Deviation | 0.5% | −0.1% | 0.9% | −2% | −2% |

diamond structure. This calculation is done in the same manner as that of Ihm and Cohen (1980), but the potential used is nonlocal and required to yield more accurate wavefunctions near the core region. In other words, the calculation is based on a density-functional formalism, and because the pseudopotential is used, the "frozen-core" approximation is also assumed.

Both the equilibrium structural properties and the lattice dynamical properties of silicon were studied. For the equilibrium properties, the calculated results for the equilibrium lattice constant, cohesive energy, bulk modulus, and shear modulus are within 2% of the experimental values. The calculated and measured values are given in Table I. References for the experimental results are in Yin and Cohen (1980). There is significant improvement over earlier work (Morita *et al.*, 1972; Wendel and Martin, 1979; Ihm and Cohen, 1979; Vergés and Tejedor, 1979). The improvement in the bulk modulus results are particularly striking.

The dynamical calculation consisted of a determination of the lattice vibrational frequencies at the $\Gamma$ (0, 0, 0) and X (1, 0, 0) points in the Brillouin zone. The results are given in Table II; references to the experimental measurements are in Yin and Cohen (1980). The agreement between theory and experiment is within 3%. This is particularly impressive when one considers that the low-lying TA phonon mode at X is ordinarily very difficult to

**TABLE II**

**Comparison of the Calculated Phonon Frequencies ($f_{calc}$) with the Experimental Values ($f_{expt}$)**

|  | LTO($\Gamma$) | TA(X) | TO(X) | LOA(X) |
|---|---|---|---|---|
| $f_{calc}$ | 15.16 (−2.4%) | 4.45 (−0.9%) | 13.48 (−3.0%) | 12.16 (−1.3%) |
| $f_{expt}$ | 15.53 | 4.49 | 13.90 | 12.32 |

[a] The deviations from $f_{expt}$ values are presented in parentheses. All values are given in units of $10^{12}$ Hz.

reproduce even with adjustable parameters (Weber, 1977). The volume dependence of the phonon frequencies was also calculated and found to be in excellent agreement with experiment.

### III. CONCLUSIONS AND THE FUTURE

It would be aesthetically pleasing to report that one unique pseudopotential exists for each element and all the properties of the elements and compounds can be computed from these potentials, but this cannot be done at this time. The outstanding features of each type of potential is as follows. The empirical pseudopotentials give the best results for electronic band structures; the soft-core local pseudopotentials yield the latest definitive results for surfaces and interfaces; the hard-core pseudopotentials yield core radii and structural indicies; and the nonlocal soft-core potentials give the structural and ground state properties. However, some of the above distinctions are made because of convenience and economics. What are the real limitations connected with the goal of having one unique pseudopotential?

The division between local and nonlocal and soft-core versus hard-core is artificial and based on convention, convenience, and computational ease. Given sufficient computer power, the nonlocal HCP of the type used by Zunger and Cohen (1978) can be used for surfaces, interfaces, total energy, and all ground state properties. The same is true of the nonlocal SCP of the type developed by Hamann *et al.* (1979) and by Kerker (1980). The Appelbaum and Hamann or Berkeley local potentials are limited. They are used currently primarily because of ease of calculation. It is also not clear that it is the local versus nonlocal aspect of these potentials which is critical. Joannopoulos and coworkers have had considerable success using local potentials (e.g., Starkloff and Joannopoulos, 1977). The real distinction is between the empirical and density-functional pseudopotentials.

The empirical pseudopotentials have experimental input and produce excellent results for the excited-state properties of crystals (e.g., optical properties). These potentials generally do not give accurate ground-state or structural properties, but this aspect is unclear since no effort has been made to develop empirical potentials to reproduce both excited- and ground-state properties. The density-functional pseudopotentials accurately reproduce the results of the all-electron calculations, and these pseudopotential calculations are, in this sense, "first-principles" calculations. However, as expected, the density-functional pseudopotentials have the same limitations as the all-electron calculations based on the density-functional approach.

One specific problem in this area is "the band gap" problem. The density-functional approach yields an energy gap which is too small. For example, in Si, the empirical pseudopotentials reproduce the experimental gap of ~1.1 eV whereas the DFP calculations and all-electron calculations give values which are ~0.5 eV. In defense of the DFP approach, it should be emphasized that these calculations were designed for ground-state properties, and from this point of view, they are successful. However, at present, this limitation prevents researchers in this area from claiming that one pseudopotential "does it all." I expect significant effort in this area, and hopefully, the various approaches will coalesce.

Returning to my comments at the beginning of this chapter, can the approach be made more global? Will the calculations always be limited to simple structures? It is difficult to expect any quantum-mechanical scheme of this complexity to be as flexible as the more empirical approaches like the use of Pauling's rules. For 50 years, researchers have used Pauling's rules, radii, and bond lengths with success, and there is good reason to believe that this approach will continue to be valuable. However, the pseudopotential should begin to contribute more significantly. The techniques are not difficult, but they need to be standardized. Some empirical rules and insights should emerge once more complex systems are studied.

Therefore, on the empirical and qualitative side, it is likely that more extensive calculations will reveal the regularities which researchers can absorb and use to make qualitative predictions. On the detailed, precise, calculational end, there should be significant progress. We are told that in the next half dozen years the computing power of a CDC 7600 will be available to individuals in desk-top-sized containers. If these extrapolations are even approximately correct, calculations of the type described in this paper could be easily done by a variety of researchers.

In the near future, it is reasonable to expect more studies of the type done by Ihm and Cohen (1979) and Yin and Cohen (1980). Materials other than Si should be tested. It should also be possible to calculate pressure phase diagrams. If the structures studied are not too close in energy, then current models should be capable of reproducing the measured results. Surface reconstructions are also within the reach of the current calculations. A good prototype would be the Si (111) buckled surface. It may be possible to compute the magnitudes of the parameters determining the buckled geometry. These calculations, if successful, could augment LEED measurements or even replace them in some cases.

## ACKNOWLEDGMENTS

This work was supported by the National Science Foundation Grant No. DMR7822465 and Division of Materials Sciences, Office of Basic Energy Sciences, U.S. Department of Energy Grant No. W-7405-ENG-48.

**REFERENCES**

Andreoni, W., Baldereschi, A., Meloni, F., and Phillips, J. C. (1978). *Solid State Commun.* **25**, 245.
Appelbaum, J. A., and Hamann, D. R. (1976). *Rev. Mod. Phys.* **48**, 479.
Chadi, D. J. (1979). *Phys. Rev. Lett.* **43**, 43.
Chelikowsky, J. R., and Cohen, M. L. (1976). *Phys. Rev. B* **14**, 556.
Chelikowsky, J. R., and Phillips, J. C. (1978). *Phys. Rev. B* **17**, 2453.
Cohen, M. L. (1973). *Science* **179**, 1189.
Cohen, M. L., and Bergstresser, T. K. (1966). *Phys. Rev.* **141**, 739.
Cohen, M. L., and Heine, V. (1970). *Solid State Phys* **24**, 38.
Hamann, D. R., Schlüter, M., and Chiang, C. (1979). *Phys. Rev. Lett.* **43**, 1494.
Harrison, W. A. (1980). "Electronic Structure and the Properties of Solids." Freeman, San Francisco, California.
Hohenberg, P. C., and Kohn, W. (1964). *Phys. Rev.* **136**, 864.
Ihm, J., and Cohen, J. L. (1979). *Solid State Commun.* **29**, 711.
Ihm, J., and Cohen, M. L. (1980). *Phys. Rev. B* **21**, 1527.
Ihm, J., Zunger, A., and Cohen, M. L. (1979). *J. Phys. C* **12**, 4409.
Ihm, J., Cohen, M. L., and Chadi, D. J. (1980). *Phys. Rev.* **B21**, 4592.
Kerker, G. P. (1980). *J. Phys. C* **13**, L189.
Kohn, W., and Sham, L. J. (1965). *Phys. Rev.* **140**, 1133.
Lam, P. K., Cohen, M. L., and Zunger, A. (1980). *Phys. Rev.* **B22**, 1698.
Littlewood, P. (1979). *J. Phys. C* **12**, 4441.
Louie, S. G., Ho, K. M., and Cohen, M. L. (1979). *Phys. Rev. B* **19**, 1774.
Machlin, E. S., Chow, T. P., and Phillips, J. C. (1977). *Phys. Rev. Lett.* **38**, 1292.
Miedema, A. R. (1976). *Phillips Tech. Rev.* **36**, 217.
Morita, A., Soma, T., and Takeda, J. (1972). *J. Phys. Soc. Jpn.* **32**, 29.
Moruzzi, B. L., Janak, J. F., and Williams, A. R. (1978). "Calculated Electronic Properties of Metals." Pergamon, Oxford.
Penn, D. R. (1962). *Phys. Rev.* **128**, 2093.
Phillips, J. C. (1970). *Rev. Mod. Phys.* **42**, 317.
Phillips, J. C. (1973). "Bonds and Bands in Semiconductors." Academic Press, New York.
Phillips, J. C., and Kleinman, L. (1959). *Phys. Rev.* **116**, 287.
St. John, J., and Bloch, A. N. (1974). *Phys. Rev. Lett.* **33**, 1095.
Schlüter, M., and.Chelikowsky, J. R., Louie, S. G., and Cohen, M. L. (1975). *Phys. Rev. B* **12**, 4200.
Simons, G., and Bloch, A. N. (1973). *Phys. Rev. B* **7**, 2754.
Starkloff, T., and Joannopoulos, J. D. (1977). *Phys. Rev. B* **16**, 5212.
Vergés, J. A., and Tejedor, C. (1979). *Phys. Rev.* **20**, 4251.
Weber, W. (1977). *Phys. Rev.* **15**, 4789.
Wendel, H., and Martin, R. M. (1979). *Phys. Rev. B* **19**, 5251.
Yang, Y. W., and Coppens, P. (1974). *Solid State Commun.* **15**, 1555.
Yin, M. T., and Cohen, M. L. (1980). *Phys. Rev. Lett.* **45**, 1004.
Zunger, A., and Cohen, M. L. (1978). *Phys. Rev. B* **18**, 5449.
Zunger, A., and Cohen, M. L. (1979). *Phys. Rev. B* **20**, 4082.

# 4

# Quantum-Defect Orbital Radii and the Structural Chemistry of Simple Solids

## AARON N. BLOCH and GINA C. SCHATTEMAN

| | | |
|---|---|---:|
| I. | Introduction . . . . . . . . . . . . . . . . . . . . . . . . | 49 |
| II. | Review of Some Fundamental Concepts . . . . . . . . . . . . | 50 |
| | A. The Simons Potential, Orbital Quantum Defects, and Orbital Core Radii . . . . . . . . . . . . . . . . . . . . . . | 50 |
| | B. Chemical Content of the Orbital Radii . . . . . . . . . | 53 |
| III. | Relationship of the Orbital Radii to Hartree–Fock Wavefunctions | 55 |
| IV. | Orbital Electronegativities and Renormalized Orbital Radii . . . | 57 |
| | A. Quantum-Defect Electronegativity Scale . . . . . . . . . | 57 |
| | B. Extrapolation and Renormalization of the Orbital Radii . . . | 59 |
| V. | The Problem of the Octet Binary Compounds . . . . . . . . . | 63 |
| | A. Structural Coordinates . . . . . . . . . . . . . . . . . | 63 |
| | B. Elastic Constants in Tetrahedral Structures . . . . . . . | 64 |
| | C. Distortions of the Wurtzite Structure . . . . . . . . . . | 68 |
| | D. Structural Maps . . . . . . . . . . . . . . . . . . . . | 69 |
| VI. | Conclusions . . . . . . . . . . . . . . . . . . . . . . . . | 70 |
| | References . . . . . . . . . . . . . . . . . . . . . . . . | 71 |

## I. INTRODUCTION

Some time ago, we observed that chemical trends in the structural properties of many simple solids are monitored with remarkable precision by a set of purely atomic parameters (Bloch and Simons, 1972; St. John and Bloch, 1974). These are the orbital core radii $r_l$, defined for a one-electron ion by the model pseudopotential of Simons (1971a,b), and determined directly from spectroscopic quantum defects through simple algebraic relations (Simons, 1971a; Simons and Bloch, 1973). Among other results, we found that the orbital radii (in various combinations) established an index of hybrid character, a reasonable scale of orbital electronegativities, a measure of certain structural distortions, and a set of coordinates for maps which quantitatively delineate the most stable structures of 93 nontransition elements and octet binary compounds.

Structure and Bonding in Crystals, Vol. I

Although these findings are consistent with intuition and with qualitative physical reasoning, their microscopic basis has remained poorly understood. More recently, however, a series of extensions, modifications, and critiques of our work have appeared which have begun to contribute toward new insight. First, Phillips and co-workers (for example, Machlin *et al.*, 1977; Chelikowsky and Phillips, 1978) have applied the orbital radii to new problems and offered some new interpretations. Second, Andreoni *et al.* (1978, 1980) have raised important questions concerning the validity of the whole scheme in light of discrepancies between the Simons and Hartree–Fock wavefunctions. Finally, Zunger and Cohen (1978) have introduced an analogous set of orbital radii derived from their density-functional pseudopotentials for neutral atoms. These lack the computational and physical simplicity of the originals, but they do bring the transition elements into the scheme for the first time. Zunger (1980; also this volume) has thus been able to use them to extend the embrace of our simple ideas to nearly 500 compounds.

In light of these developments the time appears right for a reexamination and refinement of our original concepts. This we attempt in the following sections. In Sec. II we review the basic reasoning which leads from the Simons potential through orbital quantum defects to the core radii, and we sample some of the chemical trends in $r_l$. Section III reexamines the core radii in light of the critique by Andreoni *et al.* (1978, 1980), and shows that they are indeed a valid measure of core size and strength as reflected in the structure of the Hartree–Fock wavefunction. From this perspective $r_l^{-1}$ emerges in Sec. IV as the $l$th orbital component of a Gordy-type electronegativity. By using this orbital electronegativity scale to correct the radii for the energy dependence of the pseudopotential in the post-transition elements, we remove the chemically unrealistic irregularities encountered by Chelikowsky and Phillips (1978). The corrected radii are used in Sec. V to generate a new set of structural coordinates, loosely based upon heteropolar and homopolar contributions to the binding energies of simple binary compounds. We find these coordinates to be accurate indicators of structural distortions, elastic constants, bond charges, and the relative stabilities of 12 crystal structures among more than 110 binary semiconductors. Finally, Sec. VI summarizes our conclusions.

## II.   REVIEW OF SOME FUNDAMENTAL CONCEPTS

### A.   The Simons Potential, Orbital Quantum Defects, and Orbital Core Radii

For simplicity, we confine our initial attention to compounds of nontransition elements whose core and valence electronic states are well separated in energy. Within a series of such compounds whose valence systems are isoelectronic (for example, the $A^N B^{8-N}$ octet binary semiconductors), chemical trends in structural or physical properties must ultimately originate from underlying trends in the character of the constituent atomic cores (see, for example, Simons, 1971b). Qualitative discussions of these trends are often framed in terms of traditional constructs such as core size and electronegativity. These in turn must bear a close relationship to the range and strength of the potential, $V(r)$, between an isolated core and a single valence electron.

The chemical content of $V(r)$ becomes particularly transparent when the potential is written in the approximate form suggested by Simons (1971a,b):

$$V(r) = -\frac{Z}{r} + \sum_l \frac{l'(l'+1) - l(l+1)}{2r^2} \mathscr{P}_l \tag{1}$$

The first term on the right is the coulomb attraction between the electron and an ionic core of net charge $Z$, as in a hydrogenic system. The second term expresses the departure of the core from simple hydrogenic character, and consists of distinct contributions for each orbital quantum number $l$, with $\mathscr{P}_l$ the $l$-projection operator and $l'$ a dimensionless $l$-dependent parameter.

The accuracy of Eq. (1) is not unreasonable (Simons, 1971a; Simons and Bloch, 1973), but its important virtue lies in rendering the one-electron radial Schrodinger equation exactly solvable. If $R_l(r)$ is the radial part of the electronic wavefunction and $\mu_l(r) \equiv rR_l(r)$, that equation becomes:

$$-\frac{1}{2}\frac{d^2\mu}{dr^2} + \left[\frac{l'(l'+1)}{2r^2} - \frac{Z}{r} - E\right]\mu = 0 \tag{2}$$

Equation (2) is hydrogenic except that $l$ has been replaced with $l'$; the same will be true of many of the expressions characterizing its solutions. The eigenvalues can be written in the quantum-defect form:

$$E[n, l'(l), Z] = \frac{-Z^2}{2(n'+\delta)^2}, \tag{3}$$

where $n' \geqslant l+1$ is an integer and the quantum defect $\delta$ is simply:

$$\delta[n', l'(l)] = l' - l \tag{4}$$

The chemical information contained in the Simons potential clearly resides in the core charge $Z$ and the quantum defect. According to Eqs. (1) and (4), a positive quantum defect implies a repulsive core, and a negative quantum defect an attractive core, for electrons of given $l$. Indeed, the potential parameters for Eq. (1) can be obtained directly from experimental atomic term values using Eq. (3) and (4) (Simons, 1971a; Simons and Bloch, 1973).

This determination requires a choice of the integer $n'$. To appreciate its significance, we consider the solutions of Eq. (2). These are hydrogenic radial wavefunctions generalized for nonintegral $l'$:

$$\mu_l(r) = N^{-\frac{1}{2}}r^{l'+1}F(1 + l - n'|2l' + 2|2\gamma r)\exp(-\gamma r) \tag{5}$$

Here $N$ is a normalization constant, $F$ is the confluent hypergeometric function of the first kind, and the screening constant is:

$$\gamma = \sqrt{-2E} = \frac{Z}{n' + \delta} \tag{6}$$

The wavefunctions of Eq. (5) contain $n' - l - 1$ nodes, and we are free to choose $n' \geqslant l + 1$ for the ground state as we wish (see, for example, Simons, 1974). One obvious choice is $n' = n$, the true principal quantum number. In that case $\mu_l(r)$ becomes an approximation to the true one-electron radial wavefunction, with some instructive consequences that we shall discuss in Sec. III. For the present, however, we concentrate upon the original choice (Simons, 1971a; Simons and Bloch, 1973):

$$n' = l + 1 \tag{7}$$

Where the core does not contain orbitals of quantum number $l$, Eq. (7) is equivalent to the choice $n' = n$, and Eq. (5) again approximates the true (nodeless) radial wavefunction. Where orbitals of quantum number $l$ do appear in the core, however, Eq. (7) has two important consequences. First, according to Eqs. (3) and (4) the value of $l'$, and hence the repulsive barrier of Eqs. (1) and (2), are maximized. Second, $\mu_l(r)$ [Eq. (5)] becomes nodeless, and in fact takes the form of a simple Slater atomic orbital. Clearly, no such function can be orthogonal to the core orbitals of quantum number $l$. Hence $\mu_l(r)$ cannot be the radial part of a true valence wavefunction: it is instead a radial pseudowavefunction, and the corresponding $V(r)$ [Eq. (1)] is a model pseudopotential. It mimics the physical effects of the core-valence ortho-gonality requirement with an artificial "Pauli force" (Simons, 1971a; Simons and Bloch, 1973) that repels valence electrons from the core region. In this sense the Simons potential was probably the first of what have since come to be known as "hard-core" pseudopotentials (see, for example, Simons and Bloch, 1973; Hamann, 1979; Cohen, this volume; Zunger, 1980; and references therein).

The strength of the Pauli-force repulsion is effectively measured by the position $r_l$ of the radial maximum of the nodeless pseudowavefunction $R_l(r)$.

In keeping with the hydrogenic analogy, this is just

$$r_l = \frac{l'(l' + 1)}{Z} \tag{8}$$

The radius $r_l$ is also the position of the classical balance point at which the sum of the repulsive centrifugal and "Pauli" forces exactly cancels the attractive coulomb force.* It is in this sense a core radius.†

## B.  Chemical Content of the Orbital Radii

The quantum-defect orbital radius $r_l$ is a length scale which reflects both the size of the core and the strength and sign of the core-valence interactions. It therefore serves as an index of core effects upon structural chemistry in ways that conventional energy scales cannot (Bloch and Simons, 1972; St. John and Bloch, 1974). To illustrate, consider the variation of $r_l$ with $l$ over the alkaline-earth series of elements ($Z = 2$) as displayed in Fig. 1. [Much of our discussion parallels that of Simons (1971b).]

For reference, Fig. 1 includes the orbital radii for the $He^+$ ion, which of course follow the hydrogenic sequence $r_l = 0, 1, 3, 6 \ldots$ for $l = 0, 1, 2, 3 \ldots$. The behavior of the other elements in the series is rather different. For $l = 0$, the "Pauli force" of the 1s core orbitals of $Be^+$ produces a positive quantum defect and a substantial $r_0$, and the increase of core size over the rest of the series is reflected in sequence. For $l = 1$, however, this effect cannot occur in $Be^+$, whose core contains no p electrons. Here the most important departure from hydrogenicity is the attractive charge-induced dipole interaction (Simons, 1971a; Simons and Bloch, 1973) between a valence p-electron and the $1s^2$ core; this is a weak effect but enough to reverse the sign of the quantum defect and reduce $r_l$ for $Be^+$ below the $He^+$ value. In contrast, the $Mg^+$ core with its $2p^6$ configuration is again repulsive, and the rest of the series again follows in order.

A dramatic departure from this behavior occurs for $l = 2$. As expected, the $r_2$ radii for $Be^+$ and $Mg^+$, with no d electrons in their cores, are slightly less than the hydrogenic value, but still much larger than $r_1$ or $r_0$. This suggests

---

* It is therefore the ratio of moments $\langle r^{-2} \rangle / \langle r^{-3} \rangle$, as may be ascertained by differentiating the terms in brackets in Eq. (2) with respect to $r$ and recalling that the expectation values of the repulsive and attractive forces must be equal for equilibrium to occur.

† Chelikowsky and Phillips (1978) have suggested that the classical turning point for a particle of zero energy, $r_l/2$, is a more appropriate core radius, largely on the grounds that the smaller value falls inside any reasonable solid-state or molecular screening radius. This is an attractive argument, but we continue to prefer our original definition [Eq. (8)], for reasons that will become apparent in Secs. III and IV.

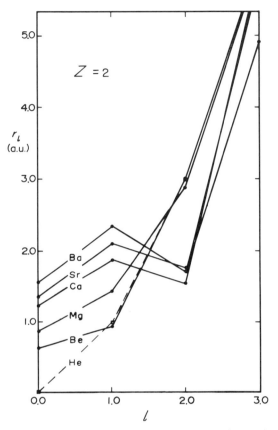

**Fig. 1.**  Quantum-defect orbital radii $r_l$ vs. $l$ for alkaline earth metals.

that d-states in these elements are too diffuse to play an effective role in bonding (Simons, 1971b; Bloch and Simons, 1972), consistent with their known propensity for s and s–p hybrid bonding. The elements Ca, Sr, and Ba, on the other hand, stand at the threshold of the transition series, and possess low-lying d-excited states: their cores are but one charge short of capturing a d-electron in the ground state. Because of the large resulting core polarizabilities, $r_2$ for these elements is reduced below $r_1$, and in the case of $Ba^+$ it approaches $r_0$. Such behavior suggests that the s–d promotion energies are small and that the s and d wavefunctions are comparable in spatial extent; it therefore implies a strong tendency toward s–d hybridization. This tendency is evident, for example, in the geometries of the molecular dihydrides of these elements (Simons, 1971b).

Based upon reasoning such as this, Bloch and Simons (1972) and St. John and Bloch (1974) were able to construct algebraic combinations of the $r_l$ which distinguished among the crystal structures of a substantial number of simple solids, and also described certain classes of crystal distortions. Since that time, other work has raised some questions and underscored some deficiencies in the scheme which have had to be dealt with before it could properly be extended. We consider these in the next two sections.

### III.  RELATIONSHIP OF THE ORBITAL RADII TO HARTREE–FOCK WAVEFUNCTIONS

The nodeless pseudowavefunction obtained by applying Eq. (7) to Eq. (5) does not necessarily resemble the best Hartree–Fock wavefunction for the same one-electron problem (Fig. 2). This discrepancy has been emphasized by Andreoni et al. (1978, 1980), who rectified the situation by multiplying the second term of $V(r)$ [Eq. (1)] by an exponential damping factor. They thereby sacrificed the analyticity of the original Simons potential, but with the aid of additional parameters and some computational labor they were able simultaneously to generate wavefunctions in close agreement with Hartree–Fock and a set of orbital radii which produced structural separations comparable with those of St. John and Bloch.

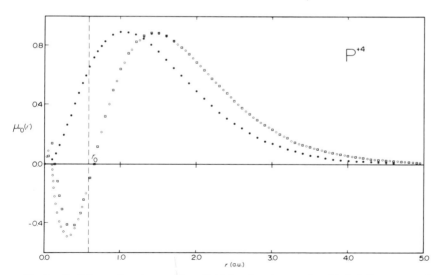

**Fig. 2.**   Radial wavefunctions $\mu_0(r) = rR_0(r)$ for phosphorous 3s orbital: ●-Simons pseudo-wavefunction, $n' = 1$;  ○-Simons wavefunction, $n' = 3$;  □-Hartree–Fock wavefunction (Clementi).

Except for the simplicity of the original Simons–Bloch radii and their present availability for a wider range of elements, we have found no practical reason to prefer one set of radii over the other, although we have not pursued the point in detail. More interesting, in our view, is an intriguing question which follows from the criticism by Andreoni et al.: since the Simons–Bloch orbital radii are derived from such a poor wavefunction, what is the source of their proven chemical utility?

The roots of the answer lie in an additional paper by Simons (1974). In a study of atomic oscillator strengths, he found that when the choice $n' = n$ is made in place of Eq. (7), the resulting wavefunction [Eq. (5)] is in excellent agreement with Hartree–Fock, without any adjustment of parameters.* The agreement is illustrated for the 3s wavefunction of $P^{+4}$ (a randomly chosen case) in Fig. 2.

Now, the $n' = 3$ wavefunction and $n' = 1$ pseudowavefunction of Fig. 2 are constructed using exactly the same experimental information, through Eqs. (3)–(5). It follows that $r_l$, which displays this same information, may be a measure of the accurate wavefunction as well as the inaccurate one. In order to understand how, we ask what quantities with the units of length are carried over unchanged from the "good" wavefunction to the nodeless pseudowavefunction, independent of the choice of $n'$. We find that there are two. The first is the reciprocal of the expectation value of $r^{-1}$:

$$\left\langle \frac{1}{r} \right\rangle^{-1} = -\left(\frac{\partial E}{\partial Z}\right)^{-1} = -\frac{Z}{2E} \tag{9}$$

as can be verified by applying either the Virial Theorem or the Hellmann-Feynman Theorem to Eq. (2). The second is the inverse of the orbital screening constant [Eq. (6)]:

$$\gamma^{-1} = (-2E)^{-\frac{1}{2}} \tag{10}$$

The first length is related to the expectation value of the coulomb potential energy; the second to the total energy. Their difference, which reflects in part the kinetic energy (or the "Pauli-force" repulsion) is exactly $r_l$:

$$r_l = \left\langle \frac{1}{r} \right\rangle^{-1} - \gamma^{-1} \tag{11}$$

according to Eqs. (3), (4), and (7)–(10).

---

* For many of the heavier elements the solution with $n' = n$ diverges at the origin, but by choosing the largest value of $n'$ for which the solution remains regular, one still obtains a pseudowavefunction in good agreement with Hartree–Fock in the chemically important region (Simons 1974).

The orbital radius $r_l$ thereby emerges as the difference between two $n'$-invariant lengths which characterize the "good" wavefunction $n' = n$ as well as the nodeless pseudowavefunction of Eq. (7). In this sense it combines something of the accuracy of the former with the simplicity of the latter.

We also note that, consistent with this discussion and that of the previous section, $r_l$ is related to the nodal structure of the $n' = n$ wavefunction. For $n - l = 1$ it is of course the radial maximum of the nodeless radial wavefunction, and for $n - l = 2$ it coincides precisely with the position of the single node. For larger $n$ the relationship is more complex, but $r_l$ is always close to the outermost node, as illustrated in Fig. 2. A similar observation has been made by W. Andreoni (private communication). It is natural to inquire whether the position of this node may not define a still more useful core radius, but we defer that question for further study.

## IV.   ORBITAL ELECTRONEGATIVITIES AND RENORMALIZED ORBITAL RADII

### A.   Quantum-Defect Electronegativity Scale

The classical work required to remove an electron from the classical balance point $r_l$ to infinity is:

$$- \int_{r_l}^{\infty} \frac{\partial}{\partial r} \left[ -\frac{Z}{r} + \frac{l'(l' + 1)}{2r^2} \right] dr = \frac{1}{2} \frac{Z}{r_l} \tag{12}$$

Then the quantity $1/r_l$ has the units of work per unit charge, or electronegativity (see, for example, Hinze and Jaffé, 1962). Since it is defined per unit core charge rather than electronic charge, $1/r_l$ is not a true chemical potential, but it nevertheless measures the depth of the $l$th potential well. Accordingly it was defined by St. John and Bloch (1974) as an orbital electronegativity:

$$X_l \equiv 1/r_l \tag{13}$$

Equation (12) identifies $X_l$ with electronegativities of the Gordy type.

The introduction of electronegativities consisting of explicit orbital components represents a degree of sophistication over more traditional scales. The differences between the $X_l$ are useful in discussions of hybridization (St. John and Bloch, 1974), while their sum defines a total electronegativity for the element:

$$X \equiv a \sum_{l=0}^{\infty} X_l + b \tag{14}$$

The constants $a = 0.328$ and $b = 0.302$ are chosen so as to reproduce the values arbitrarily assigned by Pauling to Li and Be. We prefer Eq. (14) to the

**TABLE I**

**Electronegativities of Nontransition Elements**

|  | Pauling (1960) | Quantum defect (present work) | Miedema (Boom et al., 1976) |
|---|---|---|---|
| Li | 1.0 | 1.00 | 1.00 |
| Na | 0.9 | 0.926 | 0.908 |
| K | 0.8 | 0.832 | 0.782 |
| Rb | 0.8 | 0.815 | 0.755 |
| Cs | 0.7 | 0.799 | 0.715 |
| Be | 1.5 | 1.50 | 1.50 |
| Mg | 1.2 | 1.24 | 1.14 |
| Ca | 1.0 | 1.18 | 0.939 |
| Sr | 1.0 | 1.11 | 0.871 |
| Ba | 0.9 | 1.06 | 0.810 |
| Cu | 1.9 | 1.43 | 1.63 |
| Ag | 1.9 | 1.38 | 1.59 |
| Au | 2.4 | 1.96 | 1.85 |
| Zn | 1.6 | 1.40 | 1.46 |
| Cd | 1.7 | 1.33 | 1.44 |
| Hg | 1.9 | 1.43 | 1.50 |
| B | 2.0 | 2.01 | |
| Al | 1.5 | 1.54 | 1.50 |
| Ga | 1.6 | 1.59 | 1.46 |
| In | 1.7 | 1.50 | 1.39 |
| Tl | 1.8 | 1.52 | 1.39 |
| C | 2.5 | 2.51 | |
| Si | 1.8 | 1.84 | 1.68 |
| Ge | 1.8 | 1.80 | 1.63 |
| Sn | 1.8 | 1.65 | 1.48 |
| Pb | 1.8 | 1.66 | 1.46 |
| N | 3.0 | 3.00 | |
| P | 2.1 | 2.13 | |
| As | 2.0 | 2.00 | 1.72 |
| Sb | 1.9 | 1.85 | 1.57 |
| Bi | 1.9 | 1.81 | 1.48 |
| O | 3.5 | 3.50 | |
| S | 2.5 | 2.42 | |
| Se | 2.4 | 2.21 | |
| Te | 2.1 | 2.04 | |
| Po | 2.0 | 1.95 | |
| F | 4.0 | 4.00 | |
| Cl | 3.0 | 2.70 | |
| Br | 2.8 | 2.39 | |
| I | 2.5 | 2.19 | |
| At | 2.2 | 2.11 | |

original definition of St. John and Bloch (1974) who truncated the summation after $l = 2$.

The quantum-defect electronegativity scale of Eq. (14) is compared with that of Pauling (1960) in Table I. The table also lists the currently fashionable electronegativity parameters (appropriately scaled) of Miedema and co-workers (Boom *et al.*, 1976), who have treated the heats of formation of metal alloys with remarkable success. In general, the quantum-defect scale agrees rather well with that of Pauling, and tends to fall between the Pauling and Miedema scales where those two disagree. Chelikowsky and Phillips (1978) have achieved a somewhat closer fit to the Miedema scale by introducing a set of adjustable weighting factors for the $X_l$, but we see no reason for such an exercise in the context of the present discussion.

## B. Extrapolation and Renormalization of the Orbital Radii

Rather, our principal concern is with the utility of the orbital components $X_l$ in another connection. We observe empirically that a plot of $X_l$ vs. $Z$ across any row of the periodic table is very nearly linear, so long as the definition of the core remains constant. This linearity extends from the left through stripped transition metal ions like $Mn^{+7}$ and from the right (except for $l = 0$) through noble metal atoms like $Cu^\circ$. Two important applications immediately follow.

First, by simple linear extrapolation we can assign values of $X_l$, and hence $r_l$, to elements (such as the heavy halogens) for which the requisite spectroscopic data for a direct determination are not available. The results may be somewhat more reliable than those of the more complicated curve-fitting procedures used by Chelikowsky and Phillips (1978) to extrapolate from plots of $r_l$ vs. $Z$.

Second, we are able to resolve an apparent anomaly in the systematics of the orbital radii. The linearity of the $X_l$ holds over the entire periodic table for all $l > 0$, and on the left side of the table for $l = 0$. To the immediate right of the transition series, however, $X_0$ appears anomalously large, as shown in Fig. 3. This behavior is not difficult to understand. The core of $Cu^\circ$, for example, contains a loosely bound 3d shell, which is penetrated significantly by the valence 4s electron. Under these circumstances the pseudopotential is strongly energy-dependent, and it is no longer reasonable to assert that the 4s electron experiences only the nominal net core charge $Z = 1$ (compare Van Vechten, 1969). Indeed, the use of $Z = 1$ in Eqs. (3) and (8) leads to an artificially small value of $r_0$*. As we move toward the right of

---

* It was for this reason that compounds of the noble metals were excluded from the work of St. John and Bloch (1974), and appeared anomalous in the treatment by Chelikowsky and Phillips (1978).

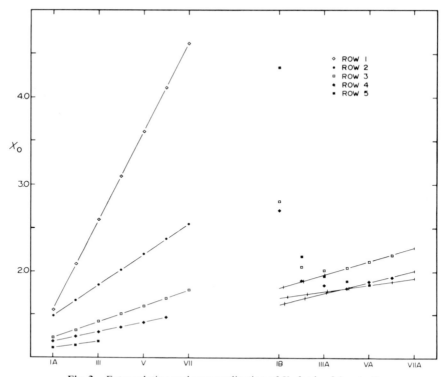

**Fig. 3.** Extrapolation and renormalization of $X_0$ for $l = 0$ (see text).

the periodic table away from the transition series, the inner d-shell becomes more tightly bound with increasing nuclear charge, and the effect is diminished.

We compensate for the difficulty by substituting for $Z$ in Eqs. (3) and (8) an $l$-dependent effective core charge $Z_l$. For $l \geqslant 1$, $Z$ remains equal to $Z$, but for s-states $Z_0$ must be determined using Fig. 3. We exploit the linearity of the $X_l$ by constructing the best straight line through the $X_0$ values on the far right of a given row, and by choosing $Z_0$ for the remainder of the row so as to force $X_0$ to fall on this line. The resulting values of $Z_0$ are displayed in Table II; for the noble metals they are somewhat less severe departures from $Z$ than proposed, for example, by Laewitz (1972).

Through Eq. (13), the new $X_0$ correspond to a renormalized set of $r_0$. For the noble metals these are a factor of 2–3 larger than the original values listed by Chelikowsky and Phillips (1978), and they restore to our scheme much of the chemical regularity of this part of the periodic table. For example,

**TABLE II**

**Effective $l = 0$ Core Charges $Z_0$ for Posttransition Elements**

| Element | $Z_0$ | Element | $Z_0$ | Element | $Z_0$ |
|---------|-------|---------|-------|---------|-------|
| Cu | 1.19 | Ag | 1.24 | Au | 1.38 |
| Zn | 2.10 | Cd | 2.15 | Hg | 2.27 |
| Ga | 3.05 | In | 3.12 | Tl | 3.17 |
| Ge | 4.02 | Sn | 4.01 | Pb | 4.10 |
| As | 5.00 | Sb | 5.00 | Bi | 5.03 |

the renormalized $r_0$ correlate closely with the tetrahedral radii deduced by Pauling (1960), as shown in Fig. 4.

The full set of available orbital radii, including extrapolated values and the renormalized $r_0$, appears in Table III. The radii listed for the transition elements are intended for use only in cases, such as ScN and Hf C, in which the d-electrons clearly act as part of the valence system.

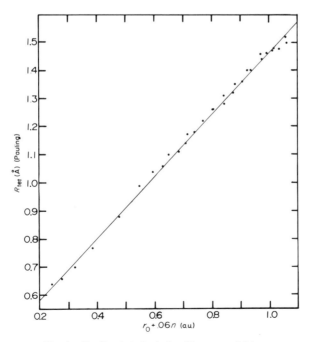

**Fig. 4.**   Pauling tetrahedral radii vs. $r_0 + 0.06n$.

**TABLE III**

**Renormalized Orbital Radii**

| | $r_0$ | $r_1$ | $r_2$ | | $r_0$ | $r_1$ | $r_2$ |
|---|---|---|---|---|---|---|---|
| Li | 0.934 | 1.879 | 5.991 | C | 0.384 | 0.472 | 1.498 |
| Na | 1.020 | 2.365 | 5.947 | Si | 0.657 | 0.888 | 1.420 |
| K | 1.363 | 2.757 | 5.291 | Ge | 0.651* | 0.919 | 1.681 |
| Rb | 1.453 | 2.947 | 4.889 | Sn | 0.761* | 1.068 | 1.868 |
| Cs | 1.624 | 3.176 | 3.961 | Pb | 0.751* | 1.086 | 1.982 |
| | | | | | | | |
| Be | 0.630 | 0.932 | 2.995 | N | 0.322 | 0.381 | 1.199 |
| Mg | 0.863 | 1.433 | 2.923 | P | 0.589 | 0.759 | 1.128 |
| Ca | 1.221 | 1.869 | 1.519 | As | 0.620 | 0.834 | 1.408 |
| Sr | 1.356 | 2.088 | 1.741 | Sb | 0.722 | 0.966 | 1.557 |
| Ba | 1.553 | 2.341 | 1.707 | Bi | 0.730* | 1.025 | 1.784 |
| | | | | | | | |
| Sc | 1.097 | 1.504 | 0.908 | O | 0.277 | 0.319 | 0.999 |
| Y | 1.252 | 1.732 | 1.181 | S | 0.534 | 0.665 | 0.936 |
| La | 1.452 | 1.982 | 1.297 | Se | 0.591 | 0.764 | 1.212 |
| | | | | Te | 0.695 | 0.888 | 1.347 |
| Ti | 0.996 | 1.286 | 0.698 | Po | 0.710[†] | 0.980[†] | 1.670[†] |
| Zr | 1.179 | 1.546 | 0.972 | | | | |
| Hf | 1.370 | 1.800 | 1.100 | F | 0.243 | 0.274 | 0.856 |
| | | | | Cl | 0.488 | 0.592 | 0.799 |
| Cu | 0.761* | 1.607 | 5.888 | Br | 0.565[†] | 0.705[†] | 1.064[†] |
| Ag | 0.877* | 1.663 | 5.903 | I | 0.666[†] | 0.822[†] | 1.188[†] |
| Au | 0.820* | 1.396 | 5.867 | At | 0.690[†] | 0.80 | |
| | | | | | | | |
| Zn | 0.722* | 1.220 | 3.064 | | | | |
| Cd | 0.835* | 1.322 | 3.176 | | | | |
| Hg | 0.796* | 1.260 | 3.239 | | | | |
| | | | | | | | |
| B | 0.477 | 0.625 | 1.996 | | | | |
| Al | 0.743 | 1.084 | 1.916 | | | | |
| Ga | 0.685* | 1.037 | 2.149 | | | | |
| In | 0.795* | 1.170 | 2.315 | | | | |
| Tl | 0.773* | 1.165 | 2.407 | | | | |

* Calculated using effective $Z_0$ values from Table 2 (see Fig. 3 and text).
† Calculated from linear extrapolation of $X_l$ vs. $Z$ (see Fig. 3 and text).

## V.  THE PROBLEM OF THE OCTET BINARY COMPOUNDS

### A.  Structural Coordinates

Rather than sample the range of problems available for study using the orbital radii, we examine a single class of materials in detail. For simplicity and for comparison with earlier work, we choose the familiar octet binary compounds $A^N B^{8-N}$.

We seek linear combinations of the $r_l$ which bear upon the structural and physical properties of the materials. One set of such combinations was introduced by St. John and Bloch (1974) on inituitive grounds, and has been used routinely since (e.g., Machlin *et al.*, 1977; Chelikowsky and Phillips, 1978; Zunger, 1980). Here we adopt a slightly different approach.

Realizing that perturbation theory is not adequate for the problem, we nevertheless consider the weighting given $r_l$ in second-order structural energies. These are (e.g., Heine and Weaire, 1970):

$$U_2 = \sum_g V_g^2 S_g^2 \chi_g \tag{15}$$

where $g$ is a reciprocal lattice vector, $V_g$ the Fourier transform of $V(r)$ at wavevector $g$, $S_g$ the structure factor, and $\chi_g$ the static density–density response function of the electron gas. Now, for an octet binary compound AB in a simple structure, the important reciprocal lattice vectors fall near the Fermi wavenumber $k_F$ and near $2k_F$, and the structure factors are such that, roughly speaking (Heine and Weaire, 1970):

$$V_g S_g \sim V_g(A) + V_g(B), \qquad g \sim 2k_F \tag{16a}$$
$$V_g S_g \sim V_g(A) - V_g(B), \qquad g \sim k_F \tag{16b}$$

Clearly Eq. (16a), which averages the contributions of atoms A and B, represents a homopolar contribution to the bonding, while the contribution of Eq. (16b) is heteropolar and vanishes when A = B.

We now apply the form factor $V_g$ for the Simons potential, Eq. (1). This can be written (Simons and Bloch, 1973):

$$V_g(A) - V_g(A)^{\text{coulomb}} = C \sum_l [r_l(A) - r_l(H)] P_l \left(1 - \frac{g^2}{2k_F^2}\right)$$
$$\equiv C[v_g(A) - v_g(H)] \tag{17}$$

where $C$ depends only on the valence electron density and $P_l$ is a Legendre polynomial.

The chemically sensitive portion of Eq. (17) is the quantity $V_g(A)$. Near the important values of $g$, this becomes:

$$v_{2k_F} = r_0 - r_1 + \cdots \tag{18a}$$
$$v_{k_F} = r_0 + \tfrac{1}{2} r_1 - \tfrac{1}{8} r_2 + \cdots \tag{18b}$$

We now combine Eqs. (16), (17), and (18) to define the chemical coordinates

$$\bar{R}_H \equiv \tfrac{1}{2}\{(r_1 - r_0)_A + (r_1 - r_0)_B\} \tag{19}$$

and

$$\Delta R_{CT} \equiv (r_0 + \tfrac{1}{2}r_1 - \tfrac{1}{8}r_2)_A - (r_0 + \tfrac{1}{2}r_1 - \tfrac{1}{8}r_2)_B \tag{20}$$

The first coordinate derives from the homopolar contribution, and is identical with one of the St. John–Bloch (1974) coordinates, later dubbed $R_\pi$ by Phillips (e.g. Chelikowsky and Phillips, 1978). According to the arguments of Bloch and Simons (1972), St. John and Bloch (1974), and Sec. II,B, measures the average propensity of the two atoms for s–p hybridization. It has therefore been truncated after $l = 1$ for systems in which only s–p bonding is important.

We expect the heteropolar coordinate, $\Delta R_{CT}$, to bear upon the degree of charge transfer between atoms A and B. It is less simple than the corresponding St. John–Bloch coordinate in part because we have arbitrarily retained the d-term so as to reflect the d-contributions to core polarizabilities and electronegativities.

The footing of the new coordinates is of course by no means rigorous, but it does reinforce the physical interpretations of St. John and Bloch. In the next two subsections we test these interpretations for each coordinate respectively, and in Sec. V,D we combine the two to form a structural map.

## B. Elastic Constants in Tetrahedral Structures

To the extent that the homopolar parameter $\bar{R}_H$ does measure a propensity for s–p hybridization (Bloch and Simons, 1972), it should index the effectiveness of the $sp^3$ covalent bonding in tetrahedral semiconductors. Martin (1970, 1972) has discussed how such bonding entails a nonvanishing ratio of the noncentral (bond-bending) to central (bond-stretching) short-range force constants, and how the growth of this ratio with effectiveness of the bonding is reflected in the shear elastic constants of the material.

In a preliminary study, J. St. John (unpublished work) compared these elastic constants with $\bar{R}_H$. She found that a strong correlation exists, but that it is skewed slightly by a size effect as one moves down an anion column of the periodic table. To compensate for anion size, St. John introduced a weighting of the components of $\bar{R}_H$ according to principal quantum number:

$$\bar{R}_H^* \equiv \frac{2}{n_A + n_B}\{n_A[r_1(A) - r_0(A)] + n_B[r_1(B) - r_0(B)]\} \tag{19}$$

The reduced shear constants $C_S^*$ and $C_{44}^*$ defined by Martin (1970) for zinc blende structures are plotted against $\bar{R}_H^*$ in Fig. 5. Also included are the theoretical cubic constants which he deduced (Martin, 1972) by a unitary

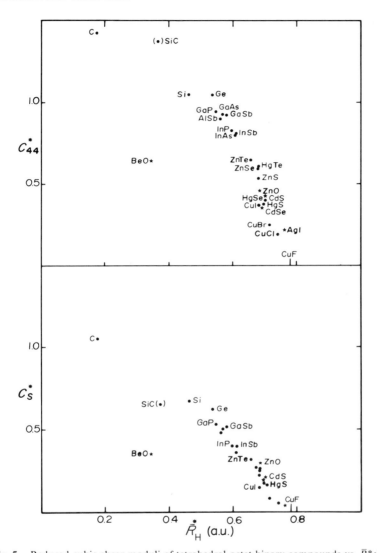

**Fig. 5.**   Reduced cubic shear moduli of tetrahedral octet binary compounds vs. $\bar{R}_H^*$: ●-zinc blende structures; ★-theoretical cubic shear moduli for wurtzite structures (Martin, 1972).

transformation from the wurtzite structure. With increasing $\bar{R}_H^*$, the effectiveness of the $sp^3$ hybridization is diminished (Bloch and Simons, 1972), and the lattice softens. Beginning with diamond, the rate of softening is gradual at first, but accelerates rapidly as $\bar{R}_H^*$ approaches what appears to be a critical value where the elastic constants vanish near 0.77. Just below this value, tetrahedral structures are very soft and are usually ionic conductors, such as

AgI; above the critical value the structure is unstable, in effect too soft to exist.

The only apparent exceptions to the otherwise smooth trends of Fig. 5 are SiC in the $C_S^*$ plot and BeO in both plots. The former was deduced theoretically from a polytype (see Martin, 1972), and may be less certain than the others. The latter are also theoretical cubic constants, deduced from the BeO wurtzite structure. It is possible to draw a smooth curve, separate from the zinc blende curve but vanishing at the same critical value, through BeO and the other wurtzite structures in Fig. 5.

Except that it gives a better account of the first-row elements, Fig. 5 is comparable with plots prepared by Martin (1970, 1972) on the basis of Phillips ionicities (e.g., Van Vechten, 1969). The Phillips ionicity, however, is deduced *a posteriori* from the solid-state properties of materials which already exist, whereas the orbital radii are atomic properties which can be used to help predict whether a given material will exist or not. For example, we note that tetrahedral CuF would lie just beyond the critical $\bar{R}_H^*$ in Fig. 5. As discussed by Chelikowsky and Phillips (1978), CuF probably does not exist as an iso-

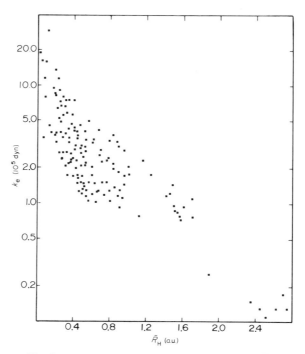

**Fig. 6.** Force constants of diatomic molecules vs. $\bar{R}_H$.

lated phase under ordinary conditions. On the other hand, since it is known that CuCl softens under pressure, Fig. 5 suggests that solid solutions of CuCl and CuF might mimic high-pressure CuCl at ordinary pressures.

It remains for us to verify that the trend of Fig. 5 is a real quantum-chemical effect, and not simply a reflection of atomic size. For comparison, we plot in Fig. 6 the force constants of a large collection of diatomic molecules. Despite considerable scatter a softening with increasing size is evident, but we find no instability like the one signaled by the sharp cutoff in Fig. 5.

Finally, it also follows from our interpretation that $\bar{R}_H^*$ should be correlated with tetrahedral bond charges (e.g., Phillips, 1969). These have been estimated phenomenologically by Velikov et al. (1973), who have used their results to prepare a plot similar to Fig. 5. Since the estimated bond charges and $\bar{R}_H^*$ both correlate well with shear constants, they also correlate with each other, as shown in Fig. 7. For comparison the figure also includes bond charges

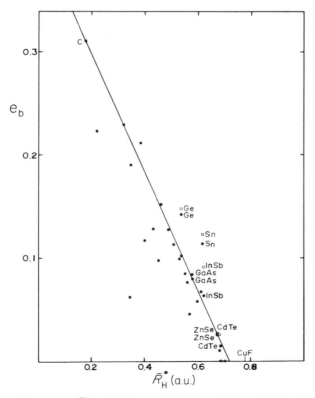

**Fig. 7.**   Bond charge $e_b$ vs. $\bar{R}_H^*$: ●-Velikov et al. (1973); ○-Walter and Cohen (1971); ★-MgS, MgSe.

determined by the pseudopotential calculations of Walter and Cohen (1971). The expected trend—a decrease of the bond charge with increasing $\bar{R}_{\mathrm{H}}^{*}$ until it vanishes at the critical value—is evident.

## C.  Distortions of the Wurtzite Structure

We now seek a test of the second heteropolar coordinate, $\Delta R_{\mathrm{CT}}$. According to our discussion in Sec. V,A, this should be related to the degree of charge transfer from one atom to the other in a simple binary compound. Now, Laewitz (1972) has discussed the role of this charge transfer in determining the minute departures of the $c/a$ ratios in wurtzite structures from their ideal value of 1.633. Further, it was observed by St. John and Bloch (1974) and reiterated in detail by Chelikowsky and Phillips (1978) that the heteropolar St. John–Bloch coordinate correlates closely with these distortions. We seek to determine whether the same is true for $\Delta R_{\mathrm{CT}}$.

In contrast with the earlier studies, our use of the renormalized orbital radii of Table 3 brings to the discussion all 20 wurtzite structures treated by

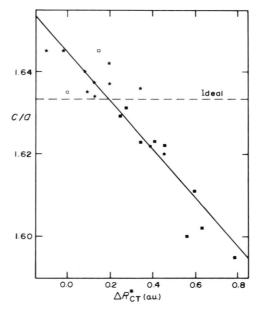

**Fig. 8.**  Axial ratios $c/a$ for wurtzite structures vs. $\Delta R_{\mathrm{CT}}^{*}$: ★-most stable as zinc blende structure; □-most stable as graphite structure: ■-most stable as wurtzite structure; ●-most stable as rock salt structure.

Laewitz (1972), plus CuH (Wyckoff, 1963) and hexagonal diamond (Kittel, 1971). A plot of their $c/a$ ratios against $\Delta R_{CT}$ is monotonic, and can be made very nearly linear if we further emphasize the charge transfer by multiplying each term in $\Delta R_{CT}$ by a factor $Z^{-1/6}$. The $c/a$ ratios are plotted against the resulting parameter, $\Delta R_{CT}^{*}$, in Fig. 8. We regard the correlation as more than satisfactory in light of the scatter and temperature variation of the data (Laewitz, 1972) and the small distortions involved. We also note that the parameter $\Delta R_{CT}^{*}$ alone is sufficient to separate from the rest of the group those dimorphous compounds which are most stable in the zinc blende form (see Laewitz, 1972; Chelikowsky and Phillips, 1978).

## D.  Structural Maps

The charge transfer and hybridization parameters $\Delta R_{CT}$ and $\bar{R}_H$ are plotted against one another in Fig. 9, which shows data points for over 100 binary

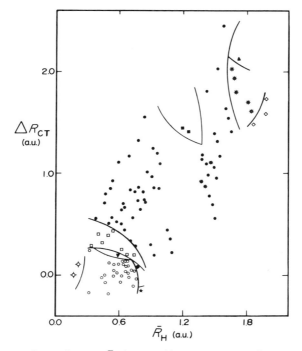

**Fig. 9.**   Structural map $\Delta R_{CT}$ vs. $\bar{R}_H$ for octet binary compounds. Structures are (Wyckoff, (1963): ✧ graphite; ○ zinc blende; □ wurtzite; ▼ HgS (cinnebar); ◆ AuCl; ★ AuI; ● NaCl; ■ PbO; ✳ GeS; ✺ CsCl; ▲ TlF; ◇ InBr.

octet and "pseudo-octet" compounds. These separate into clusters corresponding to their most stable structures at zero temperature and pressure. Dimorphous materials are assigned as discussed by St. John and Bloch (1974).

As compared with the original plot of St. John and Bloch (1974), Fig. 9 nearly doubles the data base and enlarges the number of structures included from 5 to 12. As remarked elsewhere (St. John and Bloch, 1974; Chelikowsky and Phillips, 1978; Zunger, 1980), the separation is all the more striking inasmuch as the differences in binding energy among some of these structures are often smaller than the standard errors in the best cohesive energy calculations available.

Of the 12 structures, all are clearly separated with the exception of the GeS family, which overlaps the NaCl region. J. K. Burdett (private communication) has pointed out to us that the GeS structure may be viewed as a distortion of NaCl, but we suspect that the real reason for the discrepancy is an excess of ambition on our part. Along with roughly 20 other IV–VI and III–VII compounds on the plot, GeS is really a 10-valence-electron system, but has been treated as a pseudo-octet using radii derived by treating the Ge $4s^2$ configuration as part of the core. If these 10-electron systems are removed to a separate plot, the structural separations on both plots are perfect.

It is interesting to compare Fig. 9 with the corresponding plot presented by Zunger (1980) using a data base of roughly the same size and orbital radii derived from a density-functional pseudopotential (Zunger and Cohen, 1978). The two data bases do not entirely coincide, but where they do, the Zunger plot makes six errors and Fig. 9 makes none. The difference cannot be traced to Zunger's use of St. John–Bloch combinations of radii rather than those of Fig. 9. Indeed, the hybridization parameter $\bar{R}_H$ obtained from Zunger–Cohen radii is sometimes negative, to the detriment of the structural plots and contrary to physical intuition. It is curious that the crude spectroscopic orbital radii should in this sense appear more "accurate" than radii based on a much more sophisticated pseudopotential, and the point surely invites further scrutiny.

## VI. CONCLUSIONS

The science (if it be that) of the orbital radii is still in a rudimentary stage. It is true that the radii themselves are now on a firmer physical and conceptual footing, and that they continue to delineate solid-state properties with a precision that must be taken seriously. The actual connections between these atomic parameters and their crystalline manifestations, however, are without any real theory and continue to be largely a matter of education empirical correlation rather than deduction. Thus we do not yet know, for example, how to extend our results to more sophisticated problems involving more complex structures or multiple valencies.

These remarks apply equally to alternative sets of radii such as those proposed by Zunger and Cohen (1978) or Andreoni *et al.* (1980). In this connection we remark that it is probably misleading to regard one of these sets of radii as somehow more "fundamental" than the others. The important point is that all of them are atomic parameters which bear an intriguing relation to chemical trends in solids that remains to be understood. From this perspective it is surely a matter of taste whether one prefers to obtain one's atomic parameters from computations without resort to experiment or from experiments without resort to computation. As a practical matter an approach like that of Zunger and Cohen, because it can include the transition metals and their enormous range of compounds, may eventually prove the most fruitful. Nevertheless, the quantum-defect orbital radii, by virtue of their algebraic simplicity, chemical transparency, and empirical success, continue to fascinate as vehicles toward a systematic understanding.

## ACKNOWLEDGMENTS

We have benefitted from conversations with J. C. Phillips, G. Simons, and A. Zunger, and from the unpublished work of J. St. John.

## REFERENCES

Andreoni, W., Baldereschi, A., Meloni, F., and Phillips, J. C. (1978). *Solid State Commun.* **24**, 245.
Andreoni, W., Baldereschi, A., Biemont, E., and Phillips, J. C. (1980). *Phys. Rev. B* **20**, 4814.
Bloch, A. N., and Simons, G. (1972). *J. Am. Chem. Soc.* **94**, 8611.
Boom, R., deBoer, F. R., and Miedema, A. R. (1976). *J. Less-Common Met* **46**, 271.
Chelikowsky, J. R., and Phillips, J. C. (1978). *Phys. Rev. B* **17**, 2453.
Hamann, D. R. (1979). *Phys. Rev. Lett.* **42**, 662.
Heine, V., and Weaire, D. (1970). *In* "Solid State Physics" (H. Ehrenreich, F. Seitz, and D. Turnbull, eds.), Vol. 24, p. 250. Academic Press, New York.
Hinze, J., and Jaffé, H. H. (1962). *J. Am. Chem. Soc.* **85**, 148.
Kittel, C. (1971). "Introduction to Solid State Physics," 4th ed., p. 33. Wiley, New York.
Laewitz, P. (1972). *Phys. Rev. B* **10**, 4039.
Machlin, E. S., Chow, T. P., and Phillips, J. C. (1977). *Phys. Rev. Lett.* **38**, 1292.
Martin R. (1970). *Phys. Rev. B* **1**, 4005.
Martin R. (1972). *Phys. Rev. B* **2**, 4546.
Pauling, L. (1960). "Nature of the Chemical Bond." Cornell Univ. Press, Ithaca, New York.
Phillips, J. C. (1969). "Covalent Bonding in Crystals, Molecules, and Polymers." Univ. of Chicago Press, Chicago.
St. John, J., and Bloch, A. N. (1974). *Phys. Rev. Lett.* **33**, 1095.

Simons, G. (1971a). *J. Chem. Phys.* **55**, 756.
Simons, G. (1971b). *Chem. Phys. Lett.* **12**, 404.
Simons, G. (1974). *J. Chem. Phys.* **60**, 645.
Simons, G., and Bloch, A. N. (1973). *Phys. Rev. B* **7**, 2754.
Van Vechten, J. (1969). *Phys. Rev.* **182**, 981.
Velikov, Y. K., Kadyshevich, A. E., and Rusakov, A. P. (1973). *Fiz. Tverd. Tela* (*Leningrad*) **15**, 3093.
Walter, J. P., and Cohen, M. L. (1971). *Phys. Rev. B* **4**, 1877.
Wyckoff, R. W. G. (1963). " Crystal Structures." Wiley, New York.
Zunger, A. (1980). *Phys. Rev. Lett.* **44**, 582.
Zunger, A., and Cohen, M. L. (1978). *Phys. Rev. Lett.* **41**, 53.

# 5

# A Pseudopotential Viewpoint of the Electronic and Structural Properties of Crystals

## ALEX ZUNGER

|      |                                                              |     |
|------|--------------------------------------------------------------|-----|
| I.   | Introduction . . . . . . . . . . . . . . . . . . . . . . . .  | 73  |
| II.  | Pseudopotentials and Structural Scales . . . . . . . . . . .  | 77  |
| III. | First-Principles Density-Functional Pseudopotentials . . . . . | 83  |
|      | A. Construction of Density-Functional Atomic Pseudopotentials | 83  |
|      | B. Simple Universal Form of the Density-Functional Pseudopotential . . . . . . . . . . . . . . . . . | 97  |
|      | C. Application to Electronic Structure Calculations . . . . . | 98  |
| IV.  | Trends in Orbital Radii . . . . . . . . . . . . . . . . . .   | 102 |
|      | A. Chemical Regularities . . . . . . . . . . . . . . . . .    | 102 |
|      | B. Screening Length and Orbital Radii . . . . . . . . . . .   | 113 |
|      | C. Comparison with Other Orbital Radii . . . . . . . . . .    | 113 |
| V.   | Separation of the Crystal Structure of 565 Binary AB Compounds | 117 |
| VI.  | Summary . . . . . . . . . . . . . . . . . . . . . . . . . .   | 131 |
|      | References . . . . . . . . . . . . . . . . . . . . . . . . .  | 132 |

## I. INTRODUCTION

In 1956, at the American Society of Metals meeting on alloy phases, J. C. Slater commented: "I don't understand why you metallurgists are so busy in working out experimentally the constitution of polynary metal systems. We know the structure of the atoms, we have laws of quantum mechanics, and we have calculating machines, which can solve the pertinent equations rather quickly" (Slater, 1956). Today, almost 25 years later, our computing ability has increased by two to three orders of magnitude, yet no complex alloy structure has been predicted by such a variational quantum-mechanical approach. At the same time, the semiclassical notions of Pauling, Hume-Rothery, Pearson, and others have continued to guide metallurgists, crystallographers, and crystal chemists in rationalizing even very complex crystal structures.

Our experience in understanding the occurrence of a large variety of crystal structures in nature has been traditionally expressed in two general frameworks: variational quantum mechanics and a semiclassical approach. The

Structure and Bonding in Crystals, Vol. I

bulk of our experience in understanding the structural properties of molecules and solids from the quantum-mechanical viewpoint is expressed in terms of constructs originating from the calculus of variation: total energy minimization, optimum subspaces of basis functions, etc. In this approach, one constructs a quantum mechanical energy functional representing the Born–Oppenheimer surface of a compound; its variational minimum in configuration space $\{\mathbf{R}\}$ is then sought, usually by first reducing the problem to a single-particle Schrödinger equation. The elementary constructs defining this energy functional—the interelectronic effective potential $V_{ee}(\mathbf{r}, \mathbf{r}')$ and the electron-core potential $V_{ec}(\mathbf{r}, \mathbf{R})$—can be treated at different levels of sophistication (e.g., semiempirical tight-binding, Thomas–Fermi, Hartree–Fock, density-functional, pseudopotential, etc.). Similarly, a number of choices exist for the wavefunction representation (e.g., the Bloch and molecular-orbital representations or the Wannier and valence-bond models, etc.). This approach has become increasingly refined recently, producing considerable detailed information and insight into the electronic structure of molecules (e.g., Schaefer, 1977a,b) and simple solids (e.g., Moruzzi et al., 1978).

The semiclassical approach to crystal and molecular structure, on the other hand, involves the construction of phenomenological scales ("factors") on which various aspects of bonding and structural characteristics are measured. These include chemical, crystallographic, and metallurgical constructs, such as the electronegativity, the geometry and size factors, the coordination number factor, the average electron number factor, the orbital promotion energy factor, etc. (e.g., Pearson, 1969). These factors are then represented by various quantitative scales (bond order, elemental work function, ionic, metallic, and covalent radii, electronegativity scales, etc.) that are used to deductively systematize a variety of structural properties. Such intuitive and often heuristic scales have had enormous success in rationalizing a large body of chemical and structural phenomena, often in an ingenious way (Pauling, 1960; Hume-Rothery and Raynor, 1954; Pearson, 1972; Darken and Gurry, 1953; Miedema, 1976). More recently, these semiclassical scales have been used in *quantitative* models, such as the semiempirical valence force field method (Pawley, 1968; Warshel, 1977) and Miedema's heat of formation model (Miedema, 1973, 1976; Miedema et al., 1975), where the remarkable predictive power of these approaches has been demonstrated over large data bases (literally hundreds of molecules and solids).

Even before the pioneering studies of Goldschmidt, Pauling, and others, it was known thermodynamically that the structure-determining energy $\Delta E_s$ of most ordered solids is small compared to the total cohesive energy $\Delta E_0$. Measured heats of transformation and formation data (Hultgren et al., 1973; Kubaschewski and Alcock, 1979), as well as quantum-mechanical calculations of stable and hypothetical structures, indicate that $\Delta E_s/\Delta E_0$ can be as small as $10^{-3}$–$10^{-4}$. This poses an acute difficulty for variational quantum-mechanical models. The elementary constructs of the quantum-mechanical approach $V_{ee}(\mathbf{r}, \mathbf{r}')$ and $V_{ec}(\mathbf{r}, \mathbf{R})$, are highly nonlinear functions of the individual atomic orbitals that interact to form the crystalline wavefunctions (due to both the operator nonlocality of $V_{ee} + V_{ec}$ and their self-consistent dependence on the system's wavefunctions). Consequently, the structural energies $\Delta E_s$ become analytically inseparable from the total energies $\Delta E_0$. One is then faced with the situation that the complex *weak interactions*, responsible for stabilizing one crystal structure rather than another, are often masked by errors and uncertainties in the calculation of the *strong Coulombic interactions* in the total interaction potentials $V_{ee}(\mathbf{r}, \mathbf{r}')$ and $V_{ec}(\mathbf{r}, \mathbf{R})$. Even though $\Delta E_s$ can be calculated quantum-mechanically with the aid of large computers (for sufficiently simple systems), it is notable that the extent and complexity of the information included in $V_{ee}(\mathbf{r}, \mathbf{r}')$ and $V_{ec}(\mathbf{r}, \mathbf{R})$ far exceeds that required to characterize a crystal structure. For example, although the 12 transition metals Sc, Ti, V, Cr, Fe, Y, Zr, Nb, Mo, Hf, Ta, and W have distinctly different quantum-mechanical effective potentials and are characterized by systematically varying cohesive energies $\Delta E_0$, all of them appear in the same body-centered cubic (bcc) crystal form as elemental metals. Hence, at present, the quantum-mechanical approach seems to lack the simple *transferability* of structural constructs from one system to the other, as well as the *physical transparency* required to assess the origin of structural regularities. The semiclassical approach, on the other hand, concentrates on the construction of physically simple and transferable coordinates that may systematize directly the trends underlying the structural energies $\Delta E_s$. The major limitations of the semiclassical approach seem to lie in the occurrence of internal linear dependencies among the various structural factors (e.g., orbital electronegativity and orbital promotion energy), as well as in the appearance of a large number of crystalline structures placed within narrow domains of the phenomenological structural parameters (e.g., Mooser–Pearson plots for non-octet AB compounds or diagrams of the frequency of occurrence of a given structure versus average electron per atom ratio). Even so, the semiclassical approaches provide valuable insight into the problem because they point to the underlying importance of establishing system-invariant *energy scales* (e.g., electronegativity, promotion energy) as well as *length scales* (e.g., covalent, metallic, and ionic radii).

For the 50–60 nontransition-metal binary octet compounds, the problem of systematizing the five crystal structures (NaCl, CsCl, diamond, zinc blende, and wurtzite) has been solved through the use of the optical dielectric electronegativity concept of Phillips and Van Vechten (1970; also Phillips, 1970).

This concept diagrammatically displays periodic trends when transferable elemental coordinates are used. Such diagrammatic Pauling-esque approaches are extended here to include intermetallic transition-metal compounds (a total of 565 compounds).

In this paper, I show that the recently developed first-principles nonlocal atomic pseudopotentials provide nonempirical elementary energy and length scales. By using a dual and transferable coordinate system derived from these scales, one is able to topologically separate the crystal structures of 565 binary compounds (including simple and transition-metal atoms) with a surprising accuracy. At the same time, these quantum-mechanically derived pseudopotentials allow one to conveniently define the elementary constructs $V_{ee}(\mathbf{r}, \mathbf{r}')$ and $V_{ec}(\mathbf{r}, \mathbf{R})$ and use them in detailed electronic structure calculations for molecules, solids, and surfaces. As such, this approach may provide a step in bridging the gap between the quantum-mechanical and semiclassical approaches to electronic and crystal structure.

The theoretical prediction of stable crystal structures is given diagrammatically much as in Mooser–Pearson plots. We show that each element A in the periodic table is characterized by three core radii $r_s^A$, $r_p^A$, and $r_d^A$, which measure the effective size of the atomic cores as experienced by valence electrons with angular momentum $l = 0$, 1, and 2, respectively. These radii are derived in Sec. III from the pseudopotential theory and tabulated in Table I for 70 elements. For each binary compound AB, we then construct the dual coordinates $R_\pi^{AB} = \left| r_p^A - r_s^A \right| + \left| r_p^B - r_s^B \right|$ and $R_\sigma^{AB} = \left| (r_p^A + r_s^A) - (r_p^B + r_s^B) \right|$. On an $R_\pi^{AB}$ vs. $R_\sigma^{AB}$ plot, we then find that the different groups of crystal structures of the 565 binary compounds occupy different regions. If one lumps together some of the crystallographically related structures, the accuracy of this prediction is better than 93%. The radii given in Table I can be used to systematize and analyze a large number of structural properties of crystals.

The success of this approach in correctly predicting the structural regularities of as many as 565 binary compounds using elemental coordinates that pertain directly only to the s and p electrons (and only indirectly to the d electrons through the screening potential produced by them) presents a striking result: it suggests that the *structural part* $\Delta E_s$ of the cohesive energy may be dominated by the s–p electrons. This points to the possibility that, while the relatively localized d electrons determine both *central cell effects* (such as octahedral ligand field and Jahn–Teller stabilizations) and the regularities in the structure-insensitive cohesive energy $\Delta E_0$ of crystalline and liquid alloys, the longer range s–p wavefunctions are responsible for stabilizing one complex space group arrangement rather than another. There is a striking resemblance between this result and the semiclassical ideas indicating a remarkable correlation between the stable crystal structure of transition-metal systems and the number of s and p electrons, put forward by Engel in

1939 (also 1967) and subsequently greatly refined by Brewer (1963, 1967, 1968; Brewer and Wengent, 1973). In the Engel–Brewer approach, the d electrons play an important but indirect role in determining the energy required for exciting the ground atomic configuration to one that has available for bonding a larger number of unpaired s and p electrons. The Engel–Brewer approach has enabled the extension of the Hume-Rothery rules to transition metal systems simply by counting only the contributions of s and p electrons, and at the same time it has explained the stabilities of the bcc, hcp (hexagonal close-packed), and fcc (face-centered cubic) structures of the 33 elemental transition metals, the effects of alloying in multicomponent phase diagrams, as well as pressure effects on crystal structure stabilities, phenomena yet to be tackled by variational quantum-mechanical approaches. These conclusions on the crucial *structural* roles played by the s and p coordinates should be contrasted with the contemporary quantum-mechanical resonant tight-binding approaches suggested first by Friedel (1969) for elemental transition metals and recently extended to compounds by Pettifor (1979), Varma (1979), and others. These approaches emphasize the exclusive role of d electrons in determining *cohesive properties*. This approach explains the periodic trends in $\Delta E_0$ and the bcc–fcc structural transitions of both elemental and alloyed transition metal systems by considering changes in the one-electron d energy levels, assumed to have a rectangular density of states.

The plan of this paper is as follows: in Sec. II, we introduce the pseudo-potential concept and show how it can be used in general to define atomic parameters that correlate with crystal structures. In Sec. III, we derive the first-principles atomic pseudopotentials within the density-functional theory of electronic structure. In Sec. IV, we then show how these atomic pseudo-potentials can be used to define intrinsic core radii that correlate with a large number of electronic and structural properties of crystals. These radii are used to separate diagrammatically the stable crystal structure of 565 binary AB compounds.

## II.  PSEUDOPOTENTIALS AND STRUCTURAL SCALES

Although traditionally the inner core orbitals and the outer valence orbitals are often treated on an equal footing in variational calculations of the electronic structure of atoms, molecules, and solids, it was recognized quite early that a large number of bonding characteristics are rather in-sensitive to the details of the core states (Hellman, 1935a,b, 1936; Gombas, 1935, 1967; Fock *et al.*, 1940). This relative insensitivity is a manifestation of the fact that the interaction energies involved in chemical bond formation $(10^{-1}–10 \text{ eV})$, banding in solids $(1–25 \text{ eV})$, or scattering events near the

Fermi energy ($10^{-2}$–$10^{-3}$ eV) are often much smaller than the energies associated with the polarization or overlap of core states. Hence, the core orbitals with their nearly spherical symmetry and high binding energies are nearly unresponsive to many of the scattering phenomena that determine "valence-like" properties. Many methods treating the quantum structure of bound electrons, nucleons, and general Fermions have consequently omitted any reference to the core states, variationally treating only "valence" states (Hückel, CNDO, tight-binding, Hubbard models, optical potentials in nuclear physics, effective potentials in Fermi-liquid theories, and field-theory models of the Lamb shift, empirical valence potentials in atomic physics, etc.). Clearly, however, if no constraints are placed, such a variational treatment will result in an unphysical lowering of the energy of the valence states into the empty core ("variational collapse"). Much of the empirical parametrization characteristic of such methods is implicitly directed to avoid such a pathology. It was first recognized however by Phillips and Kleinman (1959) that the price for reducing the orbital space to valence states alone can be represented by an additional nonlocal potential term (pseudopotential) in the Hamiltonian.

Although the pseudopotential concept has offered great insight into the nature of bonding states in polymers and solids (e.g., Phillips, 1973), its calculation in practical electronic structure application has generally been avoided (Cohen and Bergstresser, 1966; Brust, 1968; Cohen and Heine, 1970). Instead, it has been replaced by a local form with disposable parameters adjusted to fit selected sets of data (semiconductor band structures, Fermi surface of metals, atomic term values, etc.). Since the valence electronic energies near the Fermi level are determined (to within a constant) by relatively low-momentum transfer electron-core scattering events ($|q| \approx 2k_F$), it has been possible in the past to successfully describe the one-electron optical spectra and Fermi surface of many solids assuming core pseudopotentials that are truncated to include only small momentum components (i.e., smoothly varying in the core region in configuration space). The freedom offered by the insensitivity of the electronic band structure dispersion relation $\varepsilon_j(\mathbf{k})$ to the variations of the pseudopotential in the core region has been exploited to obtain empirical potentials converging rapidly in momentum space and hence amenable to electron-gas perturbative theories (Harrison, 1966) and plane-wave-based band structure calculations (Cohen and Heine, 1970; Brust, 1968).

These soft-core *empirical* pseudopotentials have produced the best fits to date for the observed semiconductor band structures (e.g., see Cohen and Bergstresser, 1966), and their descendants, the soft-core self-consistent pseudopotentials, have yielded the most detailed information on semiconductor surface states (e.g., see Appelbaum and Hamann, 1976). The insensitivity of $\varepsilon_j(\mathbf{k})$ to the high-momentum components of the pseudopotential has prompted an enormous literature in which different forms for the potential have been suggested (empty cores, square wells, Gaussian-shaped, etc.). Since, however, these pseudopotentials were fitted predominantly to energy levels in atoms and solids (and were not constrained to produce physically correct wavefunctions) they often yielded systematic discrepancies with experiment or all-electron calculations of the bonding charge density in molecules and solids (Yang and Coppens, 1974; Harris and Jones, 1978; Hamann, 1979): while correctly predicting a build-up of covalent charge on the bonds, such empirical pseudopotentials incorrectly suggested a *bond-perpendicular* charge density, rather than a *bond-elongated* density as envisioned by Coulson and Pauling and subsequently measured experimentally and supported by more refined calculation (Zunger, 1980). Such discrepancies result from the fact that higher momentum components (e.g., $|q| \gtrsim 6k_F$ in crystalline silicon), not included in energy-level-fitted soft-core pseudopotentials, are of importance in determining the directional distribution of the bonding charge density. It is such systematic omissions which make the soft-core empirical potentials inappropriate for predicting stable structures. The striking success of the empirical pseudopotential is that it made it possible to reduce the informational content of the often complex electronic spectra of semiconductors to a few (usually three to five) nearly transferable elemental parameters (empirical pseudopotential form factors). The assumed locality of the pseudopotential, as well as its truncation to low-momentum components, however, has limited its chemical content to reflect predominantly the low-energy electronic excitation spectrum rather than explicit structural and chemical regularities.

Recently, Simons (1971a,b) and Simons and Bloch (1973) have observed that there exists at least one class of structurally significant empirical pseudopotentials containing *very high momentum components* (i.e., $|q| \gg 2k_F$, or hard-core pseudopotentials). The general form of a screened pseudopotential is:

$$V_{\text{eff}}^{(l)}(\mathbf{r}) = V_{\text{ps}}^{(l)}(\mathbf{r}) + V_{\text{scr}}[n(\mathbf{r})] \tag{1}$$

(We use a capital $V(\mathbf{r})$ to denote solid-state potentials, while $v(r)$ will denote atomic or ionic potentials.) Here $V_{\text{ps}}^{(l)}(\mathbf{r})$ is the bare pseudopotential acting on the $l$th angular momentum wavefunction component, and $V_{\text{scr}}[n(\mathbf{r})]$ is the Coulomb, exchange, and correlation screening due to the pseudo charge density $n(\mathbf{r})$. The conventional core attraction Coulomb term $-Z/r$ is replaced by an angular momentum-dependent and spatially varying effective charge $Z_{\text{eff}}^{(l)}(\mathbf{r}) = r V_{\text{ps}}^{(l)}(\mathbf{r})$, while $V_{\text{scr}}[n]$ continues to represent interelectronic (valence–valence) interactions. For the simple case of *one-electron ions*, chosen by Simons and Bloch, the screening potential reduces to zero. The bare atomic pseudopotential $v_{\text{ps}}^{(l)}(r)$ was then assumed to take a simple hard-core form:

$$v_{\text{eff}}^{(l)}(r) = v_{\text{ps}}^{(l)}(r) = \frac{B_l}{r^2} - \frac{Z_v}{r} \tag{2}$$

where $Z_v$ is the valence charge and the parameter $B_l$ is adjusted such that the negative of the orbital energies $\varepsilon_{nl}$ obtained from the pseudopotential equation

$$\{-\tfrac{1}{2}\nabla_r^2 + v_{\text{eff}}^{(l)}(r)\}\psi_{nl}(r) = \varepsilon_{nl}\psi_{nl}(r) \tag{3}$$

match the observed ionization energies of one-electron ions such as $Be^{+1}$, $C^{+3}$, $O^{+5}$, etc. These hard-core pseudopotentials are characterized by an orbital-dependent crossing point $r_l^0$ at which $v_{\text{eff}}^{(l)}(r_l^0) = 0$. These orbital radii then possess the same periodic trends underlying the observed single-electron ionization energies through the periodic table. The remarkable feature of these radii is that they form powerful structural indices, capable of systematizing the various crystal phases of the octet $A^N B^{8-N}$ nontransition-metal compounds (St. John and Bloch, 1979). Such structural plots have been extended by Machlin et al. (1977) very successfully to some 45 nonoctet (nontransition-metal) compounds. More details on this approach are available in the cited literature.

The realization that these empirical orbital radii are characteristic of the atomic cores, and as such are approximately transferable to atoms in various bonding situations, has led to the construction of a number of new phenomenological relations of the form $G = f(r_l^0)$, correlating physical observables $G$ in *condensed phases* with the orbital radii of the constituent *free atoms*. Some examples of $G$ are the elemental work functions, the melting points of binary compounds, and the Miedema coordinates treated by Chelikowsky and Phillips (1978). What has been realized is that the characteristics of an isolated atomic core, reflected in the spectroscopically determined $l$-dependent turning points $r_l^0$, contain the fundamental constructs describing structural regularities in polyatomic systems. This can be contrasted with phenomenological electronegativity scales that are based on observables pertaining to the polyatomic systems themselves, such as the thermochemical Pauling scale, the dielectric Phillips–Van Vechten scale, and the Walsh scale.

While one normally considers structural and bonding characteristics to be predominantly determined by the atomic valence orbitals, these are not amenable to an analysis that reveals structural regularities in a simple manner. For instance, the different chemistries associated with carbon and silicon compounds are not transparently reflected by contrasting the carbon 2s and 2p with the silicon 3s and 3p orbitals, simply because the qualitative difference in their nodal structure precludes the construction of a simple quantitative scale. In the pseudopotential representation, the nodal valence orbitals are transformed into nodeless valence pseudo wavefunctions such that, for example, the relevant differences in the new carbon 2s and silicon 3s orbitals can be measured on a simple quantum mechanical scale. Such a scale is provided by the orbital radii. The chemically pertinent information of these nodeless valence orbitals is coded in the pseudopotential. By building into the simple pseudopotentials of Eq. (2) the experimentally observed regularities of the ionization energies in the periodic table, Simons, Bloch, Chelikowsky, and Phillips (e.g., Phillips, 1977, 1978) have achieved an orbital radii scale that deciphers this core code.

It is not surprising that, although the orbital radii have typical dimensions of the core states, they do reflect structural regularities characteristic of the outer valence orbitals. This should be contrasted with the classical definitions of ionic, tetrahedral, covalent, or metallic radii: these definitions attempt to reproduce observed bond distances as *sums of single-site radii*. This bond additivity constraint forces these radii to have dimensions typical of valence orbitals, and as such these radii depend on the chemical environment (ionicity, coordination number, valency, spin state, etc.) rather strongly (e.g., Shannon and Prewitt, 1969). Even so, these classical radii constitute a very important reduction in the informational degrees of freedom required to specify chemical bonds: using, typically, 5000 measured bond distances, one has deduced about 250 ionic radii. The orbital radii approach takes, however, a different viewpoint: it assumes that the valence properties that an atom will take in bonded situations are coded in its effective core. Using the orbital radii as the characteristic fingerprint of the atomic cores, one achieves a further reduction of the structural information to a single set of transferable elemental radii.

This empirical Simons–Bloch radii have, however, few obvious shortcomings. Since the general atomic pseudopotential $v_{\text{eff}}^{(l)}(\mathbf{r})$ of Eq. (1) can be reduced to a simple form with $v_{\text{scr}} = 0$ only for single-electron-stripped ions, the empirical Simons–Bloch orbital radii can only be invoked for atoms for which stripped-ion spectroscopic data exists. This excludes most transition elements, which form a wealth of interesting intermetallic structures. Yet, even so, the extraction of a bonding scale from data on ions that lack any valence–valence interactions (e.g., $C^{+3}$ and $O^{+5}$, representing chemical affinities of neutral C and O) may distort the underlying chemical regularities. In addition, the restriction to single-electron species means that the post-transition-series atoms (e.g., Cu, Ag, Au or Zn, Cd, Hg) are treated as having only one and two valence electrons, respectively, much like the alkali and alkaline earth elements, respectively. However, the increase in melting points and heats of atomization and the decrease in nearest-neighbor distances in going from Group IIB to IB metals (e.g., Zn → Cu, Cd → Ag, and Hg → Au), as compared with the *opposite trend* in going from Group IIA to IA metals (Ca → K, Sr → Rb, and Ba → Cs), completely eliminates any possibility of Cu, Ag, and Au having effectively a single-bonding electron. Similar indications on the extensive s–d and p–d hybridization are given by the large bulk of photoemission data on Cu and Ag halides (Goldman, 1977). In keeping with the single-valence-electron restriction, one is also forced to define the d-orbital coordinate of the post-transition elements from the lowest *unoccupied* rather than occupied d orbital (i.e., 4d for Cu and Zn, 5d for Ag and Cd). This may be reasonably faithful to the chemical tendencies of post-transition elements with sufficiently deep occupied d orbitals and sufficiently low unoccupied d orbitals (e.g., Br, Te, I), but it is questionable for the elements with occupied semi core d shells in the vicinity of the upper valence band (e.g., CdS and ZnS). These pathologies can be corrected by empirically adjusting the valence charge $Z_v$ in Eq. (2) for these elements (A. N. Bloch, unpublished results, 1980). Finally, the simple pseudopotential of Eq. (2) is not suitable for electronic structure studies, as indicated by Andreoni *et al.* (1978, 1979), because the wavefunctions of Eq. (3) are severely distorted relative to true valence orbitals by the unphysically long-range $r^{-2}$ tail (similarly, the total energy of a solid described by this potential is divergent!). This has been corrected by Andreoni *et al.* by replacing the long-range $B_1/r^2$ term in Eq. (2) with an $A_1 \exp(-\gamma_1 r/r^2)$ term, with the additional parameter $\gamma_1$ fixed to fit the orbital maxima. This leads to a new set of renormalized orbital radii differing considerably from the Simons–Bloch set.

One is hence faced with the situation that the soft-core empirical pseudopotential (e.g., Cohen and Heine, 1970) can be used to successfully fit the low-energy electronic band structure of solids, but it lacks the structurally significant turning points [i.e., $v_{eff}(r) = 0$ only at $r = \infty$]; whereas the empirical Simons–Bloch potentials do not yield a quantitatively satisfactory description of the electronic structure but do yield the correct structural regularities. The approach that we have taken to remedy this situation is to construct a pseudopotential theory from first-principles. The first-principles approach allows for the regularities of energy levels and wavefunctions to be systematically built into the atomic pseudopotentials, without appealing to any experimental data. Because no resort is made to simple, single-electron models, transition elements can be treated as easily as other elements, without neglecting the interactions between valence electrons or assuming that the highest occupied d levels belong to a chemically passive core. Furthermore, since the bare pseudopotential $v_{ps}^{(l)}(r)$ and the screening $v_{scr}[n(r)]$ are described in terms of well-defined quantum-mechanical constructs (such as Coulomb, exchange, and correlation interactions, Pauli forces, and orthogonality holes) both the failures and the successes of the theory could be appreciated. This defines a link between the semiclassical length scale and the quantum-mechanical approach to structure.

## III. FIRST-PRINCIPLES DENSITY-FUNCTIONAL PSEUDOPOTENTIALS

### A. Construction of Density-Functional Atomic Pseudopotentials

These section describes the construction of atomic pseudopotentials from the density-functional formalism (Hohenberg and Kohn, 1964; Kohn and Sham, 1965). These potentials were first derived by Topiol et al. (1977) and Zunger and Ratner (1978) and subsequently refined by Zunger and Cohen (1978b, 1979a,b). Technical details are given elsewhere (Zunger et al., 1979b; Zunger and Ratner, 1978; Zunger and Cohen, 1978b, 1979b; Zunger, 1979).

We first start with the *all-electron* approach. Consider a many-electron system with an electronic density matrix $\rho(\mathbf{r}, \mathbf{r}')$ interacting with an external potential $V_{\text{ext}}(\mathbf{r})$. In the conventional all-electron (ae) approach, both the core (c) and valence (v) wavefunctions, $\psi_j^c(\mathbf{r})$ and $\psi_j^v(\mathbf{r})$, respectively, are treated on the same footing. The effective single-particle potential appearing in the Schrödinger equation is then written as a sum of the external potential and the interelectronic response (screening):

$$V_{\text{eff}}^{\text{ae}}(\mathbf{r}) = V_{\text{ext}}(\mathbf{r}) + V_{\text{scr}}^{c,v}[\rho(\mathbf{r}, \mathbf{r}')] \tag{4}$$

Here $V_{\text{scr}}^{c,v}[\rho(\mathbf{r}, \mathbf{r}')]$ is a functional of the total core plus valence charge density $\rho = \rho_c + \rho_v$ and includes the interelectronic Coulomb $V_{ee}[\rho]$ as well as exchange $V_x[\rho]$ and correlation $V_{cr}[\rho]$ terms. These screening terms take different forms in the Hartree–Fock and density-functional formalisms used here. The external potential $V_{\text{ext}}(\mathbf{r})$ may be identified in atoms with the electron-nuclear attraction term $-(Z_c + Z_v)/r$ (where $Z_c$ and $Z_v$ denote the number of core and valence electrons, respectively) or with the sum of the analogous terms and the Ewald ion–ion repulsion in infinite systems. The wavefunctions $\{\psi_j^{c,v}(\mathbf{r})\}$ of the all-electron Hamiltonian $H^{\text{ae}} = -\frac{1}{2}\nabla^2 + V_{\text{eff}}^{\text{ae}}$ have a nodal structure resulting from the orthogonality constraint. These form a basis for constructing the ground-state density matrix $\rho(\mathbf{r}, \mathbf{r}')$, which is then used to calculate self-consistently the screening potential as well as the ground-state total energy.

We now turn to the *pseudopotential* approach. In this representation, one seeks an effective potential $V_{\text{eff}}^{\text{ps}}$ that will produce in a variational Schrödinger equation the *valence* wavefunctions $\chi_j(\mathbf{r})$ and orbital energies $\lambda_j$ as its lowest-lying solutions. As, by construction, no core states occur, $\chi_j(\mathbf{r})$ does not have to be core-orthogonal and, hence, may be constructed as nodeless for each of the lowest angular symmetries. One therefore replaces the all-electron effective potential of Eq. (4) by the pseudopotential effective potential:

$$V_{eff}^{ps} = V_{ps}(\mathbf{r}) + V_{scr}^{v}[n(\mathbf{r})]$$
$$= [V_{ext}^{v}(\mathbf{r}) + W_{R}(\mathbf{r})] + V_{scr}^{v}[n(\mathbf{r})] \qquad (5)$$

where $V_{ext}^{v}$ is the valence-projected external potential [e.g., in an atom $-Z_v/r$ rather than $-(Z_v + Z_c)/r$], $W_R(\mathbf{r})$ is the yet-unspecified repulsive part of the pseudopotential, and $V_{scr}^{v}[n(\mathbf{r})]$ is the screening due to the valence pseudo charge density $n(\mathbf{r}) = \sum_j \chi_j^*(r)\chi_j(r)$.

Instead of constructing $W_R(\mathbf{r})$ directly for the molecules or solids of interest, one attempts first to calculate this for simple model systems such as atoms. Then, the total pseudopotential $V_{ps}(\mathbf{r})$ for a general system will be approximated by superimposing the atomic pseudopotentials $v_{ps}^{(l)}(r)$ over all atoms and angular momenta, i.e.,

$$V_{ps}^{(l)}(\mathbf{r}) = \sum_{\mathbf{R}_n} v_{ps}^{(l)}(\mathbf{r} - \mathbf{R}_n)\hat{P}_l$$

where $\hat{P}_l$ is the angular momentum projection operator, and $\mathbf{R}_n$ is the position vector of atom $n$. To build into the atomic pseudopotentials an element of transferability, one needs to construct $v_{ps}^{(l)}(r)$ such that its dependence on the chemical environment is minimal. Formally this amounts to minimizing the energy and quantum-state dependence of $v_{ps}^{(l)}(r)$. The mathematical implications of this have been previously discussed (Zunger et al., 1979). It will suffice here to say that such a minimization of the pseudopotential's energy and state dependence can be achieved by maximizing the spatial range (starting from $r = \infty$ and going inwards to a finite value $r = R_c$) of identity between the true valence orbital $\psi_{nl}^{v}(r)$ and the pseudo orbital $\chi_{nl}(r)$ and, at the same time, minimizing to the extent possible the amplitude and lowest derivatives of $\chi_{nl}(r)$ in the core region ($0 \leqslant r \leqslant R_c$). The formulation of *transferable* atomic pseudopotentials, through the imposition of certain physically motivated constrains on the pseudo wavefunctions, is central to the present approach. No such explicit considerations have been undertaken in the development of previous pseudopotentials.

To the extent that the construction of the pseudopotentials $V_{ps}^{(l)}(\mathbf{r})$ can be made simple, the study of valence-related properties of solids through the solution of the *pseudopotential* single-particle problem is both computationally and conceptually simpler than that study via the solution of the *all-electron* problem. This relative simplicity is not only because the pseudopotential approach treats fewer ("reactive") electrons and permits nodeless and spatially smooth wavefunctions, but it is also because, to within a good approximation, the atomic pseudopotential can be constructed from nearly system-invariant transferable quantities. Such transferable atomic pseudo-potentials $v_{ps}^{(1)}(r)$ can then be used, through Eq. (5), to construct self-consistently the effective potential for arbitrary molecules and solids and obtain their electronic structure at a fraction of the complexity and computational effort required in a comparable all-electron calculation.

Specializing Eqs. (4) and (5) for atoms, the all-electron and pseudopotential single-particle equations are:

$$\left\{ -\tfrac{1}{2} \nabla_r^2 - \frac{Z_c + Z_v}{r} + \frac{l(l+1)}{2r^2} + v_{ee}[\rho_c + \rho_v] + v_x[\rho_c + \rho_v] \right.$$

$$\left. + v_{cr}[\rho_c + \rho_v] \right\} \psi_{nl}(r) = \varepsilon_{nl} \psi_{nl}(r) \tag{6}$$

and

$$\left\{ -\tfrac{1}{2} \nabla_r^2 - \frac{Z_v}{r} + v_{ps}^{(1)}(r) + \frac{l(l+1)}{2r^2} + v_{ee}[n] + v_x[n] \right.$$

$$\left. + v_{cr}[n] \right\} \chi_{nl}(r) = \lambda_{nl} \chi_{nl}(r) \tag{7}$$

respectively, where $v_{ee}$, $v_x$, and $v_{cr}$ denote the density-functional interelec-tronic Coulomb, exchange, and correlation potentials and $\nabla_r$ is the radial Laplacien. In contrast to the empirical pseudopotential method (e.g., Cohen and Heine, 1970), $v_{ps}^{(1)}(r)$ in Eq. (7) is not determined by fitting the energies $\lambda_{nl}$ to experiment, leaving the wavefunction $\chi_{nl}$ to be implicitly and arbitrarily fixed by such a process. Instead, we first construct physically desirable pseudo wavefunctions $\chi_{nl}$ and then solve for the pseudopotential $v_{ps}^{(1)}(r)$ that will produce these wavefunctions together with the theoretically correct orbital energies $\lambda_{nl} = \varepsilon_{nl}$ from the single-particle equation, Eq. (7).

To construct such pseudo wavefunctions, we postulate a number of con-straints. We will first require that the pseudo wavefunction $\chi_{nl}(r)$ be given as a linear combination of the "true" all-electron core and valence orbitals of Eq. (6):

$$\chi_{nl}(r) = \sum_{n'} C_{n,n'}^{(1)} \psi_{n'l}^{c,v}(r) \tag{8}$$

Since the pseudo wavefunctions $\{\chi_{nl}(r)\}$ are now the lowest solutions to the pseudo-Hamiltonian, they will be nodeless for each of the lowest angular symmetries (e.g., while the all-electron 4s orbital of Cu has three nodes, the pseudo 4s orbital will have zero nodes, the 5s one node, etc.). The coeffi-cients $\{C_{n,n'}^{(1)}\}$ will be, hence, chosen below to satisfy this condition. Note that in a single-determinental representation, such a mixing of rows and

columns leaves the energy invariant. We then require that the orbital energies $\lambda_{nl}$ of the pseudopotential problem equal the "true" valence orbital
energies $\varepsilon_{nl}$. The first constraint assures us that the pseudo wavefunctions
are contained in the same core-plus-valence orbital space defined by the
underlying density-functional theory; the second ensures that the spectral
properties derived from the pseudopotential single-particle equation match
those of the valence electrons in the all-electron problem.

Without specifying at this stage the choice of the unitary rotation coefficients $\{C_{n,n'}^{(l)}\}$, Eqs. (6)–(8) can already be solved to obtain the atomic pseudopotential $v_{ps}^{(l)}(r)$ in terms of the latter and the known quantities defining the
all-electron atomic equation, Eq. (6):

$$v_{ps}^{(l)}(r) = \left\{ U_1(r) - \frac{Z_v}{r} \right\} + \left\{ -\frac{Z_c}{r} + v_{ee}[\rho_c] + v_x[\rho_c] + v_{cr}[\rho_c] \right\}$$

$$+ \{ v_x[\rho_c + \rho_v] - v_x[\rho_c] - v_x[\rho_v] \} + \{ v_{cr}[\rho_c + \rho_v] - v_{cr}[\rho_c] - v_{cr}[\rho_v] \}$$

$$+ \{ v_{ee}[\rho_c] - v_{ee}[n] \} + \{ v_x[\rho_v] - v_x[n] \} + \{ v_{cr}[\rho_v] - v_{cr}[n] \} \tag{9}$$

where the "Pauli potential" $U_1(r)$ is given by

$$U_1(r) = \frac{\sum_{n'} C_{n,n'}^{(l)} [\varepsilon_{nl} - \varepsilon_{n'l}] \psi_{n'l}(r)}{\sum_{n'} C_{n,n'}^{(l)} \psi_{n,l}(r)} \tag{10}$$

and the core, valence, and pseudo charge densities are given as

$$\rho_c(r) = \sum_{nl}^{c} |\psi_{nl}^{c}(r)|^2$$

$$\rho_v(r) = \sum_{nl}^{v} |\psi_{nl}^{v}(r)|^2 \tag{11}$$

$$n(r) = \sum_{nl}^{v} |\chi_{nl}(r)|^2$$

One notices that in the pseudopotential representation each angular component of the system's wavefunction is experiencing a different external
potential $v_{ps}^{(l)}(\mathbf{r})$, whereas in the regular, all-electron representation, $v_{ext}(\mathbf{r})$
was local [Eq. (4)]. This is a direct consequence of eliminating the subspace
of core orbitals from explicit consideration, replacing thereby the dynamical
effects of the core electrons by a static potential. Such a pseudopotential
transformation allows us to conveniently decompose the chemically coded
characteristics of the core into *orbital contributions*.

The atomic pseudopotential in Eq. (9) has a simple physical interpretation. The "Pauli potential" $U_1(r)$ is the only term in $v_{ps}^{(l)}(r)$ that depends on

the wavefunction it operates on (i.e, "nonlocal"), whereas all other terms in Eq. (9) are common to all angular momenta (i.e., "local"). Note that for atomic valence orbitals that lack a matching $l$-component in the core (e.g., carbon 2p or silicon 3d, lacking $l = 1$ and $l = 2$ core states, respectively), the all-electron valence orbital $\psi_{nl}^{v}(r)$ is nodeless—no mixing of other orbitals in Eq. (8) is needed for elimination of nodes. Hence, $\chi_{nl} = \psi_{nl}$ and, from Eq. (10), $U_{1}(r) = 0$ for such states. In these cases, the pseudopotential is local and purely attractive due to the dominance of the all-electron term, $-(Z_{c} + Z_{v})/r$. In all other cases, $U_{1}(r)$ is positive and strongly repulsive, but confined to the atomic core region [for distances from the origin at which all core orbitals $\psi_{nl}^{c}(r)$ are small relative to the valence orbital $\psi_{nl}^{v}(r)$, the energy difference in the numerator causes $U_{1}(r)$ to be zero]. $U_{1}(r)$ replaces the core-valence orthogonality constraint and is a realization in coordinate space of Pauli's exclusion principle. Its precise form depends on the choice of the mixing coefficients $\{C_{n,n'}^{(1)}\}$ and is discussed below. We see that the pseudopotential nonlocality, often neglected in the empirical pseudopotential approach (Cohen and Heine, 1970; but compare Chelikowsky and Cohen, 1976; Pandey and Phillips, 1974) emerges naturally in this formulation from the quantum shell structure of the atom. Similarly, Phillips's pseudopotential kinetic energy cancellation theorem (e.g., Cohen and Heine, 1961) is simply represented as a cancellation (or over-cancellation) between the non-classical repulsive Pauli potential and the core-valence Coulomb attraction $-Z_{v}/r$.

The second term in Eq. (9) represents the total screened potential set up by the core charge density $\rho_{c}(r)$. It approaches $-Z_{c}/r$ at small distances and decays to zero exponentially at the core radius (with a characteristic core screening length) due to rapid screening of the core point charge by the core electrons. The third and fourth terms in Eq. (9) represent the non-linearity of the exchange and correlation potentials, respectively, with respect to the interference of $\rho_{c}$ and $\rho_{v}$. They measure the core-valence interactions in the system and are proportional to the penetrability of the core by the valence electrons.

The fifth term in Eq. (9) is the Coulomb orthogonality hole potential. It has its origin in the charge fluctuation $\Delta(r) = \rho_{v}(r) - n(r)$ that results from the removal of the nodes in the pseudo wavefunctions [i.e., the transformation in Eq. (8)]. The electrostatic Poisson potential set up by $\Delta(r)$ is then given by the fifth term in Eq. (9). Finally, the last two terms in Eq. (9) represent, respectively, the exchange and correlation potentials set up by this orthogonality hole charge density $\Delta(r)$.

The form of the first-principles pseudopotential in Eqs. (9)–(10) makes it easy to establish contact with the successfully simplified early empirical pseudopotentials. Hence, for example, in the Abarenkov-Heine (1965) model potential it was implicitly assumed that a pseudopotential cancellation between a repulsive Pauli force and an attractive Coulomb potential $-Z_v/r$ exists, but instead of calculating the spatial details of the cancellation its net result was assumed to take the form of a constant $v_{ps}^{(l)}(r) = A_1$ for $r$ smaller than some radius $R_1$ (i.e., inside the core), with $v_{ps}^{(l)}(r) = -Z_v/r$ for $r > R_1$. Abarenkov and Heine's empirical constants $A_1$ may be identified in the present formulation with the volume integral of $[U_1(r) - Z_v/r]$ from the origin to $R_1$ [neglecting all but the first term in Eq. (9)]. Similarly, Ashcroft (1966) has suggested an empirical "empty core" pseudopotential, postulating that the net result of the cancellation between $U_1(r)$ and $-Z_v/r$ inside the core region is zero. Indeed, for a sufficiently large core radius (i.e., of the order of Pauling's ionic radius), such a simple model represents well $v_{ps}^{(l)}(r)$ in Eq. (9).

Up to this point, we have not yet specified the form of the transformation coefficients in Eq. (8) determining the precise relationship between the pseudo and "true" wavefunctions. Clearly, one would like to constrain the pseudo wavefunction in Eq. (8) to be normalized. In addition, the relaxation of the orthogonality constraint may be exploited to construct $\chi_{nl}(r)$ as nodeless for each of the lowest angular states, permitting thereby a convenient expansion of the pseudo wavefunctions in spatially simple and smooth basis functions. Even so, $\chi_{nl}(r)$ is underdetermined: there are an infinite number of choices of $\{C_{n,n'}^{(l)}\}$ leading to normalized and nodeless $\chi_{nl}(r)$. This is a manifestation of the well-known pseudopotential nonuniqueness. The resolution of this nonuniqueness is precisely the point at which one applies one's physical intuition (and physical prejudices). Note, however, that in the present approach, any of the infinite and legitimate choices of $\{C_{n,n'}^{(l)}\}$ permits a rigorous digression from the pseudo wavefunction to the true valence wavefunction: the choice of a *linear* form for $\chi_j(r)$ in Eq. (8) allows for $v_{ps}^{(l)}(r)$ to be computed from an arbitrary set $\{C_{n,n'}^{(l)}\}$ and for the resulting pseudopotential to be used to greatly simplify the calculation of the electronic structure of arbitrary molecules or solids. Upon completion, one can simply recover the true wavefunction through a core orthogonalization:

$$\psi_j(\mathbf{r}) = \frac{1}{N}\{\chi_j(\mathbf{r}) - \sum_i^{core} \langle \chi_i | \psi_i^c \rangle \psi_i^c(r)\} \tag{12}$$

given the known core states $\psi_i^c(r)$. This property is not shared by other pseudopotentials (e.g., Kerker, 1980; Chelikowsky and Cohen, 1976). The choice of $\{C_{nn'}^{(l)}\}$ has, however, a direct bearing on the *transferability* of the atomic pseudopotentials from one system to another as well as on the degree to which the true valence wavefunctions can be reproduced without resort to core orthogonalization. Our choice of wavefunction transformation coefficients (Zunger, 1979) is based simply on *maximizing the similarity between the true and pseudo orbitals* [within the form of Eq. (8)] with a minimum core amplitude, subject to the constraints that $\chi_{nl}(r)$ be normalized and nodeless. This simple choice produces highly energy-independent, and thus transferable, pseudopotentials, and at the same time the imposed wavefunction similarity leads to pseudo wavefunctions that retain the full chemical information contained in the valence region of the "true" wavefunctions. Details of the numerical procedure used to obtain $\{C_{n,n'}^{(l)}\}$ are given elsewhere (Zunger and Ratner, 1978; Zunger, 1979). The underlying principle for obtaining maximal wavefunction similarity can however be demonstrated with a simple example. Consider a first-row atom having a single 1s core state. The pseudo-orbitals according to Eq. (8) have the form:

$$\chi_{2s}(r) = C_{1s,2s}^{(0)}\psi_{1s}^{c}(r) + C_{2s,2s}^{(0)}\psi_{2s}^{v}(r)$$
$$\chi_{2p}(r) = \psi_{2p}^{v}(r) \tag{13}$$

Normalization leads to:

$$[C_{1s,2s}^{(0)}]^2 + [C_{2s,2s}^{(0)}]^2 = 1 \tag{14}$$

Imagine now starting from $C_{1s,2s}^{(0)} = 0$ and $C_{2s,2s}^{(0)} = 1$ and gradually increasing $C_{1s,2s}^{(0)}$ from zero, keeping $C_{2s,2s}^{(0)} = (1 - C_{1s,2s}^2)^{\frac{1}{2}}$. With more core character included, the node in $\psi_{2s}^{v}(r)$ shifts towards the origin. The first point at which the node coincides with the origin, giving a legitimately nodeless $\chi_{2s}(r)$, occurs at

$$\chi_{2s}(0) = 0 = C_{1s,2s}^{(0)}\psi_{1s}^{c}(0) + C_{2s,2s}^{(0)}\psi_{2s}^{v}(0) \tag{15}$$

Given the values of the all-electron orbitals $\psi_{1s}^{c}$ and $\psi_{2s}^{v}$ at the origin, Eqs. (14)–(15) determine, therefore, the expansion coefficients and hence the atomic pseudopotential in Eq. (9). One can imagine, however, a process in which one continues to mix core character over and beyond what is necessary just to eliminate the node in $\chi_{2s}(r)$. This continues to produce legitimate $\chi_{2s}(r)$ orbitals in the sense that they are nodeless. However, due to the admixture of excess $\psi_{1s}^{c}$, the similarity of $\chi_{2s}$ to the true $\psi_{2s}^{v}$ decreases. Hence, the maximum similarity criterion becomes identical in this case with the condition that $\chi_{nl}(r = 0) = 0$, or a minimal core content in $\chi_{nl}(r)$. The vanishing value of $\chi_{nl}(r)$ at the origin causes the repulsive Pauli potential $U_l(r)$ in Eq. (10) to have a singularity at the origin since $v_{ps}^{(l)} \sim U_l(r) \sim \dfrac{\nabla_r^2 \chi_l(r)}{\chi_l(r)}$. Combining such a positive and strongly repulsive $U_l(r)$ with the negative core attraction term $-Z_v/r$ in Eq. (9) leads necessarily to a hard-core-type pseudopotential with a characteristic crossing point $v_{ps}^{(l)}(r_l^0) = 0$ at $r = r_l^0$. *The occurrence of hard-core pseudopotentials with their attendant orbital radii $r_l^0$ is therefore a consequence of the chemically motivated constraint that the pseudo wavefunctions of the form in Eq. (8) have the maximum possible similarity to the true valence* *wavefunction in the chemically important tail region.* This short-range repulsive nature of $U_1(r)$ builds into the first-principles pseudopotentials the high-momentum components absent in the empirical pseudopotentials (Cohen and Heine, 1970) that are fit to experimental energies alone.

In the more general case of an atom belonging to an arbitrary row in the periodic table, obtaining maximal wavefunction similarity is formulated as a constrained minimization of the core projection of the pseudo wavefunctions (e.g., Zunger, 1979). The general small-$r$ expansion of the pseudo orbital becomes

$$\lim_{r \to 0} \chi_{nl}(r) = A_0 r^{\eta+1} + A_1 r^{\eta+l+1} + A_2 r^{\eta+l+2} \tag{16}$$

The choice of $\eta \geq 2$ leads to a minimum core amplitude pseudo wavefunction with its attendant maximum similarity to the true valence wavefunction. Inserting Eq. (16) into Eqs. (10) and (9) leads, for *any* $\eta \geq 2$, to:

$$\lim_{r \to 0} v_{ps}^{(l)}(r) = \frac{\tilde{B}_l}{r^2} - \frac{Z_v}{r} + \cdots \tag{17}$$

Hence, the Simons–Bloch (1973) empirical pseudopotential [Eq. (2)] is recovered as the small-$r$ limit of the first-principles pseudopotential. Clearly, however, at finite $r$-values, the present potential [Eq. (9)] differs substantially from the Simons–Bloch form. The choice $\eta = 0$ leads to a soft-core pseudopotential [$\lim_{r \to 0} v_{ps}^{(l)}(r) = $ constant]. The associated pseudo wavefunction is now finite at the origin, leading necessarily to a reduced similarity between the true and pseudo wavefunctions in the chemically relevant valence region. Our choice of the wavefunction transformation in Eq. (8) produces, therefore, unique pseudopotentials by going to the extreme limit of wavefunction similarity that is possible within the underlying density-functional orbital space.

Other possibilities for choosing pseudo wavefunctions exist and are discussed elsewhere (Zunger and Cohen, 1979b; Zunger, 1979; Redondo *et al.*, 1977). These procedures involve various ways of constructing pseudo wavefunctions from components lying *outside the density-functional orbital space* [unlike Eq. (8)] and do not maintain physically transparent analytical forms such as in Eqs. (9)–(10). Hence, to distinguish them from the present density-functional pseudopotentials, we refer to these as trans-density-functional (TDF) pseudopotentials. Although such procedures lead sometimes to a somewhat better numerical accuracy in the wavefunctions, we restrict ourselves in what follows to the conceptually simpler density-functional pseudopotentials.

The approach described above for constructing orbital-dependent pseudo potentials can easily be extended to spin- and orbital-dependent potentials

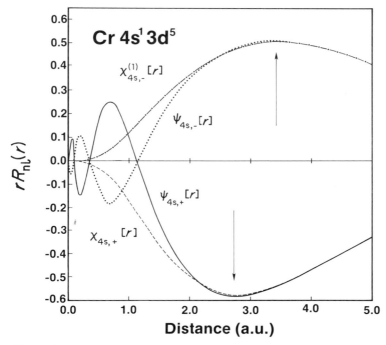

**Fig. 1.** Comparison of the 4s all-electron $[\psi_{4s\uparrow}(r), \psi_{4s\downarrow}(r)]$ and pseudo $[\chi_{4s\uparrow}(r), \chi_{4s\downarrow}(r)]$ wavefunctions of the Cr atom for spin up(↑) and down(↓).

(Zunger, 1980a). This generalization is simple, and we will not describe the details here, but rather give an illustrative example. Figure 1 compares the spin-up and spin-down pseudo wavefunction $\chi_{4s\uparrow}(r)$ and $\chi_{4s\downarrow}(r)$, which are eigenstates of the pseudo-Hamiltonian for the Cr atom, with the all-electron density-functional orbitals $\psi^v_{4s\uparrow}(r)$ and $\psi^v_{4s\downarrow}(r)$. The two sets of orbitals match very closely from $r = \infty$ up to a point $r \approx 2$ a.u., which lies inwards to the outer maximum of the true valence wavefunction. Hence, most bonding effects should be reasonably reproduced by the pseudo wavefunctions.

Figure 2 depicts various components of the $l = 0$ atomic pseudopotential in Eq. (9) for Sb. The curve labeled (1) is the Pauli term $U_1(r)$, the curve labeled (2) shows the $-Z_v/r$ term, and curve (3) represents all other terms in Eq. (9). Finally, curve (4) shows the total pseudopotential. Figure 3 shows the atomic pseudopotential for the second-row atoms, and Fig. 4 shows similar results for the transition elements. First-principles atomic pseudopotentials were generated for 70 atoms with $2 \leqslant Z \leqslant 57$ and $72 \leqslant Z \leqslant 86$ (i.e., the first five rows).

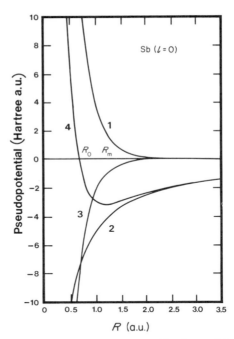

**Fig. 2.** Components of the atomic pseudopotential $v_{ps}^{(1)}(r)$ [Eq. (9)] for $l = 0$ of the Sb atom: (1) Pauli potential $U_1(r)$; (3) core screening; (2) the Coulomb attraction $-Z_v/r$; and (4) the total pseudopotential. $R_0$ and $R_m$ denote the points of crossing and minimum, respectively.

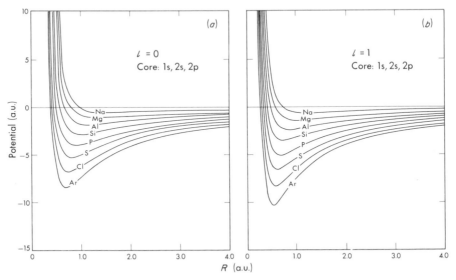

**Fig. 3.** Atomic pseudopotentials for the second-row atoms: (a) s potentials; (b) p potentials [Eq. (9)].

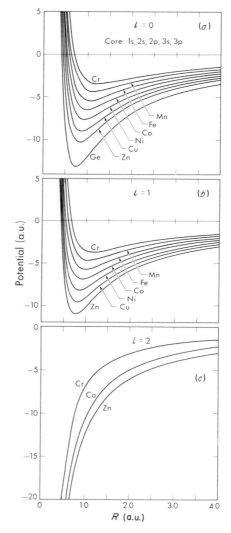

**Fig. 4.** Atomic pseudopotentials for the 3d elements: (a) s potentials; (b) p potentials; (c) d potentials [Eq. (9)]. Note that due to the absence of core states of $l = 2$ symmetry, the d potential for this row is purely attractive.

A notable feature of these "hard-core" potentials is the occurrence of a crossing point $v_{ps}^{(l)}(r_1^0) = 0$ at $r = r_1^0$. From Eqs. (9)–(10), it is seen that, physically, this point is where the repulsive Pauli potential is balanced by the Coulomb attraction $- Z_v/r$, renormalized by the screened core potential, exchange-correlation nonlinearity, and the Coulomb and exchange-correlation orthogonality hole potentials.

It seems somewhat puzzling at first sight that pseudopotentials of free-electron-like metals such as Na, Al, and K may have large momentum components or even a hard core, because the nearly-free-electron (NFE) model seems to have worked so well for these materials. However, the successes of the NFE model may have been overstated, in view of the fact that *wavefunction-related properties* of the free-electron metals, such as the shape of the optical conductivity (Bennett and Vosko, 1972), the metallic ground-state charge density and form-factors (Walter *et al.*, 1973; compare, however, with Bertoni *et al.*, 1973; Hafner, 1978), as well as the properties of impurities in metals (Hodges, 1977), are poorly reproduced by local and weak pseudopotentials. Moreover, the occurrence of rather complex crystal structures involving "simple" free-electron atoms (semimetals such as the B32 structure of LiAl, Laves phase materials such as $K_2Cs$, the existence of the compound $Na_2K$ but not NaK or $NaK_7$, etc.) as well as the existence of stable multiple valencies of these systems (e.g., AlF vs. $AlF_3$, etc.) cannot be understood in terms of local NFE pseudopotentials. Hence, although such weak and NFE pseudopotentials had to be assumed for many elements (including Groups IIIA–VIA atoms) for the very popular low-order perturbation theories to be valid, the underlying assumption—that the complex chemistry of the related compounds could be understood in terms of weak and isotropic perturbations of a homogeneous electron gas—seems naive. However, the great analytical beauty of the perturbation theory approaches (Harrison, 1966) need not be sacrificed when hard-core pseudopotentials are used. Instead, a new definition is necessary for the unperturbed system as a suitably *nonhomogeneous* form, including most of the chemically relevant potential fluctuations in zero order (Simons–Bloch pseudopotentials? square wells?).

Using the calculated atomic pseudopotentials of Eqs. (9)–(11), we now define the crossing points using the ground-state *screened* atomic pseudopotentials $v_{\text{eff}}^{(l)}(r)$:

$$v_{\text{eff}}^{(l)}(r) = v_{\text{ps}}^{(l)}(r) + \frac{l(l+1)}{2r^2} + v_{\text{ee}}[n] + v_x[n] + v_{\text{cr}}[n] \qquad (18)$$

as:

$$v_{\text{eff}}^{(l)}(r_l) = 0$$

Here $v_{\text{eff}}^{(l)}(r)$ is the total effective potential experienced in a ground-state pseudo atom by electrons with angular momentum $l$. These form the structural indices $\{r_l\}$, which we use in connection with predicting the stable crystal structure of compounds. Table I gives the $\{r_l\}$ values of the 70 elements

**TABLE I**

Classical Crossing Points of the Self-Consistently Screened Nonlocal Atomic
Pseudopotentials (Including the Centrifugal term) (a.u.)[a]

| Atom | $r_s$ | $r_p$ | $r_d$ | Atom | $r_s$ | $r_p$ | $r_d$ |
|------|-------|-------|-------|------|-------|-------|-------|
| Li | 0.985 | 0.625 | — | Rb | 1.67 | 2.43 | 0.71 |
| Be | 0.64 | 0.44 | — | Sr | 1.42 | 1.79 | 0.633 |
| B | 0.48 | 0.315 | — | Y | 1.32 | 1.62 | 0.58 |
| C | 0.39 | 0.25 | — | Zr | 1.265 | 1.56 | 0.54 |
| N | 0.33 | 0.21 | — | Nb | 1.23 | 1.53 | 0.51 |
| O | 0.285 | 0.18 | — | Mo | 1.22 | 1.50 | 0.49 |
| F | 0.25 | 0.155 | — | Tc | 1.16 | 1.49 | 0.455 |
| Ne | 0.22 | 0.14 | | Ru | 1.145 | 1.46 | 0.45 |
| | | | | Rh | 1.11 | 1.41 | 0.42 |
| Na | 1.10 | 1.55 | | Pd | 1.08 | 1.37 | 0.40 |
| Mg | 0.90 | 1.13 | — | Ag | 1.045 | 1.33 | 0.385 |
| Al | 0.77 | 0.905 | — | Cd | 0.985 | 1.23 | 0.37 |
| Si | 0.68 | 0.74 | — | In | 0.94 | 1.11 | 0.36 |
| P | 0.60 | 0.64 | — | Sn | 0.88 | 1.00 | 0.345 |
| S | 0.54 | 0.56 | — | Sb | 0.83 | 0.935 | 0.335 |
| Cl | 0.50 | 0.51 | — | Te | 0.79 | 0.88 | 0.325 |
| Ar | 0.46 | 0.46 | — | I | 0.755 | 0.83 | 0.315 |
| | | | | Xe | 0.75 | 0.81 | 0.305 |
| K | 1.54 | 2.15 | 0.37 | | | | |
| Ca | 1.32 | 1.68 | 0.34 | Cs | 1.71 | 2.60 | |
| Sc | 1.22 | 1.53 | 0.31 | Ba | 1.515 | 1.887 | 0.94 |
| Ti | 1.15 | 1.43 | 0.28 | La | 1.375 | 1.705 | 0.874 |
| V | 1.09 | 1.34 | 0.26 | Hf | 1.30 | 1.61 | 0.63 |
| Cr | 1.07 | 1.37 | 0.25 | Ta | 1.25 | 1.54 | 0.605 |
| Mn | 0.99 | 1.23 | 0.23 | W | 1.22 | 1.515 | 0.59 |
| Fe | 0.95 | 1.16 | 0.22 | Re | 1.19 | 1.49 | 0.565 |
| Co | 0.92 | 1.10 | 0.21 | Os | 1.17 | 1.48 | 0.543 |
| Ni | 0.96 | 1.22 | 0.195 | Ir | 1.16 | 1.468 | 0.526 |
| Cu | 0.88 | 1.16 | 0.185 | Pt | 1.24 | 1.46 | 0.51 |
| Zn | 0.82 | 1.06 | 0.175 | Au | 1.21 | 1.45 | 0.488 |
| Ga | 0.76 | 0.935 | 0.17 | Hg | 1.07 | 1.34 | 0.475 |
| Ge | 0.72 | 0.84 | 0.16 | Tl | 1.015 | 1.22 | 0.463 |
| As | 0.67 | 0.745 | 0.155 | Pb | 0.96 | 1.13 | 0.45 |
| Se | 0.615 | 0.67 | 0.15 | Bi | 0.92 | 1.077 | 0.438 |
| Br | 0.58 | 0.62 | 0.143 | Po | 0.88 | 1.02 | 0.425 |
| Kr | 0.56 | 0.60 | 0.138 | At | 0.85 | 0.98 | 0.475 |
| | | | | Rn | 0.84 | 0.94 | 0.405 |

[a] The core shell is defined in each case as the rare-gas configuration of the
preceding row. The Kohn and Sham exchange is used.

for which the density-functional pseudopotential equations have been solved. We have not included the heavier elements since the present pseudopotential theory is nonrelativistic. In what follows, we will hence not discuss the structural stability of lanthanide and actinide compounds.

The structure of Eq. (18) reflects our discussion in the Introduction concerning the quantum-mechanical and semiclassical viewpoints of electronic structure: $v_{ps}^{(l)}(r)$ is the realization of the quantum-mechanical electron-core potential $V_{ec}(\mathbf{r}, \mathbf{R})$, whereas the last three terms in Eq. (18) represent the interelectronic potential $V_{ee}(\mathbf{r}, \mathbf{r}')$. The semiclassical factors are then represented by the $\{r_l\}$ scale implicit in the screened effective potential.

In developing the density-functional pseudopotentials, we have tacitly assumed a specific partitioning of the atomic orbitals into core and valence. In the present theory, core orbitals are those appearing as closed-shell states in the rare gas atom of the preceding row in the periodic table. Note, however, that although we may understand the low-energy electronic excitation spectra of a compound such as ZnSe by assuming that the Zn 3d orbitals belong to in a passive core state, such an assumption may be invalid in intermetallic compounds, where the Zn 3d orbitals can be in near resonance with the d orbitals of another element (e.g., CuZn). Given the fact that any such delineation into core and valence is merely based on an arbitrary assumption on the passivity of certain selected orbitals to chemical perturbations of interest, one may ask whether structurally meaningful orbital radii can be extracted from a pseudopotential scheme.

In fact, the choice of the orbital radii from the *screened* pseudopotential [Eq. (18)], rather than from the *bare* pseudopotential $v_{ps}^{(l)}(r)$ in Eq. (9) (e.g., Simons and Bloch, 1973; Andreoni *et al.*, 1979), is based precisely on an attempt to avoid such a nonuniqueness. Although the bare pseudopotential of Eq. (9) has the form

$$v_{ps}^{(l)}(r) = U_1(r) + f(Z_c, Z_v, \rho_c, \rho_v, n) \tag{19}$$

the *screened* pseudopotential can be written as

$$V_{eff}^{(l)}(r) = U_1(r) + g(Z, \rho_c + \rho_v) \tag{20}$$

Note that whereas $U_1(r)$ [Eq. (10)] depends only on orbitals with angular momentum $l$, the valence pseudo charge density $n(r)$ [Eq. (11)] depends on all orbitals that are assigned as valence states. Consequently, if the Zn 3d orbitals are assumed to belong to the core, the *bare pseudopotential* $v_{ps}^{(l)}(r)$ for s and p electrons is different than if the d electrons were assigned to the valence. In contrast, it follows from Eq. (20) that the *screened pseudopotential* $v_{eff}^{(l)}(r)$ for $l = 0, 1$ is invariant under such a change in the assignment of the d electrons. Our definition of the structural indices $r_l$ is therefore independent of the assignment of orbitals $\psi_{nl}(r)$ from other angular shells as core or valence. Also note that the definition of orbital radii from the screened pseudopotentials of Eq. (18) permits a direct inclusion of electronic exchange and correlation effects in the structural coordinates $r_l$ (see Schubert, 1977), whereas the semiclassical electron concentration factor (Hume-Rothery and Raynor, 1954) is represented simply by $Z_v$.

## B.   Simple Universal Form of the Density-Functional Pseudopotential

The idea of atomic radii is not new in pseudopotential theory (see Sec. II). The basic thrust of the pseudopotential concept is to transform the chemical picture of the existence of an *orbital subspace* of nearly chemically inert core states into a delineation either in *configuration space* or in *momentum space* of a core region of the potential (with its attendant cancellation effects between orthogonality repulsion and Coulomb attraction) and a valence region (with its weaker effective potential). What is new in our present approach is that whereas in the empirical pseudopotential methods the radii were imposed *extraneously*, either explicitly (Abarenkov and Heine, 1965; Ashcroft, 1966; Shaw, 1968; Simons and Bloch, 1973; Natapoff, 1975, 1976, 1978) or implicitly (Cohen and Bergstresser, 1966), the present theory provides them as a natural fingerprint of the internal quantum structure of the isolated atom. Hence, although the empirical pseudopotential methods assumes the existence of opposing forces in the core region leading to the occurrence of a delineating radius, in these approaches the forces are not calculated. The radii are in turn transferred from various sources (Pauling ionic radii, the position of the last node in the valence s orbital, fitting energy eigenvalues to atomic term values, optical reflectivity of semiconductors, or the Fermi surface of metals, etc.), such that although a desired fit to selected experimental observables is achieved, the underlying electronic and structural regularities may be obscured by fitting to different data or by postulating certain arbitrary analytic forms for $v_{ps}^{(l)}(r)$.

Given that the analytic form of the pseudopotential in the present approach is not assumed but rather emerges as a consequence of requiring a maximum similarity between the all-electron and pseudo wavefunctions in the chemically important tail region, it however is possible to deduce a posteriori a universal analytic form. The density-functional atomic pseudopotentials have been calculated numerically from Eqs. (9)–(11), given the all-electron density-functional wavefunctions and orbital energies (Zunger and Cohen, 1978b). In the present study, we use the orbital radii $\{r_l\}$ determined from Eq. (18) (Table I) and these numerical pseudopotentials. However, because the limiting behavior of all terms entering the pseudopotentials in Eqs. (9)–(10) is known, one can obtain an *approximate explicit analytical form* for these pseudopotentials through fitting to the numerical results. Such a fit can be done in two different ways: either emphasizing a high numerical accuracy for the fit (and hence using rather complicated fitting functions) or by using a physically transparent fitting function, sacrificing to some extent the numerical accuracy but obtaining the correct *regularities* of the pseudopotentials. This has been attempted by Lam *et al.* (1980) using the simple form:

$$v_{ps}^{(l)}(r) \approx \frac{C_{11}}{r^2} e^{-C_{21}r} - \frac{Z_c}{r} e^{-C_3r} - \frac{Z_v}{r} \qquad (21)$$

The coefficients $\{C_{11}, C_{21}, \text{ and } C_3\}$ are tabulated by Lam *et al.* Although more complicated forms than Eq. (21) have also been used (Lam *et al.*, 1980), Eq. (21) reveals a very important characteristic of the density-functional pseudopotentials: to within a reasonable approximation, the constants $C_{11}$, $C_{21}$, and $C_3$ are *linear* functions of the atomic number, i.e.,

$$C_{11} \approx a_1 + b_1Z; \qquad C_{21} \approx c_1 + d_1Z; \qquad C_3 \approx e + fZ \qquad (22)$$

This constitutes a significant reduction in the number of degrees of freedom required to specify the potential and reveals the regularities of the periodic table through the coordinates $(Z_c, Z_v)$. This can be contrasted with the empirical pseudopotential approach in which such regularities are often obscured by fitting certain atomic pseudopotentials to optical data (Cohen and Bergstresser, 1966), whereas others are fit to metallic Fermi-surface data and the resistivity of metals (Ashcroft, 1966, 1968; Ashcroft and Langreth, 1967) or to atomic term values (Szasz and McGinn, 1967; Simons, 1971a,b; Abarenkov and Heine, 1965).

The existence of a simple linear scaling relationship in Eqs. (21)–(22) establishes a mapping of Mendeleyev's classical dual coordinates $Z_c$ and $Z_v$ characterizing the digital structure of the periodic table, into a more refined quantum-mechanical coordinate system, $r_s(Z_c, Z_v)$, $r_p(Z_c, Z_v)$, and $r_d(Z_c, Z_v)$. Given the fact that Mendeleyev's dual coordinates $(Z_c, Z_v)$ are already suggestive of broad structural trends (e.g., the AB compounds with $Z_v^A = 3$ and $Z_v^B = 5$ tend to form zinc blende structures for large $Z_c^{A,B}$ values, while compounds with $Z_v^A = 1$ and $Z_v^B = 7$ tend to form rock-salt structures, etc.), it is only reasonable to expect that with their present resolution into anisotropic orbital components, far more sensitive structural coordinates can be achieved.

## C.  Application to Electronic Structure Calculations

As discussed in Sec. II, properly constructed pseudopotentials can be used for self-consistent electronic structure calculations for molecules and solids in a way that is much simpler than the all-electron approach—or for using the periodic chemical regularities coded in the pseudopotentials to abstract structurally significant semiclassical-like scales. In this section, we briefly summarize the first application; in the following section (IV), we discuss the structural significance of the pseudopotentials.

In using pseudopotentials for electronic structure calculations, one constructs the solid-state pseudopotential $V_{ps}^{(l)}(\mathbf{r})$ as a superposition of the transferable *atomic* pseudopotentials $v_{ps}^{(l)}(r)$. The screening $V_{scr}[n(\mathbf{r})]$ is then calculated from the self-consistent response of the valence electrons in the *solid* to this external potential. Before describing applications to electronic structure, we wish to caution the reader in relation to two fundamental limitations of first-principles pseudopotentials. First, although any pseudopotential simplifies the description of many-electron systems by projecting out the often chemically passive core wavefunctions, its application is inherently limited to physical quantities that are largely unaffected by such core states. Hence, quantities such as the Fermi contact interactions, core spin polarization, Mössbauer core shifts, or core photoelectron spectra are entirely outside the realm of application of pseudopotential theories. Second, any first-principles pseudopotential theory, attempting to replace a given all-electron representation of the electronic structure, can give results that are no more refined than the physical assumptions underlying the all-electron theory it replaces. Hence, while empirical pseudopotentials attempt to directly mimic certain experimental observables through a parametrized fit, Hartree–Fock pseudopotentials (e.g., Kahn *et al.*, 1976) or the present density-functional pseudopotentials can produce results that are at best as accurate as is the respective all-electron theory. On the other hand, not only can the first-principles pseudopotentials be progressively refined as our understanding of many-electron correlation effects improves (e.g., Zunger *et al.*, 1980, also unpublished results; Zunger, 1980a), but, even more importantly, in the present approach both the successes and the failures of the theory in explaining experiment can be analyzed and understood in terms of well-defined quantum-mechanical constructs.

As an initial step in using atomic pseudopotentials, one has to establish exactly how transferable they are from one system to another. One way of testing this is to use the *atomically derived* pseudopotential to calculate a self-consistent pseudopotential band structure of a *solid* and compare the results with an all-electron band structure calculation in which no pseudopotentials are used. Such a comparison is given in Table II (D. R. Hamann, unpublished results, 1979), which shows data for crystalline silicon obtained with the TDF pseudopotential (Zunger and Cohen, 1979b; Zunger, 1979). It is seen that the pseudopotential calculation, considering only four valence

**TABLE II**

Comparison of the Band Structure of Crystalline Silicon as Obtained from an All-Electron (Core + Valence) Calculation and a Valence-Only Pseudopotential Calculation (eV)[a]

| Level | All-electron | Pseudopotential | Level | All-electron | Pseudopotential |
|-------|-------------|-----------------|-------|-------------|-----------------|
| $\Gamma_{1,v}$ | $-12.02$ | $-11.88$ | $X_{1,c}$ | $0.55$ | $0.62$ |
| $\Gamma_{25,v}$ | $0.00$ | $0.00$ | $X_{4,c}$ | $10.32$ | $10.26$ |
| $\Gamma_{15,c}$ | $2.49$ | $2.53$ | $L_{2',v}$ | $-9.64$ | $-9.55$ |
| $\Gamma_{2',c}$ | $3.18$ | $3.07$ | $L_{1,v}$ | $-7.06$ | $-6.97$ |
| $\Gamma_{1,c}$ | $7.46$ | $7.53$ | $L_{3,c}$ | $-1.16$ | $-1.14$ |
| $\Gamma_{12,c}$ | $7.86$ | $7.85$ | $L_{1,c}$ | $1.40$ | $1.39$ |
| $X_{1,v}$ | $-7.84$ | $-7.76$ | $L_{3,c}$ | $3.37$ | $3.40$ |
| $X_{4,v}$ | $-2.82$ | $-2.78$ | | | |

[a] Using the TDF pseudopotential of Zunger (1979) and Zunger and Cohen (1979b). Results are obtained by a self-consistent linear-augmented-plane-wave method (D. R. Hamann, unpublished results, 1979) using the Wigner exchange and correlation potential.

electrons per Si atom, and the all-electron calculations, which include the full 14 electrons per Si atom, match within an average deviation of 0.06 eV over an energy range of valence and conduction bands of 20 eV. Another way of testing the pseudopotential energy dependence involves using an atomic pseudopotential derived from the *ground electronic configuration* in atomic self-consistent calculations for *excited configurations*. By means of exciting the atom (or ionizing a few electrons), a very wide range of wavefunction localization and orbital energies can be probed. This can be used to test whether the pseudopotential results continue to mimic the all-electron results away from the ground-state electronic configuration used to construct the pseudopotential. Extensive work (Zunger and Cohen, 1978b; Zunger, 1979) on many atoms indicates that the typical errors involved in orbital energies, total energy differences (i.e., excitation energies), and wavefunction moments are within $10^{-2}$–$10^{-4}$ eV, $10^{-3}$–$10^{-4}$ eV, and 0.1%–2%, respectively, over a range of about 20 eV of excitation energies. This satisfies our initial constraint that atomic pseudopotentials be constructed in a way that is approximately independent of their chemical environment.

The first-principles atomic pseudopotentials have been applied to self-consistent electronic structure calculations of polyatomic systems such as diatomic molecules—$O_2$ and $Si_2$ (Kerker *et al.*, 1979; Schlüter *et al.*, 1979); tetrahedrally bonded semiconductors—silicon (D. R. Hamann, unpublished results, 1979; Zunger and Cohen, 1979b), Ge (Zunger and Cohen, 1979b), and GaAs (Zunger, 1980b); as well as elemental transition-metal solids—Mo and W (Zunger *et al.*, 1979a; Zunger and Cohen, 1979a). In addition, these

**TABLE III**

Calculated and Observed Equilibrium Lattice Constant $a_{eq}$ (Å), Total Valence Energy $E_t$ (Rydberg), or Cohesive Energy $\Delta E_0$ (eV), and Bulk Modulus $B$ (dyn/cm$^2$) for Crystalline Silicon, bcc Tungsten, and Molybdenum[a]

|  |  | $a_{eq}$ | $E_t$ or $\Delta E_0$ | $B$ ($10^{12}$ dyn/cm$^2$) |
|---|---|---|---|---|
| Si | Observed | 5.43 | $-7.919^b$ | 0.99 |
|  | Calculated | 5.44 | $-7.959^b$ | 0.94 |
| Mo | Observed | 3.147 | $6.82^c$ | 3.23 |
|  | Calculated | 3.15 | $6.68^c$ | 3.45 |
| W | Observed | 3.165 | $8.90^c$ | 2.73 |
|  | Calculated | 3.17 | $7.90^c$ | 3.05 |

[a] Calculated results are obtained using the first-principles atomic pseudopotentials and a self-consistent mixed-basis approach (Zunger, 1980c; Zunger and Cohen, 1979a).

[b] Total valence energy (Ryd).

[c] Cohesive energy (eV).

pseudopotentials were used in the first nonlocal calculations dealing with a semiconductor surface—GaAs (110) (Zunger, 1980b), transition-metal impurities in crystalline silicon (U. Lindefelt and A. Zunger, unpublished, 1981), and the Ge–GaAs semiconductor interface (A. Zunger, unpublished results, 1980). The scope of this article does not warrant the description of detailed results, and the interested reader is referred to the original papers. Here, we will give two examples illustrating the level of agreement between these calculations and experiment.

Table III shows the predicted and observed equilibrium lattice constant $a_{eq}$, bulk modulus $B$, and cohesive energy $\Delta E_0$ (or total valence energy) for Mo, W, and Si in their observed stable crystal structure. These are obtained by solving the Schrödinger equation self-consistently with the effective potential of Eq. (5) and calculating the variational ground-state total energy using a mixed basis set of Gaussian orbitals and plane waves. The second example is given in Fig. 5, where the calculated pseudopotential charge density of bulk Si (Zunger, 1980c) is compared with the experimental bonding charge density (Yang and Coppens, 1974). Keeping in mind that no empirical input other than the crystal structure and the atomic number is used in all of these calculations, the level of agreement with experiment is striking. What has been demonstrated here is that the first-principles pseudopotential method is capable of producing transferable atomic potentials that in quantitative electronic structure calculations yield a chemical type of accuracy.

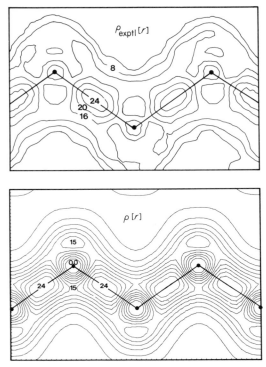

**Fig. 5.** Experimental $[\rho_{\mathrm{exptl}}(\mathbf{r})$, Yang and Coppens (1974)] and calculated valence charge density of silicon in the (110) plane, in units of electron per cell. The calculation is based on the first-principles pseudopotential using a self-consistent mixed-basis method. Full dots indicate atomic positions.

## IV.  TRENDS IN ORBITAL RADII

### A.  Chemical Regularities

We have argued that the classical turning points $r_l$ of the screened density-functional *atomic* pseudopotentials form a useful elemental distance scale for *solids*. One may then ask if indeed such atomic quantities retain their significance in the solid state. To answer this, we have performed a self-consistent band-structure calculation for bcc tungsten using our atomic pseudopotentials. This is done by assuming that the crystalline pseudopotential $V_{\mathrm{ps}}^{(l)}(\mathbf{r})$ is a superposition of the atomic pseudopotentials $v_{\mathrm{ps}}^{(l)}(r)$; but the screening $V_{\mathrm{scr}}^{v}[n]$ is calculated from the self-consistent Bloch wavefunctions of the solid (Zunger

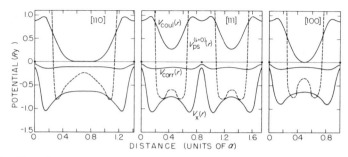

**Fig. 6.**   The self-consistent screening potential [interelectronic Coulomb $V_{coul}(\mathbf{r})$, exchange $V_x(\mathbf{r})$, and correlation $V_{corr}(\mathbf{r})$] and pseudopotential $V_{ps}^{(l=0)}(\mathbf{r})$ for bcc tungsten in different directions in the crystal.

and Cohen, 1979a), rather than from atomic orbitals. The resulting band-structure, Fermi-surface, and optical spectra are in very good agreement with previously published experimental data. One can now use the self-consistent crystalline charge density $n(\mathbf{r})$, calculate the Coulomb, exchange, and correlation screening in the solid, and extract from that the screened solid-state pseudopotentials $V_{eff}^{(1)}(\mathbf{r})$ [Eq. (1)] their classical turning points. Obviously, such a solid-state screened potential has a different form in the different crystalline directions $[h, k, l]$, resulting in spatially anisotropic orbital radii $r_l[h, k, l]$. Figure 6 shows the solid-state tungsten pseudopotential (dashed lines) as well as the three components of the screening (evaluated with respect to the Fermi energy) in the solid. Although the screened pseudopotentials show a pronounced directional character, *the solid-state radii, lying in the core region of the atoms, show only a small anisotropy*: $r_0[111] = 1.279 \pm 0.002$ a.u., $r_0[001] = 1.214 \pm 0.002$ a.u., and $r_0[110] = 1.256 \pm 0.002$ a.u., compared with the isotropic atomic value $r_0 = 1.225$ a.u. and the average crystalline value of 1.25 a.u. The near invariance of these radii with respect to the chemical environment should be contrasted with the pronounced dependence of the classical crystallographic radii (e.g., Shannon and Prewitt, 1969) on chemical factors.

Inspection of the atomic pseudopotentials depicted in Figs. 3 and 4 immediately reveals clear regularities. This may be appreciated from Fig. 7, which shows the radius $r_l^{min}$ at which the $l = 0$ pseudopotential has its minimum, plotted against the depth of the minimum $W_l$. The column structure of the periodic table is immediately apparent. At the upper left corner of the figure, we see elements such as Cs and Rb, characterized by a very shallow and extended pseudopotential; these elements are indeed the least electronegative in the first five rows of the periodic table. In the lower right corner, we find

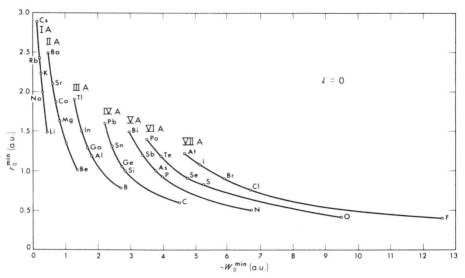

**Fig. 7.** The correlation between the radius $r_0^{min}$ at which the s pseudopotential has its minimum and the depth of the minimum $-W_0^{min}$, for the nontransition elements.

elements such as F and O, which are characterized by very deep and localized pseudopotentials; these are indeed the most electronegative elements. Clearly, as the electronegativity is a measure of the power of an atom to gain extra electrons from its environment and at the same time keep its own electrons, such a propensity is reflected in the potential-well structure of $v_{ps}^{(l)}(r)$. In contrast with the thermochemical or dielectric electronegativity scales, however, the present orbital radii define an *anisotropic* (or *l*-dependent) electronegativity scale.

We see in Fig. 7 that the first-row elements are somewhat separated from the other elements—the former having deeper potentials than might have been expected from extrapolating the data for other rows. This phenomenon, resulting from a weak pseudopotential kinetic energy cancellation for the first-row elements, is also clearly reflected in the thermochemistry of the corresponding compounds. As we move from the right to the left of the periodic table, one sees in Fig. 7 that the elements belonging to a given column can be characterized solely by their potential radii, the potential depth being nearly constant. This seems to be the basis for the success of the "empty core" pseudopotentials (Ashcroft, 1966) postulated for simple metals, in which $v_{ps}(r)$ is assumed to be zero within a sphere of radius $R_{ec}$. Only the variation of $R_{ec}$ within a column in the periodic table is used to characterize a large

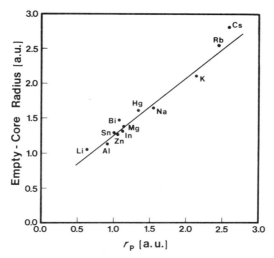

**Fig. 8.** Correlation between Ashcroft's empty-core pseudopotential radius and the p orbital radius of the present density-functional screened pseudopotential (Table I).

variety of transport and structural data for the corresponding metals (e.g., Ashcroft, 1966, 1968; Ashcroft and Langreth, 1967; Stroud and Ashcroft, 1971). In fact, one finds that these empirical empty core radii used to fit resistivity data may be identified, within a linear scale factor, with our $r_p$ screened pseudopotential coordinate (Fig. 8). Whereas the alkali elements are characterized predominantly by a single coordinate (Fig. 7), in line with their free-electron properties associated with a shallow pseudopotential, the elements to their left are characterized by a dual coordinate system. The regularities in these dual coordinates also reflect well-known chemical trends: for example, the tendency towards metalization in the C, Si, Ge, Sn, and Pb series is represented by the increased delocalization and reduced depth in their pseudopotentials, etc.

To illustrate the significance of the angular momentum dependence of the atomic pseudopotentials, Fig. 9 shows the elements ordered by their $(r_l^{min}, W_l)$ coordinates. The elements are clearly grouped according to their rows in the periodic table. At small $|W_l|$ and large $r_l^{min}$ (i.e., shallow and delocalized, or weak, potentials), we find the classical free-electron-like metals; at large $|W_l|$ and small $r_l^{min}$ (deep and localized, or strong, potentials) we find the atoms that form covalent structures and the transition elements. Each row in the periodic table is represented here by at least two lines—one connecting the full circles passing through the $l = 0$ coordinates and one connecting open

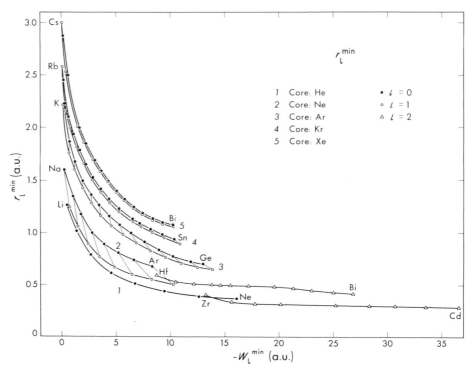

**Fig. 9.** The correlation between the radius $r_l^{min}$ at which the pseudopotential $v_{ps}^{(l)}(r)$[Eq. (9))] has its minimum and the depth of the minimum $-W_l^{min}$, arranged according to rows in the periodic table. The first and last elements of each row are denoted by their chemical symbol.

circles passing through the $l = 1$ coordinates. The full triangles denote the $l = 2$ coordinates. The $l = 1$ components of the first-row atoms as well as the $l = 2$ components of the second and third rows are purely attractive (cf. Fig. 4c), and all have a minimum of negative infinity at the origin. For clarity of display, we have connected the $l = 0$ and $l = 1$ coordinate of each atom by a straight line. Clearly, the length and slope of these lines measure the s–p nonlocality of the potential. Few interesting observations can be made. The s–p nonlocality decreases as one moves down the columns in the periodic table, as the ratio of the number of core states of $l = 0$ and $l = 1$ symmetry approaches unity, and $U_l(r)$ becomes approximately $l$-independent. The adequacy of the early *local* pseudopotential models to describe the low energy electronic excitations of compounds such as GaAs, AlAs, InSb, etc., is reflected in the present theory by the proximity of the $l = 0$ and $l = 1$ coordi- nates of the corresponding atoms. One further notes that while within a given row the slope of the line connecting the $(r_0^{\min}, W_0)$ and $(r_1^{\min}, W_1)$ coordinates is negative at the right side of each row (i.e., the p potentials are more localized and deeper than the s potentials), these slopes move gradually towards less-negative values and become positive eventually at the left side of the lower rows (i.e., the p potentials become more extended and shallower than the s potentials). This is directly related to the increased delocalization of the outer valence p orbitals as one moves towards the left side of the rows. One notes that the $l = 2$ coordinates are quite separated from the $l = 0$ and 1 coordinates, and vary almost linearly within each row. These localized d potentials are responsible for the relatively narrow and separated d bands in the transition-metal solids. Their variations along the rows parallels the changes in the d-band width in the respective elemental metals, and similarly, their proximity to the $l = 0$ coordinates governs the degree of s–d hybridization.

Having discussed some of the periodic trends exhibited by the atomic pseudopotentials, we now turn to their significance in the establishment of elementary *distance* and *energy* scales, which are quantum-mechanical extensions of similar semiclassical scales discussed in the Introduction.

Figures 10 and 11 display the multiplet-average experimental ionization energy $E_1$ of the atoms (Moore, 1971), plotted against the reciprocal orbital radius $r_1^{-1}$. For each group of elements, we show two lines: $E_s$ vs. $r_s^{-1}$ and $E_p$ vs. $r_p^{-1}$. *The striking result is that the theoretical $r_1^{-1}$ is seen to form an accurate measure of the experimental orbital energies and hence can be used as an elementary orbital-dependent energy scale*, much like Mulliken's electronegativity. Indeed, since $r_1^{-1}$ is a measure of the scattering power of a screened pseudopotential core towards electrons with angular momentum $l$, it naturally forms an electronegativity scale. There is an interesting relation between this picture and Slater's concept of orbital electronegativity within the density-functional formalism (Slater, 1974). In his approach, the spin-orbital electronegativity $X_i$ is defined as the orbital energy $\varepsilon_i$ of the density-functional Hamiltonian, which in turn equals the derivative of the total energy $E$ with respect to the ith orbital occupation number: $X_i = \varepsilon_i = \partial E / \partial n_i$. In the limit where $E$ is a quadratic function of $n_i$, this orbital electronegativity reduces to Mulliken's form. This definition is based on the notion that a chemical reaction takes place when electrons will flow from the highest occupied orbitals of a reactant to the lowest unoccupied orbitals with which a finite overlap occurs. Since the present $r_1^{-1}$ coordinate scales approximately with the orbital energy $\varepsilon_1$, the former coordinate is a realization of Slater's electronegativity in a pseudopotential representation.

The orbital radii $r_1$ also form an interesting distance scale (A. N. Bloch, unpublished results, 1980). Consider an all-electron valence atomic wave-function such as the 4s and 5s orbitals of V and Nb, respectively, depicted in Fig. 12. These wavefunctions have their outer maxima at the points

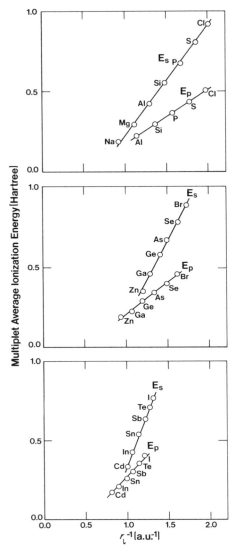

**Fig. 10.** The correlation between the observed $l$th orbital multiplet-averaged ionization energies $E_l$ and the reciprocal orbital radius $r_l^{-1}$ (Table I) for the polyvalent elements.

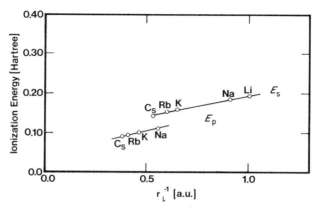

**Fig. 11.**  The correlation between the ground-state ($E_s$) and excited-state ($E_p$) orbital ionization energies of the alkali atoms and the corresponding reciprocal orbital radius $r_l^{-1}$ (Table I).

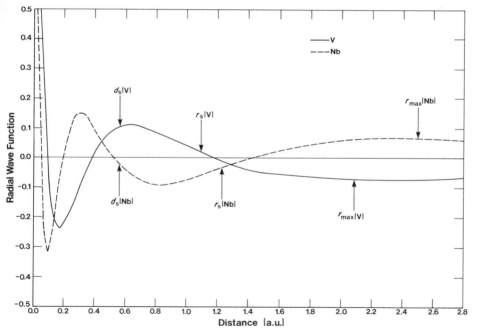

**Fig. 12.**  Radial s-type all-electron wavefunctions for V and Nb: $r_{max}$ denotes the position of the outer orbital maxima, $r_s$ is the screened pseudopotential radius, and $d_s$ is the average node position. Note that $r_s$ is pinned inwards of the last node and outwards of $d_s$.

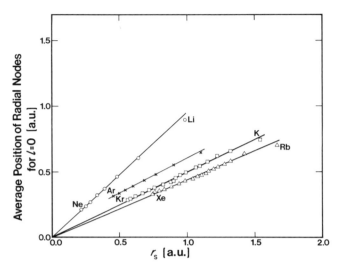

**Fig. 13.** Correlation between the average node position in the valence all-electron s wavefunction and the screened pseudopotential radius $r_s$ (Table I). The first and last atom of each row are denoted by chemical symbols.

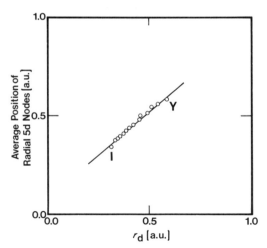

**Fig. 14.** Correlation between the average node position in the valence all-electron d wavefunction and the screened pseudopotential radius $r_d$ (Table I) for the 4d elements.

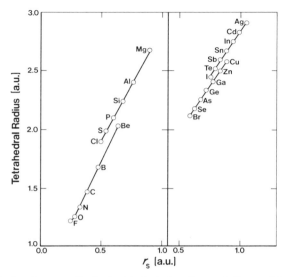

**Fig. 15.**   Relation between Pauling's tetrahedral radius and the s-orbital radius of the screened pseudopotential (Table I).

denoted by $r_{max}$ and have a number of nodes inwards to $r_{max}$. An algebraic average taken for all node positions in each wavefunction shows that these average positions (denoted by $d_1$) *are pinned at a certain distance from the orbital radius* $r_1$. Figure 13 shows the average node position $d_1$ of the outer all-electron s-type valence orbital plotted against $r_1$, and Fig. 14 shows similar results for d-orbitals (only the first and last element of each row are denoted by the chemical symbol). We find that the oribital radius $r_1$ scales linearly with the average node position, where the row-dependent scale factor increases monotonically with the position of the period in the table of elements (e.g., the scale equals 1.0, 1.5, 2.0, 2.3, and 2.7 for periods 1–5, respectively). *It is seen that the orbital radii* $r_1$ *form, therefore, an intrinsic length scale in that they carry over from the "true" wavefunctions the information on the average node position.* Hence, the dual coordinates $\{r_1, r_1^{-1}\}$ satisfy the semiclassical ideas underlying many successful structural factors (e.g., Pearson, 1972; Hume-Rothery and Raynor, 1954; Pauling, 1960) in forming elementary energy and length scales.

An additional intriguing feature of these orbital radii is their simple correlation with Pauling's tetrahedral radii (Pauling, 1960). As seen in Fig. 15, the tetrahedral radii can be identified with the $r_s$ coordinate to within a row-dependent scale factor.

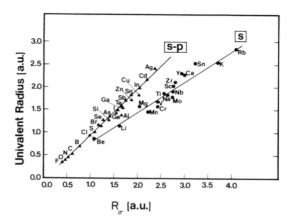

**Fig. 16.** Relation between Pauling's univalent radii and the total s–p screened pseudo-potential radius $R_\sigma = r_p + r_s$. Note that this $R_\sigma$ scale separates the univalent radii into a group of predominantly s-bonding elements (full circles) and s–p bonding elements (full triangles).

While the individual $r_s$ and $r_p$ radii measure the effective extent of the quantum cores of s and p symmetry, the sum $R_\sigma^A = r_s^A + r_p^A$ provides a measure of the total size of the effective core of atom A. Figure 16 depicts $R_\sigma^A$ versus Pauling's univalent radii for the first three periods of the table of elements. A similar correlation exists with Gordy's covalent radius (Gordy, 1946). $R_\sigma^A$ closely follows the regularities of the univalent radii, including their discontinuity at the end of the transition elements. Examination of Fig. 16 reveals that the present $R_\sigma^A$ coordinate provides a natural separation of Pauling's univalent radii into those that pertain to atoms sustaining s–p covalently bonded compounds and those in which the s-electrons largely dominate the structural properties. It is remarkable that the orbital radii derived from a pseudopotential formulation of *atomic physics* provide such a close reproduction of the length scale derived experimentally from *solid-state physics* (e.g., the empty-core radii in Fig. 8, and Pauling's tetrahedral and univalent radii in Figs. 15 and 16, respectively).

We have concentrated in this section on revealing the most significant correlations between the orbital radii and some *transferable* (rather than compound-dependent) semiclassical coordinates. We will not describe correlations with compound-dependent physical properties (e.g., melting points, deviations from ideal $c/a$ ratio in Wurtzite structures, elastic constants, etc.) not only because this may be too excessive, but also because we believe that many more such interesting correlations are likely to be discovered in the future. Such correlations between atomic $\{r_l\}$ values and physical properties $G_{AB}^{(1)}$, $G_{AB}^{(2)}$, etc., may not only serve to systematize those properties but could also point to the underlying dependencies between the seemingly unrelated physical observables $G_{AB}^{(1)}$, $G_{AB}^{(2)}$, etc.

## B.  Screening Length and Orbital Radii

One can view the quantity $r_l^{-1}$ as being an orbital-dependent screening constant pertinent to the scattering of valence electrons from an effective core. For the nontransition elements, one finds, as expected, that $r_l^{-1}$ falls off monotonically with decreasing valence charge $Z_v$, reflecting a more effective screening. However, for the 3d, 4d, and 5d transition series (Fig. 17), one finds two distinct behaviors: although $r_d^{-1}$ is a simple, monotonic function, both $r_s^{-1}$ and $r_p^{-1}$ show a break at the point where the d shell is filled. This is intimately related to a similar trend in the orbital shielding constants $Z_l^*$ calculated by Clementi and Roetti (1974) as a rigorous extension of Slater's screening rules. As seen in Fig. 18, the reciprocal screening lengths $(Z_l^*)^{-1}$ for the 3d transition series show a characteristic break around Cu–Zn, much like the corresponding reciprocal radii $r_l^{-1}$. In constrast, $(Z_l^*)^{-1}$ for the nontransition elements follow a linear trend.

This dual behavior of $r_l^{-1}$ separates the predominantly d-screening domain of the transition elements from the s–p screening domain of the posttransition elements. Note that $r_l^{-1}$ and $Z_l^*$ show uniquely this dual behavior, whereas most chemical and physical quantities are simple monotonic functions of the atomic position in these rows. It is interesting to note that such effects are clearly manifested by s and p coordinates rather than by the d coordinate. This has a central role in the structural significance of the s–p coordinates even for compounds containing transition elements.

## C.  Comparison with Other Orbital Radii

Figure 19 compares the empirical stripped-ion radii of Simons and Bloch (SB) with the present theoretical values of the density-functional orbital radii for the 41 nontransition elements calculated by SB. The latter set has recently been corrected for the posttransition elements (A. N. Bloch, unpublished results, 1980) relative to the set used by Chelikowsky and Phillips (1978) and Machlin et al. (1977). Figure 19 includes the corrected values (e.g., the $l = 0$ crossing-point radii for Cu, Ag, and Au are 0.38, 0.44, and 0.41 a.u., instead of 0.21, 0.22, and 0.13 a.u., respectively). These large corrections change quantitatively some of the results of these previous authors in analyzing the non-octet crystal structures and regularities of melting

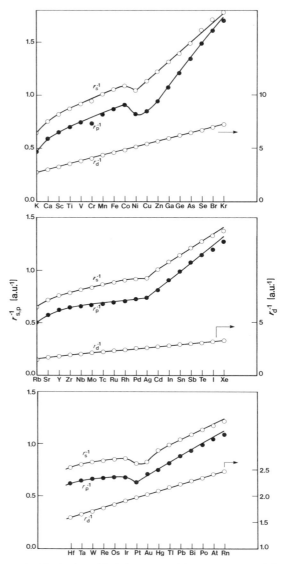

**Fig. 17.** The regularities in the orbital electronegativity parameters $r_l^{-1}$ (Table I) for the 3d, 4d, and 5d transition series. Note the linearity of the $r_d^{-1}$ scale compared with the break in the $r_s^{-1}$ and $r_p^{-1}$ scales.

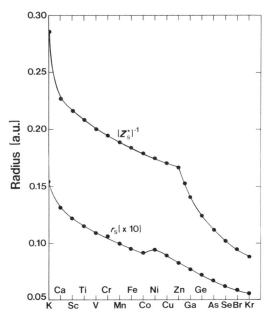

**Fig. 18.** The regularities in the reciprocal orbital shielding constants $(Z_s^*)^{-1}$ [Clementi and Roetti (1974)] for the third-row elements, compared with the screened pseudopotential radii $r_s$. Note the break at the end of the transition series.

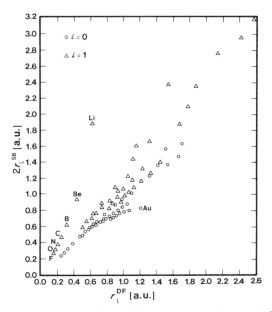

**Fig. 19.** Correlation between the Simons–Bloch empirical orbital radii $r_l^{SB}$ and the present screened pseudopotential density-functional radii $r_l^{DF}$ (Table I) for the 41 nontransition elements given by Simons and Bloch. The $l = 1$ coordinates of the first-row elements and the $l = 0$ coordinate of Au showing the largest spread, are denoted by their chemical symbol.

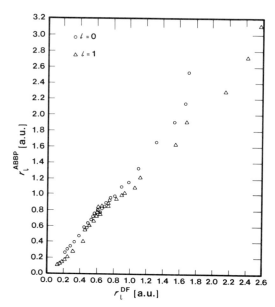

**Fig. 20.** Correlation between the Anderoni *et al.* (1979) orbital radii $r_l^{ABBP}$ and the present screened pseudopotential density-functional radii $r_l^{DF}$ (Table I) for the 27 nontransition elements given by ABBP.

temperatures, as well as the decomposition of Miedema's heat of formation model into elemental orbital radii.

Figures 20 compares the recent orbital radii developed by Andreoni, Baldereschi, Biemont, and Phillips (ABBP; see Andreoni *et al.*, 1978, 1979) with the present orbital radii, for the 27 nontransition elements calculated by ABBP. The ABBP radii are obtained from a two-parameter fit of both the Hartree–Fock stripped-ion orbital energies as well as the peak position of the orbital wavefunctions. The values for the eight 3d transition elements given by ABBP are not included in Fig. 20 since, as indicated by these authors, and as we confirm, they are not as reliable.

It can be seen that although the empirical SB radii correlate overall with the present radii, the scatter is fairly large. In particular, the SB scheme predicts $r_s \ll r_p$ for the first-row elements, whereas the present and the ABBP scheme, which attempt to reproduce both energies and wavefunctions, show $r_s > r_p$. The ABBP radii correlate well with our radii (Andreoni *et al.*, 1979) for the 27 nontransition elements. Other plots (e.g., correlation of $|r_s^A - r_p^A|$ or $|r_s^A + r_p^A|$ between the various schemes) lead to similar conclusions.

## V.  SEPARATION OF CRYSTAL STRUCTURAL OF 565 BINARY AB COMPOUNDS

The orbital radii $\{r_l\}$ derived here can be applied to predict the stable crystal structure of compounds in the same was as discussed by St. John and Bloch (1974), Machlin *et al.* (1977), and Zunger and Cohen (1978a). Having, however, the orbital radii of all atoms belonging to the first five rows in the periodic table, this theory can be applied to a far larger data base of crystals (565) than has been attempted previously (50–80).

Our first step was to compile a list of binary AB compounds whose atoms belong to the first five rows of the periodic table. We were interested in the most stable crystal form of each compound and in a structure that appears in the phase diagram at (or close to) a 50%–50% composition. We started the compilation by reviewing standard tables: Pearson's "Handbook of Lattice Spacings and Structures of Metals and Alloys" (1967); Hultgren *et al.*, "Selected Values of Thermodynamic Properties of Binary Alloys" (1973); Wykoff, "Crystal Structure" (1963); Schubert, "Kristallstrukturen Zweikomponentiger Phasen" (1964); Landolt- Börnstein,"Structure Data of Elements and Intermetallic Phases" (1971); Hansen, "Constitution of Binary Alloys" (1958); Rudman *et al.*, "Phase Stability in Metals and Alloys" (1966); Pearson, "The Crystal Chemistry and Physics of Metal and Alloys" (1972); and Parthé, "Crystal Chemistry of Tetrahedral Structures" (1964); as well as a number of basic papers, such as Rieger and Parthé (1966), Schob and Parthé (1964), and Schubert and Eslinger (1957), which give useful tables for particular structures. Whenever we have identified in this literature either a conflict in assigning a crystal structure or expressions of doubt as to the identification of the structure, other structures at somewhat different pressures or temperatures, substantial deviation from 1:1 stoichiometry, etc., we have made use of a computer-assisted literature search to find the original papers for the compounds in question. In this way, we have surveyed some 180 references. We have identified from standard sources as well as from an extended computer search a total of about 565 binary AB compounds that are near-stoichiometric, ordered, and formed from atoms belonging to the first five rows of the periodic table. Their distribution among the various crystal structures is given in Table IV.

This data base of 565 binary compounds exhibits an enormous range of physical, structural, and chemical properties. Using the terminology of the semiclassical structural factors, one notes the large range of conductivity properties spanned by these compounds (insulators, semiconductors, semi-metals, metals, superconductors), the electronegativity difference (covalent vs. large ionicity), coordination numbers (12 to 2), relative ionic sizes of the

**TABLE IV**

**AB Crystal Structures Used in the Structural Plots**[a]

| Strukturberichte or Pearson symbols | Space group | Unit cell | Prototype | Number of compounds |
|---|---|---|---|---|
| | | Octet | | |
| B1 | $Fm3m$ | cubic | NaCl | 65 |
| B2 | $Pm3m$ | cubic | CsCl | 3 |
| B3 | $F\bar{4}3m$ | cubic | ZnS | 29 |
| B4 | $P6_3mc$ | hexagonal | ZnO | 11 |
| A4 | $Fd3m$ | cubic | diamond | 4 |
| Total | | | | 112 |
| | | Non-octet | | |
| B1 | $Fm3m$ | cubic | NaCl | 33 + 8 |
| B2 | $Pm3m$ | cubic | CsCl | 122 + 2 |
| $B8_1$ | $P6_3/mmc$ | hexagonal | NiAs | 62 + 3 |
| B10 | $P4/nmm$ | tetragonal | PbO | 0 + 2 |
| B11 | $P4/nmm$ | tetragonal | CuTi | 7 |
| B16 | $Pnma$ | orthorhombic | GeS | 0 + 7 |
| B19 | $Pmma$ | orthorhombic | CuCd | 11 |
| B20 | $P2_13$ | cubic | FeSi | 17 |
| B27 | $Pnma$ | orthorhombic | FeB | 16 |
| B31 | $Pnma$ | orthorhombic | MnP | 30 |
| B32 | $Fd3m$ | cubic | NaTl | 7 |
| B33 | $Cmcm$ | orthorhombic | CrB | 41 |
| B35 | $P6/mmm$ | hexagonal | CoSn | 3 |
| B37 | $I4/mcm$ | tetragonal | SeTl | 0 + 3 |
| $B_h$ | $P\bar{6}m2$ | hexagonal | MoP | 4 |
| cP64 | $P43n$ | cubic | KGe | 6 |
| hP24 | $P6_3/mmc$ | hexagonal | LiO | 3 |
| $L1_0$ | $P4/mmm$ | tetragonal | CuAu | 27 |
| mC24 | $c2/m$ | monoclinic | AsGe | 0 + 4 |
| mC32 | $C2/c$ | monoclinic | NaSi | 1 |
| mP16 | $P2_1/c$ | monoclinic | AsLi | 4 |
| mP32 | $P2_1/n$ | monoclinic | NS | 0 + 5 |
| oC16 | $Cmcm$ | orthorhombic | NaHg | 1 |
| oC16 | $Cmca$ | orthorhombic | KO | 1 |
| oC48 | $Cmc2_1$ | orthorhombic | SiP | 0 + 1 |
| oI8 | $Immm$ | orthorhombic | RbO | 4 |
| oP16 | $P2_12_12_1$ | orthorhombic | NaP | 6 |
| tI8 | $I4/mmm$ | tetragonal | HgCl | 0 + 3 |
| tI32 | $I4_1/a$ | tetragonal | LiGe | 1 |
| tI32 | $I4/mcm$ | tetragonal | TlTe | 0 + 1 |
| tI64 | $I4_1/acd$ | tetragonal | NaPb | 7 |
| Total | | | | 453 |

[a] When two entries appear for the number of compounds, the first indicates the number of suboctet compounds and the second denotes the number of non-transition-element suberoctet compounds.

A and B atom, bonding type (covalent, ionic, metallic, etc.), range of heats of formation ($\simeq 1$–150 kcal/mole), electron per atom ratios ($\simeq 1.5$ to 8–9), etc. Given this distribution of the 112 octet compounds and 453 non-octet compounds into 5 and 31 different crystal structures, respectively exhibiting a diverse range of properties, we now ask how well can the atomically derived orbital radii scheme explain such a distribution.

We construct from the s and p atomic orbital radii the dual coordinates for an AB compound as:

$$R_\sigma^{AB} = |(r_p^A + r_s^A) - (r_p^B + r_s^B)|$$
$$R_\pi^{AB} = |r_p^A - r_s^A| + |r_p^B - r_s^B| \tag{23}$$

Here, $R_\sigma^{AB}$ is a measure of the different between the total effective core radii of atoms A and B (i.e., size mismatch); whereas $R_\pi^{AB}$ measures the sum of the orbital nonlocality of the s and p electrons on each site. Using the definition [Eq. (23)] and the values of the orbital radii given in Table I, we construct $R_\sigma^{AB}$ vs. $R_\pi^{AB}$ maps for the binary compounds. Such maps are shown for 112 octet compounds in Fig. 21 and for 356 of the non-octet compounds

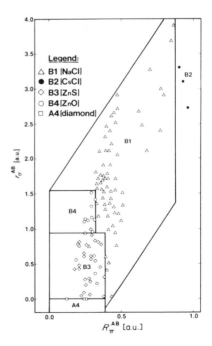

**Fig. 21.** A structural separation plot for the 112 binary octet compounds, obtained with the density-functional orbital radii, with $R_\sigma^{AB} = |(r_p^A + r_s^A) - (r_p^B + r_s^B)|$, $R_\pi^{AB} = |r_p^A - r_s^A| + |r_p^B - r_s^B|$.

Legend:
△ B1 (NaCl)
● B2 (CsCl)
◇ B3 (ZnS)
○ B4 (ZnO)
□ A4 (diamond)

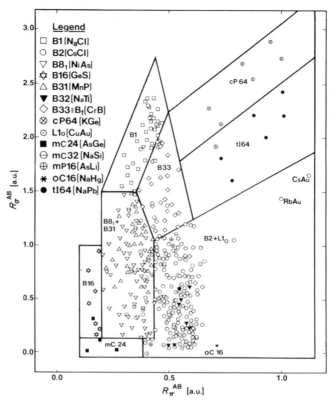

**Fig. 22.** A structural separation plot for the 356 binary non-octet compounds, obtained with the density-functional orbital radii, with $R_\sigma^{AB} = |(r_p^A + r_s^A) - (r_p^B + r_s^B)|$, $R_\pi^{AB} = |r_p^A - r_s^A| + |r_p^B - r_s^B|$.

in Fig. 22. (For example, for NiAs, we have from Table I: $r_s^{Ni} = 0.96$ a.u., $r_p^{Ni} = 1.22$ a.u., $r_s^{As} = 0.67$ a.u., and $r_p^{As} = 0.745$ a.u.; hence, $R_\sigma^{NiAs} = (1.22 + 0.96) - (0.745 + 0.67) = 0.765$ a.u. and $R_\pi^{NiAs} = (1.22 - 0.96) + (0.745 - 0.67) = 0.335$ a.u. This appears in the B8$_1$ domain in Fig. 22.)

We identify each structure by a different symbol and search for the smallest number of straight lines, enclosing minimal areas, best separating the different structures. In some cases, there exists a unique solution to this topological problem; in other cases (e.g., B33 and cP64 and tI64 structures), there are a number of permissible solutions. However, in these cases it seems to make little difference which line is chosen. While we could have lowered the number of "misplaced" compounds by using more complex lines, we feel that the more stringent criterion of using straight lines provides us with a better chance of assessing the true success of the method.

The remarkable result of these plots is that with *the same linear combination of atomic orbital radii*, most of the structures to which the 468 compounds appearing in these plots belong can be separated. The relative locations of the structural domains seems chemically reasonable. Hence, the B27–B33 (coordination number CN = 7) is intermediate between the B1 structure (CN = 6) and the B2–L1$_0$ structures (CN = 8). The mostly metallic non-octet compounds appear separate from the nonmetallic regime to the right (cP64, tI64, mC32, mP16, oC16), etc. Within single structural groups one similarly finds a chemically reasonable ordering of compounds,—e.g., polymorphic compounds (such as SiC) appear near border lines, Zinc blende–rock salt pairs that intertransform at low pressure appear along their separating line, etc. Note that even the wurtzite–zinc blende structures, which only differ starting from the third nearest-neighbors, are well separated. However, there are a number of crystallographically closely related structures that overlap: I am unable to separate the non-octet crystal type (B2) from the CuAu (L1$_0$ structure), the NiAs-type (B8$_1$) from the MnP-type (B31), and the CrB type (B33) from the FeB type (B27), etc. For charity of display, I show some of the extra overlapping structures separately in Fig. 23, *using, however, precisely the same separating lines as used for all other non-octet compounds (Fig. 22).*

It is not surprising that some of these structural pairs overlap. For instance, the B27 and B33 structures have a common structural unit consisting of a row of trigonal prisms of atom A stacked side by side and centered by a zigzag chain of B atoms (Hohnke and Parthé, 1965). The structural similarity between CsCl (B2) and CuAu (L1$_0$) has been discussed by Hume-Rothery and Raynor (1954); the relation between the NiAs (B8$_1$), MnP (B31), and the FeSi (B20) structures by Schubert and Eslinger (1957); and that between the CsCl (B2), AuCd (B19), and CuTi (B11) by Pearson (1972). In fact, examination of the thermochemical data (Hultgren *et al.*, 1973; Kubaschewski and Alcock, 1979, and references therein) indicates that if a certain compound exists in two of these related structures at somewhat different temperature, the difference in their standard heats of formation is often as small as 0.3 kcal/mole! (For example, AgCd in the B19 structure has $\Delta H = 0.094 \pm 0.004$ eV, whereas the B2 structure has a heat formation of $0.080 \pm 0.004$ eV.) Also, some of these presently unseparated structures indeed appear as mixtures when prepared from the melt [e.g., as noted by Honke and Parthé (1965), both the B27 and B33 structures are frequently found in the same arc-melted buttons of these compounds].

Since the publication of a preliminary report of this work (Zunger, 1980d), which included 495 compounds, I have been made aware of the crystal

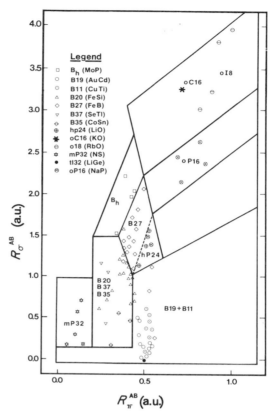

**Fig. 23.** A structural separation plot for the 81 binary non-octet compounds, obtained with the density-functional orbital radii, with $R_\sigma^{AB} = |(r_p^A + r_s^A) - (r_p^B + r_s^B)|$, $R_\pi^{AB} = |r_p^A - r_s^A| + |r_p^B - r_s^B|$.

structures of 54 more octet and suboctet compounds as well as 16 new superoctet compounds (i.e., a total of 565 compounds). I find that the lines separating the structural domains of the octet and suboctet compounds need not be changed relative to their previous assignment to incorporate the 54 new compounds. All the superoctet compounds (a total of 34) could be separated clearly as well (Fig. 24 and discussion below). This illustrates the predictive ability of the present approach.

If one is to consider the pairs of related crystal structures mentioned above as belonging to single generalized structural groups, the total number of misplaced compounds (5 octet and 32 non-octet) forms only 7% of the total

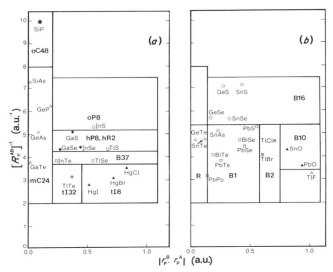

**Fig. 24.** A structural separation plot for the 34 superoctet compounds $A^N B^{(10-N)}$ with valence electron concentrations VEC = 9 (Fig. 24a) and 10 (Fig. 24b). The structural groups are defined in Table IV. The symbol R (Fig. 24) denotes a romboledral structure. In the B1 domain of Fig. 24b we have also included the compounds with VEC = 9, 11.

data base of binary compounds. In this respect the present theory is more than 90% successful.

The compounds that are "misplaced" in the present theory (i.e., their $\{R_\sigma^{AB}, R_\pi^{AB}\}$ coordinates place them in a different structural domain than that reported in the literature surveyed) are listed in Table V together with their $\{R_\sigma^{AB}, R_\pi^{AB}\}$ coordinates. In cases where a compound appears in overlapping domains or close to a border line, we indicate all the pertinent structures. Given their $\{R_\sigma^{AB}, R_\pi^{AB}\}$ coordinates, the reader can conveniently identify them on structural plots.

The list of misplaced compounds shows a number of interesting features. At least two compounds, CuF and FeC, reported to have the B3 and B1 structures, respectively, probably do not exist at all (for CuF see Barber et al., 1961; for FeC, F. Jellenek, private communication, 1980). Their "misplacement" in the present theory is hence a grafting feature.

Similarly, while OsSi (No. 11) is sometimes reported to have the B2 structure (e.g., Landolt-Bornestein, 1971) and appears in our plots in the B20 domain (No. 11 in Table 5), it is known to actually have the B20 structure and appears in B2 only with impurities. In addition, the compound PtB (No. 18 in Table V) which has been reported to have the NiAs (B8₁) structure and is placed in our plots in an entirely unrelated structural domain, has been found to have an anti-NiAs structure. NiY (No. 30) was reported in

TABLE V

**Compounds Which are Misplaced in the Present Theory**
**(out of a Total of 565)**

| Compound | Expected structure | Structural domain(s) in which it is found | $R_\sigma^{AB}$ (a.u.) | $R_\pi^{AB}$ (a.u.) |
|---|---|---|---|---|
| | | *Octet* | | |
| 1. CuF | B3 | B1 | 1.635 | 0.375 |
| 2. MgS | B1 | B3 | 0.93 | 0.25 |
| 3. BeO | B4 | B1–B3 | 0.615 | 0.305 |
| 4. MgTe | B4 | B1–B3 | 0.36 | 0.32 |
| 5. MgSe | B1 | B1–B3–B4 | 0.745 | 0.285 |
| | | *Non-Octet* | | |
| 1. CoAl | B2 | $B8_1$–B31–B20 | 0.345 | 0.315 |
| 2. FeAl | B2 | $B8_1$–B31–B20 | 0.435 | 0.345 |
| 3. NiAl | B2 | $B8_1$–B31–B20 | 0.505 | 0.395 |
| 4. CoGa | B2 | $B8_1$–B31–B20 | 0.325 | 0.355 |
| 5. FeGa | B2 | $B8_1$–B31–B20 | 0.415 | 0.385 |
| 6. NiGa | B2 | $B8_1$–B31–B20 | 0.485 | 0.435 |
| 7. NiIn | B2 | $B8_1$–B31–B20 | 0.13 | 0.43 |
| 8. MnIn | B2 | $B8_1$–B31–B20 | 0.17 | 0.41 |
| 9. CoPt | $L1_0$ | $B8_1$–B31–B20 | 0.68 | 0.40 |
| 10. TiAl | $L1_0$ | $B8_1$–B31–B20 | 0.905 | 0.415 |
| 11. OsSi | B2 | $B8_1$–B31–B20 | 1.23 | 0.37 |
| 12. CoBe | B2 | $B8_1$–B31–B20 | 0.94 | 0.38 |
| 13. PdBe | B2 | B33–B27 | 1.37 | 0.49 |
| 14. NaPb | tI64 | $B2–L1_0$–B32 | 0.56 | 0.52 |
| 15. AuBe | B20 | B1–B33–B27 | 1.58 | 0.44 |
| 16. FeC | B1 | $B1–B8_1$–B31 | 1.47 | 0.35 |
| 17. TiB | B1 | B1–B33–B27 | 1.785 | 0.445 |
| 18. PtB | $B8_1$ | B1–B33–B27 | 1.905 | 0.385 |
| 19. IrPb | $B8_1$ | $B8_1$–B31–B2 | 0.538 | 0.478 |
| 20. AgCa | B33 | $B1–L1_0$ | 0.625 | 0.645 |
| 21. HfPt | B33 | $B2–L1_0$ | 0.21 | 0.53 |
| 22. NiHf | B33 | $B2–L1_0$ | 0.73 | 0.57 |
| 23. NiLa | B33 | $B2–L1_0$ | 0.90 | 0.59 |
| 24. NiZr | B33 | $B2–L1_0$ | 0.645 | 0.555 |
| 25. PtLa | B33 | $B2–L1_0$ | 0.38 | 0.55 |
| 26. RhLa | B33 | $B2–L1_0$ | 0.56 | 0.63 |
| 27. ZrPt | B33 | $B2–L1_0$ | 0.125 | 0.515 |
| 28. PdLa | B33 | $B2–L1_0$ | 0.63 | 0.62 |
| 29. AuLa | B27–B33 | $B2–L1_0$ | 0.42 | 0.57 |
| 30. NiY | B27 | $B2–L1_0$ | 0.76 | 0.56 |
| 31. PtY | B27 | $B2–L1_0$ | 0.24 | 0.52 |
| 32. LaCu | B27 | $B2–L1_0$ | 1.04 | 0.61 |

1960 (e.g., Elliot, 1965, p. 678) to have the orthorhombic B27 structure, in conflict with the prediction of the present scheme, while in 1964 (e.g., Shunk, 1969, p. 561) it was concluded that it is actually monoclinic with a $P2_1/a$ space group. It seems that no clear identification for this structure is yet available. While HfPt (No. 21) is identified in most sources as having a B33 orthorhombic structure, a deformed B2 modification has also been reported (e.g., Shunk, 1969, p. 419). Similarly, AuBe (No. 15) has been reported in 1947 to have the B20 structure (e.g., Elliot, 1965, p. 83) while in 1962 (e.g., Shunk, 1969, p. 64) it has been identified as tetragonal. AuLa (No. 29) has been reported to transform from its high temperature B33 form to a low-temperature B27 form (both orthorhombic) at about 660°C, while in 1963 (Shunk, 1969, p. 73) it was indicated to have the cubic B2 form. It is hence clear that for some of the "misplaced" compounds, it is not yet obvious whether their misplacement is real. For the other compounds appearing in Table V their misplacement in the present phase diagrams is real and brings up a number of interesting observations.

The octet compounds MgS and MgSe have a NaCl (B1) structure but appear in our plot in the ZnS (B3) domain, near the B1 border. Experimentally (e.g., Navrotsky and Phillips, 1975) it is found that the (normalized) free energy of the B3–B1 phase transition for these compounds is nearly zero.

A large number of the other misplaced compounds have unusual properties. Two such groups of compounds show *systematic* unusual properties: the six Al and Ga compounds with the magnetic 3d transition elements (CoAl, FeAl. NiAl, CoGa, FeGa, and NiGa) and the group of ten CrB and three FeB structures (AgCa, HfPt, NiHf, NiLa, NiZr, PtLa, ZrPt, PdLa, and AuLa, and NiY, PtY, and LaCu, respectively).

The first group has the CsCl (B2) structure but appears in the present theory in the domain of the NiAs–MnP–FeSi structures. Their electric and magnetic properties have been studied intensively in the last few years (e.g., Brodsky and Brittain, 1969; Herget et al., 1970; Wertheim and Wernick, 1967; Huffman 1971; Bose et al., 1979; Müller et al., 1979; Kiewit and Brittain, 1970; Caskey et al., 1972; Sellmyer et al., 1971). It appears that these compounds are stabilized by the presence of defects, and they have a large, stable range of composition (45–55%).

Susceptibility measurements as a function of magnetic field show ferromagnetic impurities and antistructure defects in such materials. More importantly, a slight nonstoichiometry often leads to the formation of local magnetic moments. These results indicate (e.g., Sellmyer et al., 1971) that such slightly off stoichometric materials are in effect spin glasses at low temperatures. Their magnetic behavior is intermediate between that of the independent magnetic impurity problem (Kondo effect) and that characteristic of antiferromagnetic or ferromagnetic systems having long-range order due to strong magnetic interactions. It is interesting to note that this subgroup of compounds exhibiting stable intrinsic defects leading to magnetic moments, are displaced in the structural plots from the largely nonmagnetic B2 domain to the $B8_1$ region in which even the stoichiometric alloys have permanent local moments. There are indications that this subgroup of compounds have certain structural anomalies: Many authors report that the B2 aluminides could not be obtained as a single phase, and in diffraction the B2 pattern could not be separated from other diffraction lines not belonging to this structure (Schob and Parthé, 1964). It was suggested (Schob and Parthé, 1964) that many of these compounds are only *metastable* in the B2 structure. In a recent diffraction study (Gerold, 1978) it was discovered that strong distortions occur around Co sublattice sites in CoGa due to intrinsic vacancies. Similarly, a recent calculation (Bras *et al.*, 1977) of the ordering energy in FeGa, using a Bragg–Williams model, yielded very small interaction parameters of 0.049 eV and 0.03 eV (i.e., $\sim 2–1 \ kT$) for first and second neighbors, respectively. It is intriguing that the present orbital radii scheme has the ability of identifying such unusual phenomena in a few of the 122 tabulated B2 nonoctet compounds.

Our scheme suggests that similar irregularities may occur in NiIn and TiAl and perhaps even in CoBe, but to a much smaller extent, since these compounds are only marginally misplaced in the present theory. Indeed, the absorption spectra of NiIn in the 0.7–5.5 eV range (Best *et al.*, 1971) indicate almost no change with composition in its Drude regime as well as above it, suggesting constant number of electrons per cell due to the defect structure. This suggests that many of the unusual magnetic and structural properties found in the FeAl, FeGa, CoAl, CoGa, NiAl, NiGa group may may also be found in NiIn.

The second large group of misplaced compounds (numbers 20–32 in Table V) form a distinct structural group. Schob and Parthé (1964), Rieger and Parthé (1966), and Hohnke and Parthé (1965) have indicated that all CrB (B33) and FeB (B27) compounds can be separated into two groups: group I, in which a transition-metal atom combines with an s–p element (B, Si, Ge, Al, Ga, Sn, or Pb), and group II, in which a transition element from the third or fourth group combines either with another transition element from group VIII or from the Cu group. It was found that the individual trigonal prisms in both the FeB and the CrB structures have different relative dimensions in groups I and II. In particular, group I of the CrB structure shows a "normal" $a/c$ ratio greater than one, but group II compounds show a compressed prism with $a/c < 1$. Only three compounds belonging to group I (HfAl, ZrAl, and YAl) have $a/c < 1$. We find that *all of the CrB and FeB compounds that belong to group II are misplaced by our theory into the bordering CsCl domain*, whereas the three group I compounds that have $a/c < 1$

(much like group II compounds) are correctly placed. Hence, the present approach is sufficiently sensitive to separate the true physical irregularity even when simple structural factors such as the $a/c$ ratio lead to the wrong conclusion. From the results of the present approach, it would seem that group II compounds of the FeB and CrB structures should properly be identified as a separate group. As a result, if the errors made by the present theory for the latter group of compounds as well as the errors in the 1–8 (Table V) local-moment materials are regarded as *systematic irregularities*, the remaining "true" errors amount to only 2% of the total data base.

The remaining misplaced compounds may also have unusual properties; NaPb (No. 14) has an unusual structure resembling a molecular crystal with 64 atoms per cell and interacting $Pb_4$ tetrahedra (Marsh and Shoemaker, 1953; Hewaidy et al., 1964); IrPb (No. 19), according to Miedema's (1976) model, has a positive heat of formation of about 1 kcal/mole.

Similarly AgCa (No. 20) has been recently discovered (Amand and Giessen, 1978 to be one of the only known glass-forming materials that do not contain a transition element. It has also been noted (Chelikowsky and Phillips, 1978 that the ratio of anion–anion to cation–anion distances in AgCa is almost an order of magnitude smaller than in all other nontransition metal B33 compounds and that unlike the latter group of compounds it has catalytic properties in redox reactions.

The relative orientation of the structural domains in Figs. 21–23 suggest that no *single* coordinate will suffice to produce a complete topological separation between all structures. Since, however, the area of the $R_\sigma^{AB}$ vs. $R_\pi^{AB}$ plane seems to be more or less bound (e.g., $\approx 3$ a.u.$^2$ in Fig. 22) when extra compounds are added, it is likely that the two-dimensionality of this finite $R_\sigma^{AB}$–$R_\pi^{AB}$ space will eventually preclude the delineation of further structures. One may hence expect that for some critical number of structures and compounds, a third coordinate may be needed. Such a generalized multi-dimensional resolution of structural groups may also resolve some of the remaining discrepancies in the present theory. Our present approach however is aimed at demonstrating the extent of structure delineation possible with the minimum number of two coordinates using the simplest possible separating lines.

One simple example for an additional coordinate is the classical (e.g. Pearson, 1969) valence electron concentration VEC, measuring for the binary AB system the total number of valence electrons $Z_v^A + A_v^B$ [c.f. Eq. (9)] in the compound. In the semiclassical approaches to structure it is known that whereas the VEC value alone does not separate different crystal structures, compounds with the same valence electron number often belong to the same broad structural groups. One can use this additional coordinate together with our orbital radii to obtain a better structural resolution of the marginally resolved structures, and at the same time provide a clear structural delineation of all *superoctet* (i.e., $Z_v^A + Z_v^B > 8$) compounds. As with the suboctet compounds, no previous approach has succeeded in systematizing these complex crystal structures.

We find that while the definition of the structural coordinates $R_\pi^{AB}$ and $R_\sigma^{AB}$ [Eq. (23)] used for the octet (Fig. 21) and mostly suboctet (Figs. 22, 23) compounds yields an overall separation also of the superoctet compounds, a more sensitive delineation is obtained with the slightly modified coordinates $R_1 = |r_p^B - r_p^A|$ and $R_2 = (R_\pi^{AB})^{-1}$ suggested by Littlewood (1980). Here, $R_1$ is a measure of the p-orbital electronegativity difference between atoms A and B, while $R_2$ measures the s–p nonlocality on the two sites. The reason that $|r_p^B - r_p^A|$ forms a better structural coordinate for superoctet compounds is that these systems involve relatively heavy atoms (e.g., Pb, Sn, Bi, Tl, Hg) for which the s electrons are paired and strongly bound relative to the p electrons. Hence, these semicore s orbitals become chemically inactive and only the contribution of the p electons needs to be included in the electronegativity parameter $R_1$.

Figures 24a and 24b show structural plots for the superoctet compounds with 9 and 10 valence electrons respectively. Since there are only a few B1 superoctet compounds, we have included those with VEC = 9 (SnAs), VEC = 10 (PbS, PbSe, PbTe, and PbPo) and VEC = 11 (BiSe, BiTe) on the same plot in Fig. 24b. Figure 24 includes 18 compounds which have appeared in the previous nonoctet plots (Figs. 22 and 23): B37 (InTe, TlSe, TlS), mC24 (GaTe, GeAs, GeP, SiAs), B16 (GeS, SnS, GeSe, SnSe, InS), pseudo-B8$_1$ (GaS, GaSe, GeTe, SnTe, InSe), and the orthorombically distorted B1 compound TlF. It is seen that the seven different structures of the VEC = 9 compounds as well as the six different VEC = 10 structures are very clearly resolved, the only exception out of these 34 compounds being the B37 compound TlS which is marginally displaced into the neighboring hP8-hR2 domain (Fig. 24a).

It is interesting to note that the present scheme also predicts unusual *electronic properties* of compounds belonging to the same structural group. For instance, the B2 compounds CsAu and RbAu that appear in Fig. 22 as isolated from the other 147 B2 + L1$_0$ compounds have semiconducting properties (e.g., Spicer, 1962), while all other suboctet compounds belonging to these structures seem to be metals. A recent calculation of the electronic band structure of CsAu (Hasegawa and Watabe, 1977) has indicated that if relativistic corrections are neglected, CsAu appears to be a metal, which disagrees with experiment, whereas the inclusion of relativistic effects lowers the Au s valence band to form a semiconductor. It is remarkable indeed that such complicated electronic structure factors are required in quantum-mechanical band-structure calculations to reveal the unusual semiconducting behavior suggested here simply by the atomic orbital radii.

If the predictive power of the present orbital radii scheme in relation to unusual electronic properties is not accidental, it would be interesting to speculate on its consequences. One would guess, for instance, that all *sub-octet nontransition element* compounds having $R_\pi^{AB}$ larger than roughly 0.7 a.u. are nonmetals! This includes not only the known nonmetallic compounds belonging to the LiAs group (KSb, NaGe, NaSb, but not LiAs) and the KGe group (CsGe, CsSi, KGe, RbGe, and RbSi, whereas KSi is a border-line case), but also the tI64 (NaPb) group (CsPb, CsSn, KPb, KSn, RbPb, and RbSn, but not NaPb), the B2 compounds LiAu and LiHg, the $L1_0$ compound NaBi, the oC16 compound NaHg, and the mC32 compound NaSi. In the sequence of alkali–gold compounds LiAu, NaAu, KAu, RbAu, and CsAu, one would similarly predict that the transition between metallic and insulating behavior occurs between NaAu and KAu.

It is important to emphasize that the ability to separate structures shown by the present orbital radii (Figs. 21–24) is far from being trivial or accidental. This can be demonstrated by constructing structural plots using different coordinates. We have used Miedema's (1973, 1976, Miedema *et al.*, 1975) coordinates $R_2 = |\phi_A^* - \phi_B^*|$ and $R_1 = |n_A^{*\frac{1}{3}} - n_B^{*\frac{1}{3}}|$, where $\phi_A^*$ and $n_A^{*\frac{1}{3}}$ are the effective elemental work function and cell boundary density to the power of $\frac{1}{3}$. Those coordinates were extremely successful in predicting the signs (and often the magnitudes) of the heats of formation of more than 500 compounds. We have also used a Mooser-Pearson (1959) plot, where $R_2$ is the elemental electronegativity difference, and $R_1$ is the average principal quantum number. Finally, we constructed a plot using Shaw's parameters, where $R_2$ is the elemental electronegativity difference and $R_1 = \frac{1}{2}(Z_A + Z_B)/[\frac{1}{2}(n_A + n_B)]^3$, where $Z_A$ and $n_A$ are the atomic number and the principal quantum number of the outer valence orbital, respectively.

In a Miedema plot, one notices a rough separation of the B1 and $B8_1$ structures (CsAu and RbAu appear, as in our case, at high $\Delta\phi^*$, $\Delta n^{*\frac{1}{3}}$), whereas most other structures are nearly indistinguishable. This illustrates the great difficulty in carrying the success of a theory that predicts *global binding energies* $\Delta E_0$ into the prediction of *structural energies* $\Delta E_s$ (cf. Sec. I).

The Mooser-Pearson plot for these compounds appears visually as if only 104 compounds (i.e., isolated points) are plotted. In fact, it includes 360 compounds belonging to 14 different structures. This strong overlap of different structures on the same $(R_1, R_2)$ coordinate reflects the insensitivity of the scale to separating such structures. A somewhat better separation is evident using Shaw's parameters, but the overlap of different structures is still extremely large.

It is likely that one could construct first-principles atomic pseudopotentials using somewhat different procedures than have been used here (Sec. III,A) for defining the pseudo wavefunctions. While this would result in a different set of orbital radii, they will most likely scale linearly with the present set of radii. Consequently, one may expect that the systematization of the crystal structures, based on such radii, will be essentially unchanged.

The relationship between the structural stability of a polyatomic system and the degree of repulsiveness of the effective atomic cores of the constituent atoms has been discussed in 1948 by Pitzer in a remarkable paper preceeding all pseudopotential theories. While one might have thought then naively that the electron-core attraction term $-Z_v/r$ would lead to a strong penetration by valence electrons of the core regions of neighboring atoms, Pitzer has realized that the core electrons set up a repulsive potential with a characteristic radius inside which such a penetration is discouraged. Hence, the triple bond energy of $N \equiv N$ is much higher than that of $P \equiv P$ (and the bond length in $N \equiv N$ is significantly shorter than in $P \equiv P$) because the repulsive core size of nitrogen is so much smaller than that of phosphorous. Similarly, the occurance of multiple chemical bonds with first row elements as compared with the rare occurance of such bonds (with a small bond energy) with heavier atoms has been naturally explained in terms of the large repulsive core size of the latter elements. In addition, Pitzer noted that whereas Pauling (e.g., 1960) has suggested that single bond energies (e.g., N—N) should be roughly $\frac{3}{4}$ of the tetrahedral bond energy (e.g., C—C), in fact the ratio of the two is closer to $\frac{1}{2}$. This discrepancy was simply explained (Pitzer, 1948) by the fact that the change from the bond angle of $90°$ characteristic of p-type single bonds to a tetrahedral angle of $109.5°$ minimized in the latter case their overlap with the repulsive core.

In the present orbital radii approach, these ideas are realized in a simple manner. To first order, the change in energy per atom introduced by incorporating an atom in a polyatomic system is proportional to:

$$\delta E \sim \sum_l k_l \int \Delta \rho_l(r) [U_l(r) + \Delta(r)] \, dr \qquad (24)$$

where $\Delta \rho_l$ is the $l$th component of the charge redistribution, $U_l(r)$ is the Pauli repulsive potential [Eq. (10)], $\Delta(r)$ the $l$-independent part of the atomic pseudopotential [Eq. (9)] and $k_l$ are constants. The first $\Delta \rho_l(r) U_l(r)$ term in Eq. (24) leads to a repulsive and angular-momentum dependent contribution (for electron-attracting species) while the second term is isotropically attractive (for similar atoms). Neglecting, for this simple argument, the nonlinear dependence of the charge density redistribution $\Delta \rho_l(r)$ on $U_l(r)$, one notes that Pitzer's ideas on the destabilizing role of large-core atoms—as well as the relative stability of structures that minimize such repulsions through conformational changes in bond angles—are directly manifested in Eq. (24). One could further note that charge redistribution effects occuring predominantly outside the pseudopotential core [where $U_1(r) \approx 0$] do not contribute to such strongly directional repulsive terms. It would seem reasonable that the dominance of the centrifugal barrier at small distances form the origin will cause the charge redistribution effects in the high angular momentum components of the density to be confined to regions outside $U_1$ (i.e., $r \gtrsim r_1$). This simple picture clearly indicates the important structural role played by the $r_s$ and $r_p$ coordinates, as compared to higher angular-momentum orbital coordinates.

## VI.  SUMMARY

It has been demonstrated that the pseudopotential theory in its present nonempirical density-functional form is capable of providing transferable atomic pseudopotentials $v_{ps}^{(l)}(r)$ that can be used both for performing reliable quantum-mechanical electronic structure calculations and for defining semi-classical-like elementary length and energy scales. The resulting radii correlate with a large number of classical constructs that have been traditionally used to systematize structural and chemical properties of many systems.

At the same time, the orbital radii derived here are capable of predicting the stable crystal structure of the 112 octet compounds (Fig. 21), 419 suboctet compounds (Figs. 22–23) and the 34 superoctet compounds (in Fig. 24) with a remarkable success, exceeding 95%. The compounds for which the present theory does not predict the correct crystal structure are analyzed and found to be largely characterized as defect structures with many unusual electronic, magnetic and structural properties. Although I am unable at this time to provide a direct *causal* quantum-mechanical model explaining this remarkable success, I am hopeful that the use of these concepts in solving practical crystallographic, metallurgical, and chemical problems may be useful and will also eventually provide the insight needed for a microscopic theory elucidating the success of the orbital radii.

## ACKNOWLEDGMENTS

I would like to thank A. N. Bloch, M. L. Cohen, and J. C. Phillips for inspiring discussions. Special thanks are due to E. Parthé and F. Hulliger for reviewing the present compilation of crystal structures. I am grateful to L. Pauling for bringing Pitzer's work (1948) to my attention and discussing related work.

## REFERENCES

Abarenkov, J. V., and Heine, V. (1965). *Philos. Mag.* [8] **12**, 529.
Amand, R., and Glessen, B. C. (1978). *Scr. Metall.* **12**, 1021.
Andreoni, W., Baldereschi, A., Meloni, F., and Phillips, J. C. (1978). *Solid State Commun.* **24**, 245.
Andreoni, W., Baldereschi, A., Biemont, E., and Phillips, J. C. (1979). *Phys. Rev. B* **20**, 4814.
Appelbaum, J. A., and Hamann, D. R. (1976). *Rev. Mod. Phys.* **48**, 3.
Ashcroft, N. W. (1966). *Phys. Lett.* **23**, 48.
Ashcroft, N. W. (1968). *J. Phys. C* **1**, 232.
Ashcroft, N. W., and Langreth, D. C. (1967). *Phys. Rev.* **159**, 500.
Barber, M., Linnett, J. W., and Taylor, M. (1961). *J. Chem. Soc.* p. 3323.
Bennett, B. I., and Vosko, S. H. (1972). *Phys. Rev. B* **6**, 7119.
Bertoni, C. M., Bortolani, V., Calandra, C., and Nizzoli, E. (1973). *J. Phys. F* **3**, L244.
Best, K. J., Rodies, H. J., and Jacobi, H. (1971). *Z. Mettallkd.* **62**, 634.
Bose, A., Frohberg, G., and Wever, H. (1979). *Phys. Status Solidi A* **52**, 509.
Bras, J., Couderc, J. J., Fagot, M., and Ferre, J. (1977). *Acta Metall.* **25**, 1077.
Brewer, L. (1963). *In* "Electronic Structure and Alloy Chemistry of the Transition Elements" (P. Beck, ed.). Wiley (Interscience), New York.
Brewer, L. (1967). *In* "Phase Stability in Metals and Alloys" (P. Rudman, J. Stringer, and R. I. Jaffee, eds.), p. 39. McGraw-Hill, New York.
Brewer, L. (1968). *Science* **161**, 115.
Brewer, L., and Wengert, P. R. (1973). *Metall. Trans.* **4**, 2674.
Brodsky, M. B., and Brittain, J. O. (1969). *J. Appl. Phys.* **40**, 3615.
Brust, D. (1968). *Methods Comput. Phys.* **8**, 33.
Caskey, G. R., Franz, J. M., and Sellmyer, D. J. (1972). *J. Chem. Solids* **34**, 1179.
Chelikowsky, J. R., and Cohen, M. L. (1976). *Phys. Rev. B* **14**, 556.
Chelikowsky, J. R., and Phillips, J. C. (1978). *Phys. Rev. B* **17**, 2453.
Clementi, E., and Roetti, C. (1974). *At. Data Nuc. Data Tables* **14**, 1974.
Cohen, M. L., and Bergstresser, T. K. (1966). *Phys. Rev.* **141**, 739.
Cohen, M. L., and Heine, V. (1961). *Phys. Rev.* **122**, 1821.
Cohen, M. L., and Heine, V. (1970). *Solid State Phys.* **24**, 38.
Darken, L. S., and Gurry, R. W. (1953). "Physical Chemistry of Metals." McGraw-Hill, New York.
Elliot, R. P. (1965). "Constitution of Binary Alloys," 1st Suppl. McGraw-Hill, New York.
Engle, N. (1939). *Ingenioren* M101.
Engel, N. (1967). *Acta Metall.* **15**, 557.
Fock, V., Vesselow, M., and Petraschen, M. (1940). *Zh. Eksp. Teor. Fiz.* **10**, 723.
Friedel, J. (1969). *In* "The Physics of Metals" (J. M. Ziman, ed.) p. 1. Cambridge Univ. Press, London and New York.
Gerold V. (1978). *J. Appl. Crystallogr.* **11**, 153.
Goldman, A. (1977), *Phys. Status Solidi B* **81**, 9.
Gombas, P. (1935). *Z. Phys.* **94**, 473.
Gombas, P. (1967). "Pseudopotentials." Springer, Vienna (and references therein).
Gordy, W. (1946). *Phys. Rev.* **69**, 604.
Hafner, J. (1978). *Solid State Commun.* **27**, 263.
Hamann, D. R. (1979). *Phys. Rev. Lett.* **42**, 662.
Hamann, D. R., Schlüter, M., and Chiang, C. (1980). *Phys. Rev. Lett.* **43**, 1494.
Hansen, P. M. (1958). "Constitution of Binary Alloys." McGraw-Hill, New York.
Harris, J., and Jones, R. O. (1978). *Phys. Rev. Lett.* **41**, 191.

Harrison, W. A. (1966). "Pseudopotentials in the Theory of Metals." Benjamin, New York.

Hasegawa, A., and Watabe, M. (1977). *J. Phys. F* **7**, 75.

Hellman, H. (1935a). *J. Chem. Phys.* **3**, 61.

Hellman, H. (1935b). *Acta Physicochim. URSS* **1**, 913.

Hellman, H. and Kassatotschikin, W. (1936). *J. Chem. Phys.* **4**, 324.

Herget, R., Wreser, E., Gengnagel, H., and Gladun, A. (1970). *Phys. Status Solidi* **41**, 255.

Hewaidy, I. F., Busmann, E., and Klein, W. (1964). *Z. Anorg. Allg. Chem.* **328**, 283.

Hodges, C. H. (1977). *J. Phys. C* **7**, L247.

Hohenberg, P. C., and Kohn, W. (1964). *Phys. Rev.* **136**, 864.

Honke, D., and Parthé, E. (1965). *Acta Crystallogr.* **20**, 572.

Huffman, G. P. (1971). *J. Appl. Phys.* **42**, 1606.

Hultgren, R., Desai, R. D., Hawkins, D. T., Gleiser, H. G., and Kelley, K. K. (1973). "Selected Values of the Thermodynamic Properties of Binary Alloys." Am. Soc. Metals, Metals Park, Ohio.

Hume-Rothery, W., and Raynor, G. V. (1954). "The Structure of Metals and Alloys." Institute of Metals, London.

Kahn, L. R., Baybutt, P., and Truhlar, D. G. (1976). *J. Chem. Phys.* **65**, 3826.

Kerker, G. P. (1980). *J. Phys. C* **13**, L189.

Kerker, G. P., Zunger, A., Cohen, M. L., and Schlüter, M. (1979). *Solid State Commun.* **32**, 309.

Kiewit, D. A., and Brittain, J. O. (1970). *J. Appl. Phys.* **41**, 710.

Kohn, W., and Sham, L. J. (1965). *Phys. Rev.* **140**, 1133.

Kubaschewski, O., and Alcock, C. B. (1979). "Metallurgical Thermochemistry," 5th ed. Pergamon Oxford.

Lam, P., Cohen, M. L., and Zunger, A. (1980). *Phys. Rev. B* **22**, 1698.

Landolt-Börnstein (1971). *In* "Structure Data of Elements and Intermetallic Phases" (K. H. Hellwege and A. M. Hellwege, eds.), New Ser., Vol. 6. Springer-Verlag, Berlin and New York.

Littlewood, P. (1980). *J. Phys. C* **12**, 4441.

Machlin, E. S., Chow, T. P., and Phillips, J. C. (1977). *Phys. Rev. Lett.* **38**, 1292.

Marsh, R. E., and Shoemaker, D. P. (1953). *Acta Crystallogr.* **6**, 197.

Meisel, M. W., Helperin, W. P., Ochiani, Y., and Brittain, J. O. (1978). *J. Phys. F* **10**, 1105.

Miedema, A. R. (1973). *J. Less-Common Met.* **32**, 117.

Miedema, A. R. (1976). *J. Less-Common Met.* **46**, 67.

Miedema, A. R., Boom R., and De Boer, F. R. (1975). *J. Less-Common. Met.* **41**, 283.

Moore, C. E. (1971). "Atomic Energy Levels," NSRDS-NBS, Vols. 34–35. US Gov. Printing Office, Washington, D.C.

Mooser, E., and Pearson, W. B. (1959). *Acta Crystallogr.* **12**, 1015.

Moruzzi, V. L., Janak, J. F., and Williams, A. R. (1978). "Calculated Electronic Properties of Metals." Pergamon, Oxford.

Müller, C., Seifert, G., Lautenschläger, G., Wonn, H., Ziesche, P., and Mrosan, E. J. (1979). *Phys. Status Solidi B* **91**, 605.

Natapoff, M. (1975). *J. Phys. Chem. Solids* **36**, 53.

Natapoff, M. (1976). *J. Phys. Chem. Solids* **37**, 59.

Natapoff, M. (1978). *J. Phys. Chem. Solids* **39**, 1119.

Navrotsky, A., and Phillips, J. C. (1975). *Phys. Rev. B* **11**, 1583.

Pandey, K. C., and Phillips, J. C. (1974). *Phys. Rev. B* **9**, 1552.

Parthé, E. (1964). "Crystal Chemistry of Tetrahedral Structures." Gordon & Breach, New York.

Pauling, L. (1960). "The Nature of the Chemical Bond." Cornell Univ. Press, Ithaca, New York.

Pawley, G. S. (1968). *Acta Crystallogr., Sect. B* **24**, 485.

Pearson, W. B. (1967). "A Handbook of Lattice Spacings and Structures of Metals and Alloys," Vols. 1 and 2. Pergamon, Oxford.

Pearson, W. B. (1969). *In* "Development in the Structural Chemistry of Alloy Phases" (B. C. Giessen, ed.), p. 1. Plenum, New York.

Pearson, W. B. (1972). "The Crystal Chemistry and Physics of Metal Alloys." Wiley (Interscience), New York.

Pettifor, D. G. (1979). *Phys. Rev. Lett.* **42**, 846.

Phillips, J. C. (1970). *Rev. Mod. Phys.* **42**, 317.

Phillips, J. C. (1973). "Bonds and Bands in Semiconductors." Academic Press, New York.

Phillips, J. C. (1977). *Solid State Commun.* **22**, 549.

Phillips, J. C. (1978). *Comments Solid State Phys.* **9**, 11.

Phillips, J. C., and Kleinman, L. (1959). *Phys. Rev.* **116**, 287.

Phillips, J. C., and Van Vechten, J. A. (1970). *Phys. Rev. B* **2**, 2147.

Pitzer K. S. (1948). *J. Am. Chem. Soc.* **70**, 2148.

Redondo, A., Goddard, W. A., and McGill, T. C. (1977). *Phys. Rev.* **15**, 5038.

Rieger, W., and Parthé, E.. (1966). *Acta Crystallogr.* **22**, 919.

Rudman, P. S., Stringer, J., and Jaffe, R. I. (1966). "Phase Stability in Metals and Alloys." McGraw-Hill, New York.

St. John, J., and Bloch, A. N. (1974). *Phys. Rev. Lett.* **33**, 1095.

Schaefer, H. F., ed. (1977a). "Methods of Electronic Structure Theory." Plenum, New York.

Schaefer, H. F., ed. (1977b). "Application of Electronic Structure Theory." Plenum, New York.

Schlüter, M., Zunger, A., Kerker, G. P., Ho, K. M., and Cohen, M. L. (1979). *Phys. Rev. Lett.* **42**, 540.

Schob, O., and Parthé, E. (1964). *Acta Crystallogr.* **19**, 214.

Schubert, K. (1964). "Kristallstrukturen Zweikomponentiger Phasen." Springer-Verlag, Berlin and New York.

Schubert, K. (1977). *In* "Intermetallic Compounds" (J. H. Westbrook, ed.), p. 100. Robert E. Krieger Publ. Co., Huntington, New York.

Schubert, K., and Eslinger, P. (1957). *Z. Metallkd.* **48**, 126, 193.

Sellmyer, D. J., Caskey, G. R., and Franz, J. (1971). *J. Phys. Chem. Solids* **33**, 561.

Shannon, R. D., and Prewitt, C. T. (1969). *Acta Crystallogr., Sect. B* **25**, 925.

Shaw, R. W. (1968). *Phys. Rev.* **174**, 769.

Shunk F. A. (1969). "Constitution of Binary Alloys," 2nd Suppl. McGraw-Hill, New York.

Simons, G. (1971a). *J. Chem. Phys.* **55**, 756.

Simons, G. (1971b). *Chem. Phys. Lett.* **12**, 404.

Simons, G., and Bloch, A. N. (1973). *Phys. Rev. B* **7**, 2754.

Slater, J. C. (1956). "Theory of Alloy Phases," pp. 1–12. Am. Soc. Metals, Cleveland, Ohio.

Slater, J. C. (1974). "The Self-Consistent Field for Molecules and Solids," Vol. 4, p. 94. McGraw-Hill, New York.

Spicer, W. E. (1962). *Phys. Rev.* **125**, 1297.

Stroud, D., and Ashcroft, N. W. (1971). *J. Phys. F* **1**, 113.

Szasz, L., and McGinn, G. (1967). *J. Chem. Phys.* **47**, 3495.

Topiol, S., Zunger, A., and Ratner, M. (1977). *Chem. Phys. Lett.* **49**, 367.

Varma, C. M. (1979). *Solid State Commun.* **31**, 295.

Walter, J. P., Fong, C. Y., and Cohen, M. L. (1973). *Solid State Commun.* **12**, 303.

Warshel, A. (1977). *In* "Semiempirical Methods of Electronic Structure Calculations" (G. A. Segal, ed.), p. 133. Plenum, New York.

Wertheim, G. K., and Wernick, J. H. (1967). *Acta Metall.* **15**, 297.

Wykoff, R. W. G. (1963). "Crystal Structures," 2nd ed., Vols. 1–3. Wiley (Interscience), New York.

Yang, Y. W., and Coppens, P. (1974). *Solid State Commun.* **15**, 1555.

Zunger, A. (1979). *J. Vac. Sci. Technol.* **16**, 1337.

Zunger, A. (1980a). *Phys. Rev. B* **22**, 649.

Zunger, A. (1980b). *Phys. Rev. B* **22**, 959.

Zunger, A. (1980c). *Phys. Rev. B* **21**, 4785.

Zunger, A. (1980d). *Phys. Rev. Lett.* **44**, 582.

Zunger, A., and Cohen, M. L. (1978a). *Phys. Rev. Lett.* **41**, 53.

Zunger, A., and Cohen, M. L. (1978b). *Phys. Rev. B* **18**, 5449.

Zunger, A., and Ratner, M. (1978). *Chem. Phys.* **30**, 423.

Zunger, A., and Cohen, M. L. (1979a). *Phys. Rev. B* **19**, 568.

Zunger, A., and Cohen, M. L. (1979b). *Phys. Rev. B* **20**, 4082.

Zunger, A., Kerker, G. P., and Cohen, M. L. (1979a) *Phys. Rev. B* **20**, 581.

Zunger, A., Topiol, S., and Ratner, M. (1979b). *Chem. Phys.* **39**, 75.

Zunger, A., Perdew, J. P., and Oliver, G. (1980). *Solid State Commun.* **34**, 933.

# 6

# Elementary Quantitative Theory of Chemical Bonding

## WALTER A. HARRISON

|       |                                                      |     |
|-------|------------------------------------------------------|-----|
| I.    | Introduction                                         | 137 |
| II.   | The Formulation                                      | 138 |
| III.  | The Bonding Energy                                   | 141 |
| IV.   | Fourth-Order Terms                                   | 143 |
| V.    | The Chemical Grip                                    | 143 |
| VI.   | AB Compounds                                         | 145 |
| VII.  | The Role of Noble Metal d States, Ion Distortion     | 146 |
| VIII. | The s–p Hybridization Energy                         | 146 |
| IX.   | The Oxygen Bridge                                    | 151 |
| X.    | Tetrahedral Complexes                                | 152 |
| XI.   | Perovskites; The s–d Hybridization Energy            | 153 |
| XII.  | Central-Atom Hybrids and Bond Orbitals               | 153 |
|       | References                                           | 154 |

## I. INTRODUCTION

A general program has been undertaken with the goal of understanding and representing the electronic structure of solids sufficiently simply that the entire range of properties can be calculated in terms of it (Harrison, 1980). This effort was directed at the dielectric properties as well as the elastic and bonding properties. The present study addresses specifically those aspects of the bonding properties which are most relevant to crystal-structure determination.

For ionic and covalent solids the approach was based upon a minimal-basis-set LCAO formulation of the electronic structure with universal parameters. This is a first-principles, though only approximate, approach, and no additional scaling parameters are introduced. Thus quantitative estimates of the various physical effects can be obtained in order to learn which are

Structure and Bonding in Crystals, Vol. I

important. The goal here is to provide the formulation which makes some of the familiar concepts of bonding theory concrete and quantitative.

## II. THE FORMULATION

The electronic structure is formulated first for ionic systems; KCl is a suitable prototype. The crystal may be thought of as consisting of spherical closed-shell ions packed together to minimize the electrostatic energy. Such a description, in which the electronic states are simply free-ion states, gives a reasonable account of the cohesion, equilibrium spacing, and elastic properties of alkali halides. The principal approximation made in using it is the neglect of matrix elements coupling the states on neighboring ions.

The approach to be used here is based upon this representation of the electronic structure and the systematic addition of interatomic matrix elements by perturbation theory. This is a rigorously correct method for solution of the one-electron problem, and the one-electron description is adequate to give an understanding of almost all of the properties of solids. However, the calculations will be based upon a number of approximations which tremendously simplify the theory. These approximations have been explored in detail with a wide-ranging application of them to the dielectric and bonding properties of ionic and covalent solids (Harrison, 1980). The expansion based upon the polar limit is a portion of that theory which gives a particularly clear framework for understanding the bonding properties of ionic and covalent solids. The principal approximations upon which the analysis is based are given first:

(a)   The one-electron states are represented as linear combinations of atomic orbitals (LCAOs). This could be made as accurate as wished, but here the orbitals are limited to a *minimal basis set*, only those orbitals which are occupied, or partially occupied, in the free atom.

(b)   Only *nearest-neighbor interatomic matrix elements* are included. The consequences of this with respect to the electronic energy bands are unambiguous and there is no question but what it leads to some error. However, the matrix elements will be chosen in a way appropriate to this approximation and it is not correct to add second-neighbor corrections without redoing the first-neighbor problem at the same time. The inclusion of more distant neighbors is not very difficult, but within the minimal basis concept it is also not very valuable.

The orbitals on neighboring atoms are treated as orthogonal to each other. This is not an approximation since orthogonal orbitals could be constructed and we shall obtain matrix elements directly without ever constructing the orbitals in detail. Since the values are obtained to be consistent with the orthogonality assumption, it would not be correct to add nonorthogonality effects without redetermining the parameters used here.

(c)  For the energies of the electronic states in the ions (the diagonal terms in the Hamiltonian matrix), the *free-atom term values* are used; these are listed ($\varepsilon_p$ for the valence p state and $\varepsilon_s$ for the valence s state) for the nontransition elements in Table I. It might seem surprising to ignore intraatomic level shifts (due to varying level occupation) and Madelung terms, but this approximation is based upon analysis of the known energy bands (Harrison, 1980) for a variety of solids and a consideration of what adjusted values give the best account of these bands. The finding that the atomic term values

TABLE I[a]

**Term Values and Ionic Radii of the Nontransition Elements**

| 1 | 2 | 3 | 4 | 5 | 6 | 7 | 8 | 9 | 10 | 11 |
|---|---|---|---|---|---|---|---|---|---|---|
| | Be 4 | B 5 | C 6 | N 7 | O 8 | F 9 | Ne 10 | Na 11 | Mg 12 | |
| | 8.17 | 12.54 | 17.52 | 23.04 | 29.14 | 35.80 | 43.2 | 5.13 | 6.86 | |
| | 4.14 | 6.64 | 8.97 | 11.47 | 14.13 | 16.99 | 20.0 | | 2.99 | |
| | 0.30 | 0.16 | 1.59 | 1.56 | 1.46 | 1.33 | | 0.98 | 0.65 | |
| | Mg 1 | Al 13 | Si 14 | P 15 | S 16 | Cl 17 | Ar 18 | K 19 | Ca 20 | Sc 21 |
| | 6.86 | 10.11 | 13.55 | 17.10 | 20.80 | 24.63 | 28.7 | 4.19 | 5.41 | 5.85 |
| | 2.99 | 4.86 | 6.52 | 8.33 | 10.27 | 12.31 | 14.5 | | | |
| | | | | | | | | 1.20 | 0.90 | |
| | 0.65 | 0.45 | | 1.95 | 1.90 | 1.81 | | 1.33 | 0.94 | 0.68 |
| Cu 29 | Zn 30 | Ga 31 | Ge 32 | As 33 | Se 34 | Br 35 | Kr 36 | Rb 37 | Sr 38 | Y 39 |
| 6.92 | 8.40 | 11.37 | 14.38 | 17.33 | 20.32 | 23.35 | 26.5 | 3.94 | 5.00 | 5.53 |
| 1.83 | 3.38 | 4.90 | 6.36 | 7.91 | 9.53 | 11.20 | 13.0 | | | |
| 0.65 | 0.63 | 0.61 | | 2.03 | 2.01 | 1.95 | | 1.48 | 1.10 | 0.88 |
| Ag 47 | Cd 48 | In 49 | Sn 50 | Sb 51 | Te 52 | I 53 | Xe 54 | Cs 55 | Ba 56 | La 57 |
| 6.41 | 7.70 | 10.12 | 12.50 | 14.80 | 17.11 | 19.42 | 21.8 | 3.56 | 4.45 | 4.86 |
| 2.05 | 3.38 | 4.69 | 5.94 | 7.24 | 8.59 | 9.97 | 11.4 | | | |
| 1.01 | 0.89 | 0.78 | | 2.22 | 2.19 | 2.16 | | 1.67 | 1.29 | 1.04 |
| Au 79 | Hg 80 | Tl 81 | Pb 82 | Bi 83 | Po 84 | At 85 | Rn 86 | Fr 87 | Ra 88 | Ac 89 |
| 6.48 | 7.68 | 9.92 | 12.07 | 14.15 | 16.21 | 18.24 | 20.31 | 3.40 | 4.24 | 4.63 |
| 2.38 | 3.48 | 4.61 | 5.77 | 6.97 | 8.19 | 9.44 | 10.71 | | | |
| 0.70 | 0.81 | | 2.28 | | | | | 1.75 | 1.37 | 1.11 |

[a] The first two entries for each element are $-\varepsilon_s$ and $-\varepsilon_p$, respectively, in eV, from Herman and Skillman (1963). The third entry is the ionic radius given by Bending (1979). Note that the nonmetals (columns 5, 6, and 7) form close-packed ionic solids when they form AB compounds with metals to the right, but generally form tetrahedral covalent solids when they form AB compounds with metals to the left.

are suitable may be rationalized after the fact by noting *in the equilibrium crystal structure* the intraatomic level shifts very nearly cancel the Madelung terms if the equilibrium charge distribution in the solid is quite close to that of a superposition of free-atom charge densities. This is known to be true from x-ray studies and is not inconsistent with the fact that the charge density is also rather close to a superposition of free-ion densities. Thus in KCl the band gap estimate is 8.1 eV, the difference between the value $\varepsilon_p$ for Cl and the $\varepsilon_s$ value for K; the experimental value (Poole *et al.*, 1975) is 8.4 eV, in good agreement, in fact somewhat better than typical for this method.

(d) For the interatomic matrix elements of the Hamiltonian we use *universal forms*. These are written as a $V$ with two subscripts, $l$ and $l'$ giving the angular momentum quantum numbers for the two coupled atomic states (s for $l = 0$; p for $l = 1$) and a subscript $m$ giving the number of units of angular momentum around the internuclear axis ($\sigma$ for $m = 0$; $\pi$ for $m = 1$)

$$V_{ll'm} = (ll'm) = \eta_{ll'm}\hbar^2/(md^2) \tag{1}$$

with $\eta_{ss\sigma} = -1.40$, $\eta_{sp\sigma} = 1.84$, $\eta_{pp\sigma} = 3.24$, and $\eta_{pp\pi} = -0.81$. The notation $(ll'm)$ is that used by Slater and Koster (1954); $m$ is the electron mass and $d$ the internuclear distance. These equations originated as empirical forms from fitting the known energy bands of semiconductors (Harrison, 1976) but were subsequently derived (both the forms and coefficients close to those listed above) theoretically from the requirement that LCAO and free-electron bands be simultaneously valid representations (Froyen and Harrison, 1979). The signs will not enter into the calculations, only the magnitudes, so one need not distinguish $V_{ll'm}$ from $V_{l'lm}$. When transition-metal compounds are discussed, matrix elements with d states will be needed. They are given by

$$V_{ldm} = (ldm) = \eta_{ldm}\hbar^2 r_d^{\frac{3}{2}}/(md^{\frac{7}{2}}) \tag{2}$$

with $\eta_{sd\sigma} = -3.16$, $\eta_{pd\sigma} = -2.95$, and $\eta_{pd\pi} = 1.36$. Their derivation was based upon transition-metal pseudopotential theory; values for the d *state radius* $r_d$ are given for all 27 transition metals by Harrison (1980) and Harrison and Froyen (1980).

Equations (1) and (2), with $d$ the internuclear distance, give the variation matrix elements from material to material (KCl to RbBr) as well as within a single material (KCl) under pressure.

Something in accuracy has been sacrificed by not allowing these to be adjusted or recalculated for each compound, but there is advantage in having all of the parameters of the theory from the outset and therefore no opportunity to make an incorrect theory appear successful by adjusting parameters. For any given structure the only input needed is the atomic spacing itself.

This can be obtained from experiment, detailed calculation, or on the basis of empirical rules. To aid in the use of the latter, S. J. Bending (private communication, 1979) has extended the traditional ionic radii to additional elements. These are listed in Table I. These are intended to give the internuclear distances AB compounds would have if that compound formed in the rock-salt structure. His study of observed lattice distances indicated that the same cation values could be used in the tetrahedral structure but the anion radius should be reduced for that structure by a factor of 0.977 for the oxygen row, a factor 0.956 for the sulfur row, 0.930 for the selenium row, and (by extrapolation) 0.910 for the tellurium row. They are least accurate for the oxides themselves.

## III.   THE BONDING ENERGY

Since the cohesion of alkali halides is already rather well described by the independent-ion model, a large part of what would be called the bonding energy is already included in the combination of ion formation energy, Madelung energy, and overlap interaction between closed-shell ions. This model correctly suggests that alkali halides should form in structures such as the NaCl and CsCl structure. It in fact always predicts the NaCl structure, which is found to have smaller $d$ than the CsCl structure (Hund, 1925) when reasonable overlap interactions are used. However, it does not contain the physical mechanisms which are responsible for other structures (such as zinc blende, or wurtzite structures) forming in the AB compounds. The most important such contribution is from interatomic matrix elements, the contribution which is now considered. The fact that the theory without them is as descriptive as it is, suggests that their inclusion as perturbative corrections to closed-shell system is most plausible.

Since only interatomic matrix elements are included, there are no diagonal corrections and no first-order terms. The leading contribution to the bonding energy is of second order,

$$E_{\text{bond}} = 2\Sigma_{\alpha,\gamma} V_{\alpha\gamma}^2/(\varepsilon_\alpha - \varepsilon_\gamma) \tag{3}$$

The $V_{\alpha\gamma}$ are interatomic matrix elements between orbitals of energy $\varepsilon_\alpha$ and $\varepsilon_\gamma$. We shall systematically use subscripts $\alpha$ and $\beta$ to indicate orbitals which are occupied in the free ion and $\gamma$ and $\delta$ to indicate orbitals which are empty in the free ion. To obtain the energy per ion pair in an AB compound such as KCl, one sums over the four occupied valence orbitals on the chlorine [the factor of two for spin is written explicitly in Eq. (3)] and sums over s states on the neighboring six potassium ions to which they are coupled. The denominator is negative so this term lowers the energy. The coupling between neighboring orbitals is associated with covalency so this may be called the *covalent contribution to the bonding* in an ionic crystal.

There are two aspects of this form which should be noted immediately. First, if the ions in question are closed-shell systems (no partially filled s, p, nor d shells in the free ion), this covalent contribution to the bonding corresponds to central-force interactions. Consider, for example, an anion with orbitals indicated by $\alpha$ and a neighboring cation with orbitals indicated by $\gamma$. In the sum over the occupied orbitals $\alpha$ and the empty orbitals $\gamma$ there is no influence on the result from other neighbors; it is just as an interaction between two isolated spherically symmetric ions. Equation (3) gives rise to no angle-dependent forces nor to any structure dependence aside from that on internuclear distance. It does not contain the kind of angular interactions ordinarily associated with the stability of tetrahedral structures.

Second, though Eq. (3) would always favor maximizing the number of neighbors, giving additional terms all of which lower the energy, one cannot compare the energies of two structures simply by comparing Eq. (3). The internuclear distance changes and the Madelung energy and the overlap interaction are also modified when the number of nearest neighbors is changed. A full calculation including all of the various changes in energy with structure change has not been carried out but we will be able to see clearly the different influences of each term.

It is important to note that the central-force aspect has arisen directly because the shells were closed, and this will be appropriate in almost all systems considered here. If, however, there are two electrons missing from the shell of orbitals labeled $\alpha$, the situation is quite different. It is ordinarily favorable then to leave one orbital (both spins) unoccupied. Then if this is a p orbital, neighboring cations will bond more strongly in the plane perpendicular to the axis of that orbital. One could estimate the effect, for example, of coupling to a cation s state, at a position of angle $\theta$ from the axis by first calculating the full-shell, and therefore central-force, interaction and then subtracting the effect of coupling to the empty p orbital by the matrix element $V_{sp\sigma} \cos \theta$. This gives an energy $2V_{sp\sigma}^2 \cos^2 \theta/(\varepsilon_s - \varepsilon_p)$ from Eq. (3). This is minimized at $\theta = 90°$ corresponding to all cation neighbors lying in the plane perpendicular to the empty orbital.

In the expansion in matrix elements this gives the leading and dominant contribution if it is present. The effect is sometimes called "dehybridization" though that is an unfortunate term from the point of view used here, where it might better be called a *bonding anisotropy*. "Hybridization" here will denote a fourth-order term based upon full or empty shells. This bonding anisotropy arises at a silicon surface where "dangling hybrids" would be only half occupied; there it is responsible for a well-known reconstruction of the (111) surface (e.g., Harrison, 1980). It can be analyzed using the parameters given in Sec. II, but will not be discussed here.

## IV.   FOURTH-ORDER TERMS

In order to obtain angular forces and structure-dependent interactions for closed shells, one must go to higher order in the matrix elements, the leading terms being of fourth order. Going to fourth order allows the consideration of an extraordinarily large variety of effects, each of which is in some sense a different physical effect. Four orbitals enter the calculation but since only nearest-neighbor coupling between orbitals is included, each in the succession of four orbitals must be on a nearest neighbor to its predecessor. Thus we can think diagramatically of an electron jumping from orbital $\alpha$ to orbital $\delta$ on a neighbor, to another orbital $\beta$, to another $\gamma$, and finally back to $\alpha$. This enables us to keep track of the possibilities. (The order will be convenient later.) However, the orbital $\beta$ may or may not be on the same ion as that of $\alpha$; similarly, $\gamma$ and $\delta$ may or may not be on the same ion. This gives four distinct possibilities. Allowing contributions from two shells on each atom, the possibilities are multiplied by $2^4$, or a total of 64 different physical effects. Finally, in computing total energies there is a sum over occupied states and different sums are appropriate in different circumstances. With the added complexity from coupled degenerate states, it is clearly not appropriate to enumerate all possibilities. We instead consider different systems and evaluate the effects appropriate to those types of systems.

In all cases the analysis is straightforward, though frequently the algebra is complicated. In all cases the parameters given in Sec. II will be used so that explicit estimates may be made of each effect and the magnitudes of competing effects may be compared. It does not generally give high accuracy but it allows one unambiguously to find the trends favored by each effect, and this is not always obvious at the outset.

## V.   THE CHEMICAL GRIP

The evaluation of the total energy in fourth order for a collection of full states $(\alpha, \beta)$ of a single energy $\varepsilon_\alpha$ with a collection of empty states $(\gamma, \delta)$ of energy $\varepsilon_\gamma$ is intricate but straightforward. The result is plausible and is given by Harrison (1980):

$$E_{4\text{th}} = \frac{2}{(\varepsilon_\gamma - \varepsilon_\alpha)^3} \Sigma_{\alpha,\beta,\gamma,\delta} V_{\alpha\gamma} V_{\gamma\beta} V_{\beta\delta} V_{\delta\alpha} \tag{4}$$

A factor of 2 has been included for spin so the sums are over orbitals. [This factor was inadvertently omitted by Harrison (1980) and the formulas given there are in error by that factor.] There are a variety of different kinds of contributions to Eq. (4) which may be distinguished. First are those in which orbitals $\alpha$ and $\beta$ are on the same ion and orbitals $\gamma$ and $\delta$ are on another single ion. These give only central-force interactions for the same reason as given for the bonding energy, Eq. (3), and add nothing new to the problem except that they do give a repulsion proportional to $d^{-8}$. Second are terms in which one set of orbitals (perhaps $\alpha$, $\beta$) are on the same ion but the other pair $(\gamma, \delta)$ are on different ions. This gives a three-body force and we call the sum of such terms the *chemical grip*. This will be seen to give a positive term in the energy and therefore it reduces the cohesion, but it will depend upon the angles between sets of neighbors and therefore is the leading term in the angular interaction. Third, there may be terms in which all four orbitals are on different ions; they might be called *closed-loop contributions* to the chemical grip. They do not occur in zinc blende and wurtzite structures where there are no closed loops of four nearest neighbors (the minimum is six) nor in perovskites which will be discussed later. They vanish for s–p interactions in undistorted rock-salt structures, but they would appear to contribute to the shear constant $c_{44}$ in the rock-salt structure, for example. They will not be discussed further here.

We select then terms in Eq. (4) in which one ion interacts jointly with two of its neighbors. Focus for example upon a chlorine ion in KCl and its interaction with six potassium neighbors. The basis set includes only s states on the potassium so only $V_{1s\sigma}$ matrix elements enter. In fact only the $\sigma$ matrix elements will be kept in other cases treated since they are the largest. Each pair of neighbors enters twice (e.g., $\gamma = 1$, $\delta = 2$ and $\gamma = 2$, $\delta = 1$) so one may multiply by 2 and sum only over $\gamma < \delta$. Using the transformation properties for the spherical harmonics leads to (Harrison, 1980)

$$E_{\text{grip}} = \frac{4V_{1s\sigma}^4}{(\varepsilon_\gamma - \varepsilon_\alpha)^3} \Sigma_{\gamma < \delta} P_1 (\cos \theta_{\gamma\delta})^2 \tag{5}$$

where the Legendre polynomial $P_1$ has entered squared, and $\theta_{\gamma\delta}$ is the angle between the two neighbors. The same formula would apply (again with a positive energy denominator) for the contribution of an empty cation shell interacting with its nearest-neighbors pairs; then $l$ would be the angular-momentum quantum number of the cation shell and the angle $\theta_{\alpha\beta}$ would appear under a sum over $\alpha$ and $\beta$. The energy per ion pair in an AB compound is obtained by adding a cation to an anion grip.

Note that for an s shell on the central ion, $P_1 = 1$ and there is no angular force. However, for a p shell on the central ion, $P_1$ becomes $\cos \theta_{\gamma\delta}$ and interesting angular terms are obtained. Again, these are the leading contribution of covalency to angular interactions. We explore their role in a range of circumstances.

## VI.  AB COMPOUNDS

The first set of competing structures which comes to mind is that of rock-salt (ionic) and tetrahedral (covalent) structures. In the polar limit the energy is minimized, due to low Madelung energy, by ionic structures. In the AB compounds this will be in a rock-salt or cesium chloride structure with six and eight, respectively, nearest neighbors; this high coordination number is also favored by the bonding energy, Eq. (3). The chemical grip is the leading term favoring a reduction in the coordination. There is a chemical grip associated with the p states on the anion. In the rock-salt structure each anion has three pairs of neighbors at $180°$ separation ($\cos \theta_{\gamma\delta} = -1$) and 12 pairs of neighbors at $90°$ ($\cos \theta_{\gamma\delta} = 0$). Equation (5) gives immediately $E_{\text{grip}} = 12V_{\text{sp}\sigma}^4/(\varepsilon_\text{s} - \varepsilon_\text{p})^3$. In the tetrahedral structure the anion would have six pairs of neighbors at a tetrahedral angle of $109.5°$ ($\cos \theta_{\gamma\delta} = -\frac{1}{3}$) and $E_{\text{grip}} = \frac{8}{3}V_{\text{sp}\sigma}^4/(\varepsilon_\text{s} - \varepsilon_\text{p})^3$. Thus the chemical grip strongly favors the tetrahedral structure, and when the interatomic matrix elements are sufficiently large the covalent structure should occur. This is consistent with observation and with intuitive bonding arguments, but note that the anion s states have not entered; there are no $\text{sp}^3$ hybrids in the argument.

To carry out the detailed analysis it would be necessary to introduce an overlap interaction, perhaps of a form $d^{-n}$ or $e^{-\mu d}$, add the Madelung energy, and minimize with respect to internuclear distance for each structure. Setting the spacing equal to the observed spacing (note that this gives the overlap interaction in terms of the Madelung term) would lead to a total energy, based upon overlap and Madelung energies, proportional to $Z^2e^2/d$ for each structure, favoring in general the rock-salt structure. The criterion for formation of the tetrahedral structure would be that this energy difference be exceeded by the difference in grip energy, $(28/3)V_{\text{sp}\sigma}^4/(\varepsilon_\text{s} - \varepsilon_\text{p})^3$.

First, note that this gives a formal criterion, that the ratio $R = V_{\text{sp}\sigma}^4 d/[Z^2e^2(\varepsilon_\text{s} - \varepsilon_\text{p})^3]$ exceed some numerical constant. ($V_{\text{sp}\sigma}$ was given in terms of internuclear distance in Eq. (1) and a correction, such as that due to Bending and mentioned earlier, should be made for the difference in spacings for the two structures.) In contrast it is usual in empirical approaches without detailed formulations to choose simply the criterion that some covalency, perhaps defined by $V_{\text{sp}\sigma}/(\varepsilon_\text{s} - \varepsilon_\text{p})$, exceed some numerical constant. Second, note that the analysis here would predict the value of the numerical constant whereas in empirical approaches one can obtain it only from a knowledge of the observed structures. Experience would suggest that if this program were carried through it would separate structures, but much better if the numerical constant were adjusted slightly from the predicted value. Evaluation of the parameters for the AB compounds by S. J. Bending (private communication, 1979) indicated that the simple covalency defined above separated all the AB compounds which have a unique structure except MgO and that the ratio $R$ separated them all.

## VII. THE ROLE OF NOBLE METAL d STATES, ION DISTORTION

It is natural in this context to ask the role of the d states in noble metal AB compounds in both structure determination and angular rigidity. It would at first seem plausible simply to introduce a chemical grip based upon d states interacting with the p states on the neighboring halogens. This however would be incorrect. In the ionic limit both the d states and the neighboring halogen p states are occupied, and though a shift in their energies occurs, which was the basis of the chemical grip, the shifts are opposite and exactly canceled out. It is absolutely essential in bonding arguments to remember which states are occupied.

In the LCAO context the only effects which the d states can have must arise from coupling with the only empty basis states of the minimal set, the noble-metal s states. This coupling arises through the p states on the neighboring halogens and gives an energy shift

$$E_{\text{dist}} = -4(V_{\text{sp}\sigma}V_{\text{pd}\sigma})^2\Sigma_{\gamma<\sigma}P_2(\cos\theta_{\gamma\delta})/(\varepsilon_s - \varepsilon_p)^2(\varepsilon_s - \varepsilon_d) \qquad (6)$$

of fourth order as was the chemical grip, but of quite different form. (Note the Legendre polynomial $P_2$ is not squared.) Since the effects arise from addition of empty orbitals on the same atom as the d states, it is most appropriately thought of as an effect of *ion distortion*. It vanishes in the ideal tetrahedral or cubic structure and therefore does not influence the competition between those structures, but it lowers the energy when the lattice is distorted. Thus it softens the lattice against distortion, but when explicitly evaluated by S. Froyen (private communication, 1979) it was found not to give an important correction to the elastic constants.

## VIII. THE s–p HYBRIDIZATION ENERGY

It is interesting that it has been possible to discuss the stability of the tetrahedral structures without recourse to the familiar $sp^3$ hybrids which are usually used; only the p state on the anion was included and only the s state on the cation. Clearly in the less polar compounds the other states become important. They may be systematically included starting from the polar limit.

Consider then the effect of the s state, in addition to the p state, on an anion coupled to two ligand cations. In the context of Sec. VI we might focus upon just the s states on the cations, but the analysis is the same if instead $\sigma$-oriented p states or even hybrids of the two are treated; in fact it will be seen that hybrids are frequently most appropriate. Therefore the ligand states are indicated by an $|h'\rangle$ (the prime indicating a ligand ion) and the appropriate state can be substituted in the end. The calculation is described taking the central ion to be the anion and the ligands to be cations, but the final formulas are the same if these are interchanged. The analysis is particularly simple when the central ion contains only s and p states, as assumed here, and discussion of s–d hybridization is postponed to Sec. XI.

In Fig. 1 is shown the anion with two of its cation neighbors with their $h'$-states. The angular effects of s–p hybridization arise from fourth-order terms from such complexes. The $p_y$ orbital is not coupled to either $h'$-state so solutions may be obtained based upon the five orbitals of energy $\varepsilon_s$ and $\varepsilon_p$ on the anion and $\varepsilon_{h'}$ on the two cations. In fact the solutions are even or odd under reflection in the $yz$ plane and the odd solution contains only the $p_x$ orbital and an odd combination of cation $h'$ states, $(|h'_1\rangle - |h'_2\rangle)/\sqrt{2}$. The energies are immediately

$$E = \frac{\varepsilon_{h'} + \varepsilon_p}{2} \pm \left[ \left( \frac{\varepsilon_{h'} - \varepsilon_p}{2} \right)^2 + 2V_{ph\sigma}^2 \sin^2 \theta \right]^{\frac{1}{2}} \tag{7}$$

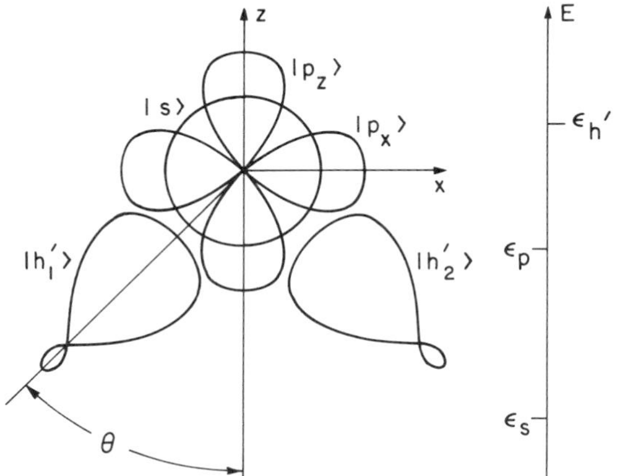

**Fig. 1.** A schematic representation of the orbitals entering the s–p hybridization energy. On the right is a diagram showing the corresponding energy levels.

(Note that the matrix element between the $p_x$ state and the $h'$ state is reduced from the $\sigma$-oriented value of $V_{ph\sigma}$ by a factor $\sin\theta$ due to the relative orientation of the p state and the internuclear distance (Slater and Koster, 1954).

The even states are obtained from the Hamiltonian matrix

$$
\begin{pmatrix}
\varepsilon_{h'} & \sqrt{2}V_{sh\sigma} & \sqrt{2}V_{ph\sigma}\cos\theta \\
\sqrt{2}V_{sh\sigma} & \varepsilon_s & 0 \\
\sqrt{2}V_{ph\sigma}\cos\theta & 0 & \varepsilon_p
\end{pmatrix}
\tag{8}
$$

One needs only the energy of the upper states near $\varepsilon_{h'}$ which, when added to the upper state of Eq. (7), gives the sum of the shifts of the states which are unoccupied. This must be exactly equal to the negative of the shifts of the three occupied states since the diagonalization leaves the trace of the $5\times 5$ Hamiltonian matrix unchanged. This energy may be obtained by writing out the secular equation for Eq. (8) and solving for $E - \varepsilon_{h'}$,

$$
E = \varepsilon_{h'} + \frac{2V_{sh\sigma}^2}{E - \varepsilon_s} + \frac{2V_{ph\sigma}^2\cos^2\theta}{E - \varepsilon_p}
\tag{9}
$$

To second order,

$$
E^{(2)} = \varepsilon_{h'} + \frac{2V_{sh\sigma}^2}{\varepsilon_{h'} - \varepsilon_s} + \frac{2V_{ph\sigma}^2\cos^2\theta}{\varepsilon_{h'} - \varepsilon_p}
\tag{10}
$$

This may be substituted in the final terms of Eq. (9) to obtain the energy to fourth order

$$
E = E^{(2)} - 2\left[\frac{V_{sh\sigma}^2}{(\varepsilon_{h'} - \varepsilon_s)^2} + \frac{V_{ph\sigma}^2\cos^2\theta}{(\varepsilon_{h'} - \varepsilon_p)^2}\right]\left[E^{(2)} - \varepsilon_{h'}\right]
\tag{11}
$$

Equation (7) may be directly expanded in $V_{ph\sigma}$ to obtain the energy to fourth order for the upper state,

$$
E = \varepsilon_{h'} + \frac{2V_{ph\sigma}^2\sin^2\theta}{\varepsilon_{h'} - \varepsilon_p} - \frac{4V_{ph\sigma}^4\sin^4\theta}{(\varepsilon_{h'} - \varepsilon_p)^3}
\tag{12}
$$

which may be added to Eq. (11). The second-order terms become independent of $\theta$ as they must and the fourth-order terms are to be reversed in sign and multiplied by 2 for spin to give the energy per ion pair.

$$
\begin{aligned}
E_{\text{tot}}^{(4)} = 8\Bigg[ &\frac{V_{sh\sigma}^4}{(\varepsilon_{h'} - \varepsilon_s)^3} + \frac{V_{sh\sigma}^2 V_{ph\sigma}^2\cos^2\theta}{(\varepsilon_{h'} - \varepsilon_s)(\varepsilon_{h'} - \varepsilon_p)}\left(\frac{1}{\varepsilon_{h'} - \varepsilon_s} + \frac{1}{\varepsilon_{h'} - \varepsilon_p}\right) \\
&+ \frac{V_{ph\sigma}^4}{(\varepsilon_{h'} - \varepsilon_p)^3}(\cos^4\theta + \sin^4\theta)\Bigg]
\end{aligned}
\tag{13}
$$

The angle-dependent part is more conveniently written

$$E_{\text{hybrid}} + E_{\text{grip}} = \frac{8V_{\text{ph}\sigma}^4}{|\varepsilon_{\text{h}'} - \varepsilon_{\text{p}}|^3}(\cos^4\theta + \sin^4\theta + \lambda\cos^2\theta) \tag{14}$$

with $\lambda$, the hybridization parameter, given by

$$\lambda = \frac{V_{\text{sh}\sigma}^2}{V_{\text{ph}\sigma}^2}\frac{\varepsilon_{\text{h}'} - \varepsilon_{\text{p}}}{\varepsilon_{\text{h}'} - \varepsilon_{\text{s}}}\left(1 + \frac{\varepsilon_{\text{p}'} - \varepsilon_{\text{p}}}{\varepsilon_{\text{h}'} - \varepsilon_{\text{s}}}\right) \tag{15}$$

Note first that if the coupling $V_{\text{sh}\sigma}$ is neglected, so $\lambda = 0$, the result may be rewritten using the identity $\cos^4\theta + \sin^4\theta = \frac{1}{2} + \frac{1}{2}\cos^2 2\theta$. Then noting that the $\theta_{\gamma\delta}$ entering the chemical-grip equation, Eq. (5), equals the $2\theta$ here, it is seen that the angle-dependent terms in Eq. (5) and Eq. (14) are identical; the chemical grip is obtained directly as it should be. In this case the minimum energy occurs at $\theta = 45°$, a perpendicular orientation of the bonds. We call the term proportional to $\lambda$ the *hybridization energy*.

It is convenient in many cases to sum the hybridization energy over all bond angles at the central ion of interest to obtain a form analogous to Eq. (5) for the grip. Writing $\cos^2\theta = (1 + \cos\theta_{\gamma\delta})/2$, Eq. (14) yields

$$E_{\text{hybrid}} = \frac{4\lambda V_{\text{ph}\sigma}^4}{|\varepsilon_{\text{h}'} - \varepsilon_{\text{p}}|^3}\Sigma_{\gamma<\delta}(1 + \cos\theta_{\gamma\delta}) \tag{16}$$

To understand the significance of the angular term, construct the sum over the vectors $\mathbf{d}_\gamma$ (all of length $d$) to the nearest neighbors Its squared length is

$$(\Sigma_\gamma\mathbf{d}_\gamma)\cdot(\Sigma_\delta\mathbf{d}_\delta) = \Sigma_\gamma d^2 + 2d^2\Sigma_{\gamma<\delta}\cos\theta_{\gamma\delta} \tag{17}$$

Then by comparison with Eq. (16) (noting that $\lambda > 0$), it is seen that the s–p hybridization energy is minimized if $\Sigma_\gamma\mathbf{d}_\gamma = 0$. *The s–p hybridization favors an arrangement of neighbors such that the central ion is at their centroid.*

Thus the regular tetrahedron of neighbors in AB compounds with four neighbors has the minimum s–p hybridization energy, but not lower than a square or rectangle of ions surrounding, and coplanar with, the central ion; it is the *grip* which specifically favors the tetrahedron. The grip energy is in general zero if all $\mathbf{d}_\gamma$ are orthogonal, but that is not possible with four neighbors and the tetrahedron appears to be lowest in grip energy.

Consider again the single pair of neighbors of Fig. 1. For this it is convenient to treat the grip and the hybridization energy together by using Eq. (14).

The energy may be minimized by setting the derivative with respect to $\theta$ of Eq. (14) equal to zero. There are two solutions corresponding to $\theta = 0°$

and $2\theta = 180°$, and one for

$$\cos 2\theta = -\lambda/2 \qquad (18)$$

Note again from Fig. 1 that $2\theta$ is the angle between bonds. In the case under consideration, $\varepsilon_{h'}$ exceeds both $\varepsilon_s$ and $\varepsilon_p$ so $\lambda$ is positive. If it is less than 2, Eq. (18) gives the angle for minimum energy and the 0° and 180° solutions are maxima. If $\lambda$ exceeds 2 the minimum occurs at $2\theta = 180°$.

We have described here the case in which $\varepsilon_{h'}$ exceeds both $\varepsilon_s$ and $\varepsilon_p$; that is the case in which the central ion is the anion. Then $\varepsilon_{h'} - \varepsilon_p$ will be less than $\varepsilon_{h'} - \varepsilon_s$. Since $V_{sh\sigma} < V_{ph\sigma}$ we may expect $\lambda$ to be less than 2 and Eq. (16) gives an equilibrium configuration with $2\theta$ less than 180°, a bent configuration.

On the other hand, for the case in which $\varepsilon_{h'}$ is less than both $\varepsilon_s$ and $\varepsilon_p$, corresponding to the central ion being the cation, the same analysis would apply except that $\varepsilon_{h'} - \varepsilon_p$ in Eq. (14) changes sign and the sign reversal should not be made in going to Eq. (13); thus both cases are covered by Eqs. (14)-(18) if $|\varepsilon_{h'} - \varepsilon_p|^3$ is used in the denominator of Eqs. (14) and (16). In this case $\varepsilon_p - \varepsilon_{h'}$ will be greater than $\varepsilon_s - \varepsilon_{h'}$ and $\lambda$ would ordinarily be expected to exceed 2; the minimum energy occurs at $2\theta = 180°$.

This difference is consistent with the colinear geometry of $CO_2$. However, in this case there is a complication which occurs also in other systems: the carbon s state lies below the oxygen p state, contrary to the assumption of the polar limit. In such a case one should really solve the Hamiltonian matrix more completely and that would be no problem in the molecule. It may in fact be done step-by-step for the molecule in a way that could also be done for the solid. Include, in addition to the carbon s and p states and oxygen p state, the oxygen s states. They are coupled to the carbon p state, but not the oxygen p state so they will tend to eliminate the crossing of levels and restore the polar limit. In perturbation theory this raising of the carbon p level is $2V_{ss\sigma}^2/(\varepsilon_s^C - \varepsilon_s^O) = 8.6$ eV, based upon an internuclear distance of $d = 1.23$ Å. This is in fact more than enough to eliminate the crossing and restore the polar limit.

Another approximation, which accomplishes the same thing and thus extends the range of the polar limit, is the introduction of hybrids on the ligand ions—e.g., hybrids $(|p\sigma\rangle + |s\rangle)2^{\frac{1}{2}}$. This is particularly appropriate in solids where hybrids are selected which are only coupled to orbitals on a single other ion. Then the problem reduces to a two-level problem and can be solved more accurately than in perturbation theory. Such ligand hybrids will be used here.

In $CO_2$, and in other cases treated here, the full solution of the problem requires the inclusion of $\pi$ bonding, which is neglected here as it was for the chemical grip. All of the needed matrix elements have been given so $\pi$ bonding could be directly included if desired.


## IX.  THE OXYGEN BRIDGE


An interesting application of Eq. (14) is to the position of an oxygen atom bonded to two neighbors, as in $SiO_2$ or in the perovskites. For this case, in contrast to $CO_2$, the central atom being considered is the anion, oxygen. In such a case the ligand states (e.g., Si hybrids) are much closer to the oxygen $\varepsilon_p$ than to the $\varepsilon_s$ and the hybridization parameter tends to be quite small, 0.14 for $SiO_2$ based upon Si $sp^3$ hybrids; thus the minimum energy comes near $2\theta = 90°$. Again the effects of the chemical grip tend to be dominant in an anion and they favor bent configurations.

However, other effects open up the angle, to $144°$ in $SiO_2$. One of these is electrostatic energy. Another, coupling between bonding units through the s–p splitting in the ligand, may be even more important (Harrison, 1980). One could proceed with the theory by calculating the change in energy as the oxygen is displaced from the position midway between two silicon neighbors. Using Eq. (14) a value proportional to $V_{ph\sigma}^4 \, \delta\theta^2/(\varepsilon_{h'} - \varepsilon_p)^3$ would be obtained, though in $SiO_2$ the matrix elements are too large for the expansion to really be good. The change in electrostatic energy under such a distortion might also be calculated, giving a value proportional to $Z^2 e^2 \, \delta\theta^2/d$. The ratio of the two is proportional to $R' = V_{ph\sigma}^4 d/[Z^2 e^2 (\varepsilon_{h'} - \varepsilon_p)^3]$. When $R'$ exceeds some numerical constant a bent configuration is predicted, assuming that the electrostatic energy is the dominant term favoring a colinear arrangement. The numerical constant has not been calculated, but $R'$ was computed for various systems to find an empirical criterion for a bent configuration.

In $SiO_2$ $R' = 2.02$ was obtained, based upon $sp^3$ hybrids on the silicon; the configuration in $SiO_2$ is bent so this is a sufficient value. For the perovskites a value may similarly be calculated replacing the ligand hybrid by the transition-metal d state. This gives values of $R'$ for three representative perovskites of 0.04, 0.03, 0.16, and 0.68 for $SrTiO_3$, $KTaO_3$, $KMoO_3$ and $ReO_3$, respectively (Harrison, 1980). All of these form straight configurations corresponding to the low values of the ratio obtained.

Note, however [see Eq. (2)], that this ratio varies as the inverse 13th power of $d$; thus under pressure the ratio $R'$ might rise above the critical value. Recent experiments by Schirber and Morosin (1979) in fact suggest that $ReO_3$ undergoes a transition to a bent configuration under pressure, suggesting that the critical value is just above 0.68. This discussion of $ABO_3$

compounds has overlooked the role of the ion A which must be important at least in some cases; $CaTiO_3$ is bent while $SrTiO_3$ is not.

## X. TETRAHEDRAL COMPLEXES

An important subgroup in many solids is the tetrahedral complex, such as form the isoelectronic series $SiO_4^{4-}$, $PO_4^{3-}$, $SO_4^{2-}$, $ClO_4^-$, $ArO_4$. In the polar limit these are thought of as a central positive ion surrounded by four $O^{2-}$ ions. One question of stability which could be asked is the energy required to remove a neutral oxygen atom from the complex. The principal contribution to this is the energy required to raise two electrons from an oxygen level ($\varepsilon_p$ for oxygen, in the absence of interatomic matrix elements) to the energy of an $sp^3$-hybrid on the central atom [$(\varepsilon_s + 3\varepsilon_p)/4$ in the absence of interatomic matrix elements]. Using the orbital energies of Table I this suggests immediately that the first three are stable but $ClO_4^-$ and $ArO_4$ are not. This confirms the expected trend of a reduction in stability to the right in the series that is intuitively expected, though $ClO_4^-$ is in fact stable.

The angular stability or rigidity of the tetrahedral complex may also be considered. Clearly the tetrahedral configuration is favored by the Madelung energy in the polar limit (with $O^{2-}$). The effects of hybridization and of the chemical grip may be added by using Eq. (14). Writing $\theta_{\alpha\beta}$ as the angle between two ligands measured from the central ion, and setting $\theta$ in Eq. (14) equal to $\theta_{\alpha\beta}/2$, one obtains

$$E_{\text{hybrid}} + E_{\text{grip}} = \frac{4V_{\text{ph}\sigma}^4}{|\varepsilon_{h'} - \varepsilon_p|^3} \sum_{\alpha<\beta}[1 + \lambda + \cos^2\theta_{\alpha\beta} + \lambda\cos\theta_{\alpha\beta}] \quad (19)$$

This is a minimum with the ligands forming a regular tetrahedron. The geometry is somewhat complicated but it is straightforward to evaluate the change in Eq. (19) when one ligand is rotated an angle $\delta\theta$ from its tetrahedral position. It is

$$\delta(E_{\text{hybrid}} + E_{\text{grip}}) = \frac{4V_{\text{ph}\sigma}^4}{|\varepsilon_{h'} - \varepsilon_p|^3}\left(\frac{2}{3} + \frac{\lambda}{2}\right)\delta\theta^2 \quad (20)$$

where $\lambda$ is positive and both terms contribute to the angular rigidity of the system. As the covalency increases, moving to the right in the series $SiO_4^{4-}$, $PO_4^{3-}$, etc., the angular rigidity increases and in this sense the stability of the structure increases. The trend is opposite to the trend in stability against an oxygen atom splitting off.

The effect of hybridization increases more rapidly than the grip with increasing covalency, and the expression for $\lambda$ diverges when the central-atom s state energy equals the oxygen hybrid energy. Presumably a more accurate calculation would remove the divergence in such a way that the trend of increasing rigidity continues beyond this point. Indeed, in the tetrahedral AB compounds the angular rigidity and elastic shear constants are found to vary as $V_2 \alpha_c^3 = V_2^4/(V_2^2 + V_3^2)^{\frac{3}{2}}$, approaching Eq. (20) in the polar limit (Harrison, 1980). Here $V_2$ and $V_3$ are defined in terms of $sp^3$ hybrids (see Sec. XII).

## XI.   PEROVSKITES; THE s–d HYBRIDIZATION ENERGY

In many compounds of transition metals, all outer electrons are stripped from the transition-metal ion in the polar limit; $TiO_2$ and $SrTiO_3$ are examples. In the latter case each Ti is surrounded by six oxygen ions in cube directions. The Madelung energy stabilizes this structure against a distortion $e_4$ changing the angles from $90°$ (Harrison, 1980, pp. 468ff.). The stability is also affected by the chemical grip associated with the Ti d state and by s–d hybridization on the Ti ion. Considering both, taking the ligand oxygen states as $\sigma$-oriented s–p hybrids, and dropping the prime on the ligand $h$ leads to

$$E_{grip} = \frac{4V_{hd\sigma}^4}{(\varepsilon_d - \varepsilon_h)^3} \Sigma_{\gamma < \delta} P_2^2 (\cos \theta_{\gamma\delta}) \tag{21}$$

and

$$E_{hybrid} = \frac{4V_{sh\sigma}^2 V_{hd\sigma}^2}{(\varepsilon_d - \varepsilon_h)(\varepsilon_s - \varepsilon_h)} \left( \frac{1}{\varepsilon_d - \varepsilon_h} + \frac{1}{\varepsilon_s - \varepsilon_h} \right) \Sigma_{\gamma < \delta} P_2 (\cos \theta_{\gamma\delta}) \tag{22}$$

Under a shear distortion $e_4$ four of the $\pi/2$ angles at each titanium are changed by $e_4$ and to second-order in $e_4$ one finds $P_2^2 = \frac{1}{4} - \frac{3}{2}e_4^2$ and $P_2 = -\frac{1}{2} + \frac{3}{2}e_4^2$. Thus the grip tends to destabilize the right angles while the s–d hybridization stabilizes it. There is some uncertainty in the choice of $\varepsilon_s$ on Ti, but if $\varepsilon_d - \varepsilon_h$ and $\varepsilon_s - \varepsilon_h$ are taken equal, the ratio of the second-order terms in $e_4$ of $E_{hybrid}/E_{grip}$ becomes $2V_{sh\sigma}^2/V_{hd\sigma}^2$. Then from Eq. (2) with $r_d = 1.08$ Å for Ti (Harrison, 1980; Harrison and Froyen, 1980) and $d = 1.95$ Å for $SrTiO_3$, one finds that the s–d hybridization term is 3.31 times as large as the grip, adding to the angular rigidity from the Madelung term.

## XII.   CENTRAL-ATOM HYBRIDS AND BOND ORBITALS

When systems become highly covalent—i.e., when the interatomic matrix elements become comparable to the term value differences—the perturbation expansion which has been made here becomes inappropriate though the effects of the grip, ion distortion, and hybridization are qualitatively the same and the LCAO formulation remains just as valid. A suitable numerical approach in such systems is again to form hybrids, now on the central ion, chosen such that each hybrid couples strongly only with one ligand; then that hybrid and the ligand states can be treated as a two-level system. For example, in the tetrahedral complexes or crystal structures, $sp^3$ hybrids are appropriate. If the matrix element between this hybrid and the orbital from the neighbor is written $-V_2$ and the energy difference between them is $2V_3$, one obtains immediately antibonding and bonding levels at energies, relative to the average of the uncoupled levels, of $\pm(V_2^2 + V_3^2)^{\frac{1}{2}}$. The lower level is occupied and the bonding properties can be treated by summing bonding energies.

Note that this approach includes the bonding energy and some of the fourth-order terms which have been considered here and terms of all higher orders since the expansion of $(V_2^2 + V_3^2)^{\frac{1}{2}}$ in $V_2$ contains terms of arbitrarily high order. However, it contains an approximation, the neglect of coupling between bond orbitals and neighboring antibonding orbitals, which was not necessary in the perturbation-theoretic approach. Corrections to this *Bond Orbital Approximation* can then be made by incorporating the neglected matrix elements in perturbation theory using the theory of Extended Bond Orbitals (Harrison, 1980; Ren, 1980). However, this approach loses the simplicity and the classification of contributions to the bonding which was provided here by the fourth-order terms in the interatomic matrix elements.

## REFERENCES

Bending, S. J., (1979) private communication.
Froyen, S., (1979) private communication.
Froyen, S., and Harrison, W. A. (1979) *Phys. Rev. B* **20**, 2420.
Harrison, W. A. (1976). *Bull. Am. Phys. Soc.* **21**, 1315.
Harrison, W. A. (1980). "Electronic Structure and the Properties of Solids." Freeman, San Francisco, California.
Harrison, W. A., and Froyen, S. (1980). *Phys. Rev.* **B 21**, 3214.
Herman, F., and Skillman, S. (1963). "Atomic Structure Calculations." Prentice-Hall, Englewood Cliffs, New Jersey.
Hund, F. (1925). *Z. Phys.* **34**, 833.
Poole, R. T., Jenkin, J. G., Liesegang, J., and Leckey, R. C. G. (1975). *Phys. Rev. B* **11**, 5159.
Ren, S.-Y., (1980). *Phys. Rev.* **B 22**, 2908.
Schirber, J. E., and Morosin, B. (1979). *Phys. Rev. Lett.* **42**, 1485.
Slater, J. C., and Koster, G. F. (1954). *Phys. Rev.* **94**, 1498.

# 7

# The Role and Significance of Empirical and Semiempirical Correlations

## LEO BREWER

| I. | Introduction | 155 |
|---|---|---|
| II. | Bonding Models | 156 |
| | A. What is the Range of Bonding Contributions per Electron? | 158 |
| | B. Promotion of Electrons | 159 |
| | C. Net Bonding per Electron | 161 |
| | D. Variations of Promotion Energies of the Lanthanides | 162 |
| | E. To What Extent Do Solid Ni to Pt and Cu to Au Require Electronic Promotion? | 163 |
| | F. Variation of Valence-State Bonding Enthalpy | 167 |
| III. | Correlation of Electronic Configuration and Crystal Structure | 170 |
| IV. | Application of a Semiempirical Correlation | 171 |
| | References | 174 |

## I. INTRODUCTION

The Schrödinger equation can, in principle, give us all the information that we need for practical applications. However, it is clear that we are not yet in the position to provide information at the needed level of accuracy except for the simplest molecules through strictly *ab initio* calculations. The application of semiempirical correlations will continue to play a most important role for the forseeable future. The term "empirical correlation" is not used in the sense of simply examining the variation of a desired property as a function of some arbitrary variable. This is sometimes done for engineering purposes when only limited groups of closely related materials are of interest and it is desirable to interpolate across a small gap. A practical correlation must have a sound theoretical framework. As a simple example, we can consider the spectra of diatomic molecules. Quantum mechanics

**155**

Structure and Bonding in Crystals, Vol. I

has characterized the main features of the electronic, vibrational and rotational contributions to the spectra, but the actual energy spacings cannot generally be calculated with adequate accuracy. It is necessary to obtain experimental data for a few key compounds to fix electronic, vibrational, and rotational parameters. Once these values are known for representative compounds, theory can be used to select appropriate parameters that vary in a relatively simple way and which can be used for predicting the properties of yet unstudied molecules. For example, we can confidently expect the radii of the atoms to increase regularly as we go down in the Periodic Table, with the size of the atom being fixed primarily by the radial distribution of the electrons in the last completed subshell. Similarly, as we move to the right in the Periodic Table, we expect the size of the atoms to decrease as the nuclear charge increases to compress the distribution of electrons in the filled subshells. Although we cannot calculate with sufficient accuracy the radii of the atoms by *ab initio* calculations, we are confident that theory does correctly predict the trends as a function of nuclear charge and the types of electronic subshells involved. Thus it only takes a few values to establish a method of predicting the values for a large number of unstudied molecules.

## II.  BONDING MODELS

There are many types of models available, such as ionic bonding, covalent bonding, hybridization, electronegativity models, acid–base models, soft acids and hard acids, etc., which scientists are using to correlate properties of materials and to predict properties for materials which have not been studied. Why are so many models necessary? It is really a matter of providing models of practical utility. A very general model that would cover a wide range of materials from covalently bonded materials to strongly ionic materials would be too complicated to be of practical use. It is generally more efficient to design a simpler model of limited applicability that is easy to use. The problem then becomes one of setting the bounds of utilization of the model, so that one knows when to stop using a given model and to switch to a more appropriate model.

Pauling (1947, 1949, 1950, 1960) has been one of the leaders in demonstrating the value of simple bonding models in evaluating thermodynamic, structural, magnetic, and many other types of measurements. As an example of the effective use of these procedures, an extension of chemical bonding models to metallic systems will be summarized. This extension is essentially an amalgamation by Engel (1939, 1940, 1945, 1949, 1954, 1964, 1967) of chemical models initiated by Pauling with metallurgical models used by

**TABLE Ia**

$\Delta H_0^\circ / R$ (kK) for Atomization of 1 Gram-Atom of the Solid Elements at 0 K

| | | | | | | | | | | | | | | | | | |
|---|---|---|---|---|---|---|---|---|---|---|---|---|---|---|---|---|---|
| H<br>26.03 | | | | | | | | | | | | | | | | | He<br>0.007 |
| Li<br>19.0 | Be<br>38.5 | B<br>67.4 | | | | | | | | | | C<br>85.5 | N<br>57.1 | O<br>30.2 | F<br>9.7 | | Ne<br>0.23 |
| Na<br>12.9 | Mg<br>17.5 | Al<br>39.3 | | | | | | | | | | Si<br>54 | P<br>39.8 | S<br>33.1 | Cl<br>16.20 | | Ar<br>0.93 |
| K<br>10.8 | Ca<br>21.4 | Sc<br>45.2 | Ti<br>56.3 | V<br>62 | Cr<br>47.6 | Mn<br>33.9 | Fe<br>49.7 | Co<br>51.0 | Ni<br>51.5 | Cu<br>40.5 | Zn<br>15.6 | Ga<br>32.6 | Ge<br>44.7 | As<br>34.3 | Se<br>28.5 | Br<br>14.19 | Kr<br>1.35 |
| Rb<br>9.9 | Sr<br>20.0 | Y<br>50.7 | Zr<br>72.6 | Nb<br>88<br>±2 | Mo<br>79.1 | Tc<br>80 | Ru<br>78.2 | Rh<br>66.7 | Pd<br>45.2 | Ag<br>34.2 | Cd<br>13.5 | In<br>29.2 | Sn<br>36.4 | Sb<br>31.9 | Te<br>25.3 | I<br>12.89 | Xe<br>1.91 |
| Cs<br>9.3 | Ba<br>22 | La<br>51.9 | Hf<br>74.7 | Ta<br>94.0 | W<br>103.3 | Re<br>93.2 | Os<br>94.8 | Ir<br>80.6 | Pt<br>67.8 | Au<br>44.3 | Hg<br>7.8 | Tl<br>21.8 | Pb<br>23.5 | Bi<br>25.3 | Po<br>17 | At<br>(11)<br>±3 | Rn<br>2.3 |
| Fr<br>(9) | Ra<br>19 | Ac<br>49<br>±2 | | | | | | | | | | | | | | | |

**TABLE I***b*

$\Delta H_0^\circ / R$ **(kK) for Atomization of 1 Gram-Atom of Solid Lanthanides and Actinides at 0 K**

| La | Ce | Pr | Nd | Pm | Sm | Eu | Gd | Tb | Dy | Ho | Er | Tm | Yb | Lu |
|------|------|------|------|------|------|------|------|------|------|------|------|------|------|------|
| 51.9 | 50.2 | 42.9 | 39.5 | (37) | 24.8 | 21.5 | 48.1 | 47.0 | 35.3 | 36.4 | 38.1 | 28.1 | 18.7 | 51.4 |
|      |      |      |      | ±2   |      |      |      |      |      |      |      |      |      |      |

| Ac | Th | Pa | U | Np | Pu | Am | Cm | Bk | Cf | Es | Fm | Md | No | Lr |
|------|------|------|------|------|------|------|------|------|------|------|------|------|------|------|
| 49 | 71.9 | (73) | 64 | 55 | 41.8 | 32 | 46 | (35) | (21) | (18) | (17) | (14) | (13) | (37) |
| ±2 |      | ±4   |      |      |      |      |      | ±2   | ±3   | ±3   | ±3   | ±3   | ±3   | ±5   |

Hume-Rothery (1931, 1936) to provide insight into the factors which fix the thermodynamic and structural properties of the various metallic phases.

During the past century, chemists have developed a detailed understanding of bonding in both organic and inorganic compounds. This knowledge can be directly applied to metallic bonding. Essentially the same glue is used to bond all materials. It is the concentration of electrons between the nuclei that is responsible for the cohesion of atoms. What limits are there to the binding contribution per electron? How does the strength of bonding vary with the type of electronic orbital used? How does the structural arrangement of the atoms depend upon the character of the electronic bonding? These are all questions that must be answered if useable thermodynamic data are to be provided by bonding models. Rather complete information is available for most of the elements, and these data can provide a check of theoretical models. Because of the wide range of melting and boiling points of the elements, it seems most appropriate to compare the enthalpies of atomization of the solid phase at 0 K, $\Delta H_0^\circ$, as a measure of the strength of cohesion. Tables I*a* and I*b* present values from Brewer (1977) for the atomization of one gram-atom solid, $\Delta H_0^\circ / R$ in kilokelvin (kK).

## A.  What Is the Range of Bonding Contributions per Electron?

Bonding can be described by several models. There is the bonding due to van der Waals or London force, which can be described in terms of a temporary polarization of the electron cloud of an atom or molecule. This interaction is predominant only for the noble gases and is too small to be significant for the other elements. Bonding can be described in terms of an ionic model in which an electron is transferred from one atom to another. The interaction is then treated as a point-charge coulombic interaction or may include dipole or higher interactions. For most elemental structures, the atoms occupy equivalent lattice positions and no charge separation takes place. Although compounds of metals and nonmetals can form strongly ionic compounds, Pauling has emphasized that intermetallic compounds cannot be expected to have substantial charge separations. The model used here for metallic systems is the valence bond model, which ascribes the cohesion to overlap of atomic orbitals from adjacent atoms to allow the electrons to occupy bonding orbitals in which they can interact with two or more nuclei. The electrons may be localized between a pair of atoms as in diamond or may be delocalized over several atoms as in benzene. The effect of increasing the number of bonding 3s and 3p electrons per atom is illustrated in Table IIa, which tabulates $\Delta H_0^\circ/R$ in kilokelvin for the atom-

**TABLE IIa**

$\Delta H_0^\circ/R$ (kK) for Atomization of 1 Gram-Atom of Solid

|  | Na | Mg | Al | Si | P | S | Cl |
|---|---|---|---|---|---|---|---|
| $n^a$ | 1 | 2 | 3 | 4 | 3 | 2 | 1 |
| $\Delta H_0^\circ/R$ | 13 | 17 | 39 | 54 | 40 | 33 | 16 |

$^a$ $n$ = Number of bonding electrons per atom.

ization of one gram-atom of the solid state for the row of elements in the Periodic Table from Na to Cl. The number of electrons used in bonding increases from 1 to 4 from Na to Si with a corresponding increase in the atomization enthalpy. The 3s and 3p orbitals can only accommodate four pairs of electrons. Thus a free P atom with five electrons has a pair of electrons in the 3s orbital and one electron in each of the p orbitals. As phosphorus atoms approach one another, the p orbitals can overlap and concentrate electrons between the nuclei so as to attract both nuclei. The filled s orbitals repel one another and are unable to concentrate electrons between the nuclei to bring about bonding. The electrons in filled orbitals are thus nonbonding electrons, and phosphorus only uses three electrons in bonding. Sulfur has only two bonding electrons since the s orbital and one of the p orbitals are filled in the separated atoms. Finally, chlorine has only one bonding electron per atom. Table IIa shows that cohesion increases with the number of bonding electrons per atom.

## B.   Promotion of Electrons

Table IIb lists the electronic configurations for the ground electronic states of the atoms from Na to Cl. For most elements with two or more valence electrons, the s orbital contains a pair of electrons which would be

**TABLE II*b***

**Ground State of Atom and Valence State of Solid**

|  | Na | Mg | Al | Si | P | S | Cl |
|---|---|---|---|---|---|---|---|
| Ground state | s | $s^2$ | $ps^2$ | $p^2s^2$ | $p^3s^2$ | $p^4s^2$ | $p^5s^2$ |
| Valence state | s | ps | $p^2s$ | $p^3s$ | $p^3s^2$ | $p^4s^2$ | $p^5s^2$ |
| $n^a$ |  | 1 | 2 | 3 | 4 | 3 | 2 | 1 |

*$^a$ $n$ = Number of bonding electrons per atom.*

nonbonding. However, for magnesium, aluminum, and silicon solids, one of the s electrons can be promoted to a p orbital, making all of the valence electrons available for bonding. The ground electronic state of the gaseous atom is promoted to the electronic configuration corresponding to the effective configuration in the solid which is termed the valence state.

The bonding ability of electrons is reduced if they are crowded together in a high-order bond (Pauling, 1960). In solid magnesium, there are many neighbors which share the bonding electrons. The net bonding enthalpy or atomization enthalpy relative to the ground atomic state, which results from the bonding of the 3p3s valence state minus the promotion energy required to convert the $3s^2$ ground state configuration to the 3p3s atomic configuration, is positive. If the bonding electrons are crowded between a single pair of atoms as in gaseous $Mg_2$, the bonding of the 3p3s valence state would be smaller than the promotion energy and promotion is not possible. $Mg_2$ is bonded only by van der Waals forces, as for the noble gas dimers. It has been demonstrated by Knight *et al.* (1975) that Mg vapor can be maintained as a permanent gas at room temperature if the container surfaces are coated to prevent nucleation of magnesium crystals large enough for the bonding enthalpy to be greater than the promotion energy. The recognition of the need to promote to obtain optimum utilization of the valence electrons is one of the most important keys to understanding the stability of metallic phases. Table II*b* lists the solid valence-state configurations for comparison with the ground state configurations for atomic Na to Cl. As will be discussed later, there is little question about the degree of promotion and the number of bonding electrons per atom for almost 80 of the elements. For the nonmetals of Groups IV to VII with four to seven valence electrons per atom, the number of bonding electrons per atom is eight minus the group number. The noble gases have no bonding electrons. For Zn, Cd, Hg, and the metals to their right in the Periodic Table, promotion from the filled d shell does not take place, but all of the outer s and p electrons are used in bonding although relativistic effects complicate the behavior of the elements from mercury to astatine. As will be discussed in detail in II, D, the lanthanide metals have three bonding electrons per atom except for divalent Eu and Yb. With the exceptions of the ferromagnetic metals Cr to Ni, the actinides beyond uranium, and the platinum group and noble metals Ru to Ag, Os to Au, and Cu, all of the remaining metals do promote sufficiently to allow use of all of their valence electrons in bonding. Before discussing the treatment of those metals for which high promotion energies limit the utilization of all the valence electrons, it is instructive to consider what limits can be placed on the net bonding contributions per electron.

## C.  Net Bonding per Electron

For a solid with one or more bonding electrons per atom, what is the maximum net cohesive enthalpy per electron? This question can be answered by reviewing the data for the almost eighty elements for which the number of bonding electrons is indisputable. In Table III, the atomization enthalpies of the solid phases from Table I have been divided by $n$, the number of

**TABLE III**

$\Delta H_0^\circ / nR$ (kK per bonding electron) for Atomization of 1 Gram-Atom of Solid at 0 K with $n$ Bonding Electrons per Atom

| $n = 1$ | $n = 2$ | $n = 3$ | $n = 4$ | $n = 5$ | $n = 6$ | $n = 7$ |
|---|---|---|---|---|---|---|
| H 26 | | | | | | |
| Li 19 | Be 19 | B 22 | C 21 | Ta 19 | W 17 | Re 13 |
| Cl 16 | S 17 | N 19 | Hf 19 | Nb 18 | Mo 13 | Tc 11 |
| Br 14 | O 15 | Ce, La, Lu, Y 17 | Zr, Th 18 | Pa (15) | U 11 | |
| Na 13 | Se 14 | Ac, Gd, Tb 16 | Ti 14 | V 12 | | |
| | Te 13 | Cm, Sc 15 | Si 13 | | | |
| | Ba 11 | | | | | |

bonding electrons, to yield $\Delta H_0^\circ/nR$ per gram-atom of bonding electrons. The first column, headed by hydrogen, lists in order of decreasing bond strength the values for elements with one valence electron per atom. The second column, headed by Be, lists atomization enthalpies divided by two for elements with two valence electrons per atom. The division by families continues to the seventh column, where the atomization enthalpies of Re and Tc divided by seven are listed. In each column, the values are tabulated in order of decreasing bond strength. With the restriction to those elements of indisputable number of bonding electrons, no elements with more cohesion per electron than the last member of each column have been omitted. It is of interest to note that metals and nonmetals are interspersed in Table III. With the exception of hydrogen, for which the electrons acting on bare nuclei can achieve an unusually small internuclear distance and unusually large cohesion, there is no element for which the net bonding enthalpy exceeds 22 kK per gram-atom of electrons. After the very tightly bound boron and carbon, no element achieves a net bonding enthalpy above 19kK per gram-atom of electrons (i.e., 1.65 e.v. per electron). This upper limit will be important in dealing with the degree of promotion of elements for which not all of the valence electrons are promoted to bonding status. For the elements of the fifth to seventh groups which have in the ground state at least one electron in each of the three p orbitals, promotion from the s to p orbital does not increase the number of bonding electrons, and the bonding configuration will be close to the gaseous atomic ground state configuration. Almost all other elements do benefit by promotion to increase the number of bonding electrons. The lanthanide metals provide a particularly striking example of the effect of rapidly varying promotion energies upon the volatilities of the metals.

### D. Variations of Promotion Energies of the Lanthanides

With the exceptions of Eu and Yb, the lanthanide metals show steadily increasing melting points from Ce to Lu. Other properties also indicate that the cohesion must be increasing from La to Lu, but the enthalpies of sublimation show a steady decrease from La to Eu, a jump to a high value for Gd followed by steady decrease to Yb, and then a jump to a high value for Lu. The valence-state bonding energies increase from La to Lu, as one would expect for increasing nuclear charge that results in decreasing internuclear distances. However, as the nuclear charge increases from La to Eu, the electrons in the 4f orbital become increasingly stabilized. The energies required to promote from the 4f orbital to 5d or 6p are tabulated by Brewer (1971). They steadily increase from La to Eu, resulting in a smaller net enthalpy of sublimation, until at Eu the promotion energy for a 4f electron is greater than the valence state bonding energy and Eu is only able to use two electrons for bonding. Thus Eu has properties similar to those of the alkaline earth metals. Beyond Eu, for which the seven 4f orbitals are half filled, additional 4f electrons are not held tightly and promotion to the trivalent state is again profitable. With increasing nuclear charge the energy for promotion of a 4f electron increases, and for Yb as for Eu, it is not possible to promote beyond the divalent state. Lutetium, which can obtain three bonding electrons without promoting any 4f electrons, again has a high enthalpy of sublimation. Thus the abnormal variation of properties of the lanthanides can be quantitatively represented through variation of the atomic spectroscopic properties and a reasonable steady increase of valence-state bonding energies from La to Lu.

### E.  To What Extent Do Solid Ni to Pt and Cu to Au Require Electronic Promotion?

In II,B, transition metals that do not use all of their valence electrons in bonding were listed. As will be discussed later, some of the 3d orbitals become so localized as the nuclear charge is increased that bonding is negligible even with a single electron in an orbital. Usually electrons become nonbonding because of pairing in the valence state. When six or more electrons go into the d orbitals, the valence state will have one or more pairs of d electrons. The electron pairs of filled orbitals repel one another as the atoms approach, and the electrons cannot concentrate between the nuclei to pull the nuclei together. The number of electron pairs in d orbitals can be reduced by promoting to p orbitals. For every d electron promoted, two bonding electrons become available. However, the bonding due to these electrons in d and p orbitals must be sufficient to offset the promotion energy required. As will be described later, when we have established tables of bonding energies, we can use the promotion energies from spectroscopic measurements tabulated by Moore (1949, 1952, 1958), Brewer (1971), and Martin $et$ $al.$ (1978) to determine possible valence states, as was done in II,D with the lanthanides. Spectroscopic data are lacking for the $d^{n-3}p^2s$ configuration ($n$ is the total number of valence electrons), for some of the metals at the right-hand side of the transition metals, and we must use other procedures to determine the degree of promotion for valence states corresponding to $d^{n-3}p^2s$ or mixtures of this configuration with $d^{n-2}ps$ yielding fractional average electron populations in a given orbital. The $d^{n-3}p^2s$ configuration is important because the ccp structure of Mn to Cu, Rh to Ag, and Ir to Au will be assigned to configurations in the range $d^{n-2.5}p^{1.5}s$

to $d^{n-3}p^2s$ in subsequent discussion. To achieve these configurations, at least 0.5 d electrons must be promoted for Mn to Ni and Ir and at least 1.5 d electrons must be promoted for Rh, Pt, and Cu to Au. Palladium, with a $4d^{10}$ ground atomic configuration, would have only weak van der Waals bonding (as for the noble gases) if it did not promote d electrons to s and p orbitals. It must promote at least 2.5 d electrons to achieve the $d^{7.5}p^{1.5}s$ to $d^7p^2s$ range that is correlated to the ccp structure. We will now consider the justification of this degree of promotion, recognizing that all of the properties of the metals that are enhanced by increasing cohesion between the atoms will depend strongly upon the degree of promotion to increase the number of bonding electrons. We will consider the properties of the metals Ni to Pt and Cu to Au.

There are a number of properties that can be used as a measure of strength of cohesion in addition to the enthalpy of atomization. Melting points and internuclear distances are not always direct measures of cohesion as substantial changes in coordination number in the melting process or substantial changes in electronic configuration can produce abnormal behavior, but the comparison made in Table IV are for similar enough metals so that variation of melting point and contraction of the lattice can be taken as a measure of degree of cohesion.

In Table IVa, the divalent elements of the first column all have the ground electronic state $s^2$, and the alkali metals of the second column have the configuration s. Ni, Pd, and Pt have the ground-state configurations $d^8s^2$, $d^{10}$, and $d^9s$, respectively. Cu, Ag, and Au have the ground-state configuration $d^{10}s$. Clearly the two electrons per atom of calcium achieved a much greater cohesion than the one electron per atom of potassium with respect to higher melting point, smaller internuclear distance, and larger atomization enthalpy. The same behavior is seen for strontium compared to rubidium and for barium compared to cesium. The trends in melting point, internuclear distance, and atomization enthalpy go hand-in-hand with the number of bonding electrons. Under each comparison of the divalent alkaline earth metal with the univalent alkali metal to its left in the Periodic Table is the similar comparison in the same period of divalent Zn, Cd, and Hg with the elements to the left. For example, divalent Cd is compared to "univalent" silver and "zerovalent" Pd. The data clearly indicate that silver does not remain univalent and palladium does not remain zerovalent and that promotion must have occurred. The same behavior is found as one moves to the left in the Periodic Table from Zn, Cd, or Hg. The melting points and cohesive enthalpies rise sharply and, most significantly, the internuclear distance decreases even though the nuclear charge is decreasing. With decrease of nuclear charge, the core of filled electronic orbitals will expand if not offset by enhanced bonding.

**TABLE IV$a$**

**Cohesion Trends: $\Delta H_0^\circ/R$ (kK) for Atomization of
1 Gram-Atom of Solid**

|                     | Ca   | K    |     |
|---------------------|------|------|-----|
| m.p. (K)            | 1113 | 336  |     |
| $R$ (Å)             | 3.9  | 4.6  |     |
| $\Delta H_0^\circ/R$ (kK) | 21.  | 11.  |     |

|                     | Zn   | Cu   | Ni   |
|---------------------|------|------|------|
| m.p. (K)            | 693  | 1358 | 1455 |
| $R$ (Å)             | 2.7  | 2.6  | 2.5  |
| $\Delta H_0^\circ/R$ (kK) | 16   | 40.5 | 51.5 |

|                     | Sr   | Rb   |     |
|---------------------|------|------|-----|
| m.p. (K)            | 1042 | 313  |     |
| $R$ (Å)             | 4.3  | 5.0  |     |
| $\Delta H_0^\circ/R$ (kK) | 20   | 10   |     |

|                     | Cd   | Ag   | Pd   |
|---------------------|------|------|------|
| m.p. (K)            | 594  | 1235 | 1827 |
| $R$ (Å)             | 3.0  | 2.9  | 2.75 |
| $\Delta H_0^\circ/R$ (kK) | 13.5 | 34   | 45   |

|                     | Ba   | Cs   |     |
|---------------------|------|------|-----|
| m.p. (K)            | 1002 | 302  |     |
| $R$ (Å)             | 4.35 | 5.25 |     |
| $\Delta H_0^\circ/R$ (kK) | 22   | 9    |     |

|                     | Hg   | Au   | Pt   |
|---------------------|------|------|------|
| m.p. (K)            | 234  | 1338 | 2045 |
| $R$ (Å)             | 3.0  | 2.9  | 2.8  |
| $\Delta H_0^\circ/R$ (kK) | 8    | 44   | 68   |

Table IV$b$ compares the atomization enthalpies as in Table IV$a$, but with extension to the trivalent metal to the left to give a better idea of the number of bonding electrons in copper and nickel, for example. The trend from Sc to Ca to K is what is expected for three to two to one bonding electrons per atom. Comparison with Table III shows that the atomization enthalpies per electron of 15, 10.5, and 11 kK for Sc, Ca, and K are in line with values

**TABLE IV$b$**

**Bonding Electrons of Copper, Silver, and Gold**

|                      | Sc   | Ca   | K    |      |
| -------------------- | ---- | ---- | ---- | ---- |
| $n$                  | 3    | 2    | 1    |      |
| $\Delta H_0^\circ/R$ (kK) | 45   | 21   | 11   |      |
|                      | Ga   | Zn   | Cu   | Ni   |
| $n$                  | 3    | 2    | 4    | 5    |
| $\Delta H_0^\circ/R$ (kK) | 33   | 16   | 40.5 | 51.5 |
|                      | Y    | Sr   | Rb   |      |
| $n$                  | 3    | 2    | 1    |      |
| $\Delta H_0^\circ/R$ (kK) | 51   | 20   | 10   |      |
|                      | In   | Cd   | Ag   | Pd   |
| $n$                  | 3    | 2    | 4    | 5    |
| $\Delta H_0^\circ/R$ (kK) | 29   | 13.5 | 34   | 45   |
|                      | La   | Ba   | Cs   |      |
| $n$                  | 3    | 2    | 1    |      |
| $\Delta H_0^\circ/R$ (kK) | 52   | 22   | 10   |      |
|                      | Tl   | Hg   | Au   | Pt   |
| $n$                  | $\leqslant 3$ | 2    | 4    | 5    |
| $\Delta H_0^\circ/R$ (kK) | 22   | 8    | 44   | 68   |

found generally throughout the Periodic Table. From Ga to Zn, the atomization enthalpy varies as expected for three to two bonding electrons, but the atomization enthalpy of 40.5 kK for Cu clearly cannot be due to only one bonding electron. Comparison with Table III shows that even hydrogen with a bare nucleus cannot obtain more than 26 kK cohesion from one electron. Below hydrogen, even the most strongly bound boron and carbon only obtain 22 and 21 kK per electron and all other elements fall below these values. In every instance, the "univalent" metals Cu, Ag, and Au achieve higher atomization enthalpies than the trivalent metals Ga, In, and Tl. The value of $n$, the number of bonding electrons, is given as 4 for Cu, Ag, and Au indicating a promotion from $d^{10}s$ of 1.5 d electrons to achieve a valence state of $d^{8.5}p^{1.5}s$ with four unpaired electrons available for bonding. To achieve five bonding electrons, Pd must promote 2.5 d electrons to achieve a valence state of $d^{7.5}p^{1.5}s$. For the same valence state, Pt must promote 1.5 d electrons and Ni must promote 0.5 d and 1.0 s electrons to p orbitals.

A more precise characterization of the number of bonding electrons and of the valence state electronic configuration for Cu to Au and for Ni to Pt requires consideration of promotion energies and valence-state bonding energies as for the lanthanides as briefly described in IV,D, and in more detail by Brewer (1967, 1971). However, it is simpler to examine the net cohesive enthalpies for which there is a clear maximum limit of cohesion per electron as shown in Table III. There can be no question upon examining Tables III and IV that Cu, Ag and Au could not attain their measured properties using only one bonding electron per atom. The comparison was made in Table IV using melting point, internuclear distance, and atomization enthalpy. One could also use bulk modulus, coefficient of expansion, compressibility, and a wide range of properties that depend upon strength of cohesion. All of these properties indicate that copper, silver, and gold are using at least four bonding electrons per atom.

## F.   Variation of Valence-State Bonding Enthalpy

The discussion of the preceeding sections has illustrated how theoretical concepts can be correlated with experimental data to arrive at semiempirical models that use meaningful parameters. The atomization enthalpies of transition metals given in Table I$a$ vary in an irregular way. A similar erratic behavior was noted for the lanthanides in Table I$b$. The discussion of the preceeding sections indicates that the net atomization enthalpy is a difference between the valence-state bonding enthalpy and the promotion energy. The promotion energy depends upon the mutual interactions of the nucleus and the electrons in the various orbitals of the ground state and valence state of the isolated atom. The valence-state bonding enthalpy has no relationship to the atomic ground state but is determined by the electrons in the valence state orbitals and their interactions not only with electrons and nucleus of a given atom but also with those of neighboring atoms. Thus the promotion energy and the valence-state bonding enthalpy vary in markedly different ways across the Periodic Table; their differences can be quite erratic. For most solid phases of the elements, spectroscopic measurements yield the promotion energies, which can be added to the observed net atomization enthalpies to obtain the valence-state bonding enthalpies. As noted in II,D in regard to the lanthanides, the resulting valence-state bonding enthalpies vary in a smooth and consistent way. This resolution of an experimental quantity into two or more factors which either are directly known or which vary in a predictable way is a common feature of well-designed semiempirical correlations. A similar example is the well-known use of the Born-Haber cycle for ionic substances, where the enthalpy of formation of a salt is separated into terms for the atomization of the elements, ionizational potentials and electron affinities, and lattice energies. In terms of this model, the irregular behavior of enthalpies of formations of salts which show increasing stability for Li to Cs for the iodides and decreasing stability from Li to Cs for the fluorides is readily understood.

### 1. Contrasting Bonding Behavior of Outer- and Inner-Shell Electrons

For the transition metals, one additional separation is required. The bonding behavior of the outer-shell p and s electrons is different from that of the inner-shell d or f electrons. This is a very important aspect of the Engel correlation which is readily understood from quantum mechanics. As can be seen from Table I$a$, or more accurately after inclusion of promotion energies, the atomization enthalpies of elements that use only outer-shell p and s electrons for bonding show a steady decrease in each family as one goes downward in the Periodic Table. The increasing size of the core of filled electronic orbitals as one goes down in the table causes an increase of internuclear distance. Over virtually all of the Periodic Table, with the exception only of the upper right-hand corner, the bonding enthalpy due to p/s electrons decreases in a steady way with increasing internuclear distance. Using this correlation for the transition metals, one can calculate the increase in bonding per electron due to the 5s or 5p electrons, for example, as one goes from Rb to Cd as the filled 4s4p core contracts with increasing nuclear charge. Subtraction of the p/s contributions from the valence-state value yields the contribution of the inner shell d electrons to the bonding enthalpy. The effect of neighboring atoms upon the d orbitals has been extensively characterized by Cotton and Wilkinson (1980). There is a splitting of the degeneracy of the d orbitals to yield, in the metal, states covering a band of energies and varying degrees of radial extension. At the beginning of each transition period, the d orbitals are quite extended. As the nuclear charge increases, they contract and become more localized. As one goes from Ca to Cr or from Sr to Mo, the average bonding enthalpy contribution per electron steadily decreases due to increased average localization as well as the utilization of more of the localized d orbitals. After more than five electrons are used in the valence state of the atom, the nonbonding electrons will be in the most localized orbitals while the bonding orbitals will utilize the most extended orbitals to achieve maximum stability. Thus, to the right of the half-filled d shell, the bonding enthalpy per d electron will increase as the bonding electrons use only the most extended orbitals. The resulting curves for the variation of bonding for either 3d, 4d, 5d, or 6d electrons can be expressed in simple analytical form and used to calculate the bonding contributions for these electrons.

A large increase in bonding enthalpy is found in going from 3d to 6d, which is readily understood from quantum mechanics. The degree of localization of the $n$d orbitals depends upon their extension beyond the high density distribution of the $n$p and $n$s orbitals. It is instructive to compare the d orbitals of Cr, Mo, and W. The 3d orbitals do not extend much beyond the filled 3s and 3p core. Some of the 3d orbitals are so localized that they can contain single electrons that do not overlap with neighboring orbitals and can remain unused in bonding. These localized 3d orbitals of Cr to Ni are responsible for the strong magnetism of these metals. On comparing Mo with Cr, the increase of nuclear charge has a much greater effect on the s and p electrons than on the d electrons. We know from quantum mechanics that the s and p electrons have a much higher probability near the nucleus than do d or f orbitals. In going from Cr to Mo, the 4d orbitals expand beyond the 4ps core to a much greater extent than for the 3d relative to 3ps orbitals. All of the 4d orbitals extend out far enough to be used in bonding and ferromagnetism is not observed for the 4d transition metals. The effect on increasing the nuclear charge from Mo to W is equally pronounced. From Table I$a$, we see that the atomization enthalpy increases from 47.6 for Cr to 79.1 for Mo and to 103.3 kK for W. A similar effect is observed for the f orbitals. The 5f orbitals are more extended than the 4f orbitals and contribute significantly to bonding for the first half of actinides. For the lanthanides, the 4f orbitals are so localized that only for cerium is there a very small bonding contribution and the lanthanides are strongly magnetic. As the nuclear charge is increased into the second half of the actinides, the 5f orbitals contract and the heavy actinides display lanthanide-like behavior in that the f electrons do not contribute to bonding and they are strongly magnetic. To understand the variation of atomization enthalpy across the Periodic Table, any useful model must recognize the different behavior of the outer-shell s and p orbitals compared to the inner-shell d and f orbitals. The outer-shell s and p electrons contribute very effectively to bonding. The contributions from inner-shell electrons decrease in the order 6d > 5d > 4d > 3d > 5f > 4f.

## 2.   *Extension of Model to Gaseous Molecules*

A bonding model of this type not only successfully correlates with the atomization enthalpies of the solid elements, but similar models can be used for a variety of covalently bonded materials as this model is based on the quantum-mechanical principles which apply to all materials. For example, the same model has been used by Brewer and Winn (1980) to predict the dissociation enthalpies of the homonuclear diatomic molecules of all of the elements through lawrencium. In the $M_2$ molecules, all of the bonding electrons must concentrate between two nuclei instead of many neighboring nuclei in the solid. Thus as emphasized by Pauling (1960), the bonding contributions per electron are substantially decreased. It was found that the degree of promotion in the $M_2$ molecules was reduced, compared to the solid, with fewer bonding electrons. In contrast to solid silicon or germanium where four electrons per atom are used in bonding with promotion to the $p^3s$ configuration, $Si_2$ or $Ge_2$ do not promote from the ground $p^2s^2$ configuration and only use two bonding electrons per atom. Similarly, none of the elements from Cu on to the right promote. Gaseous $Cu_2$ uses only one electron per atom, $Zn_2$ is a van der Waals molecule, and $Ga_2$ uses only one electron per atom. Solid germanium with four bonding electrons per atom reaches a maximum atomization enthalpy compared to gallium and arsenic. For $M_2$ gas, the maximum occurs at $As_2$. Even though the behavior of the gaseous $M_2$ molecules is considerably different from the behavior of the solid phases, the same type of model applies to both if one takes into account the variation with bond order of bonding ability of an electron, as has been well-established for single, double, and triple bonds of carbon for many years.

## III. CORRELATION OF ELECTRONIC CONFIGURATION AND CRYSTAL STRUCTURE

It is well-established in inorganic chemistry that many compounds which are isoelectronic have similar coordination numbers or structures, as noted by Cotton and Wilkinson (1980). If we move horizontally in the Periodic Table from sodium to argon, there is a change in crystal structure for each element which clearly indicates a strong correlation between electrons per atom and structure. Fifty years ago, Hume-Rothery (1931) noted that many intermetallic compounds with varying compositions had the same crystal structure if they had the same average number of valence electrons per atom. From examination of alloy systems, where the bonding is due to only p/s electrons, the cubic close-packed (ccp) structure can be correlated with a range of approximately 2.5 to 3 p/s electrons per atom (e/a). The hexagonal close-packed (hcp) structure can be correlated with 1.7 to 2.1 p/s electrons per atom and the body-centered cubic (bcc) structure is limited to less than 1.5 p/s electrons per atom. He encountered difficulties in using the correlation for transition metals as he did not know how to count the number of electrons. Using the total number of valence electrons gave too high an electron count. Engel (1939, 1940, 1945, 1949) resolved this dilemma by recognizing the difference between the outer-shell p/s electrons and the inner-shell d and f electrons. In contrast to the change from bcc to hcp to ccp to diamond structure for the adjoining nontransition elements Na, Mg, Al and Si, the transition metals from Rb to Mo all have the bcc structure as one of their structures. There is a change to hcp for Tc and Ru, and finally the fcc structure is found for Rh, Pd, and Ag. Engel, in agreement with Pauling (1960), recognized that the unfilled d shells of the first half of the transition elements act as sinks for electrons and, thus, tend to keep the p/s electron concentration below 1.5 e/a. After the $d^5s$ configuration is reached for Mo, the addition of another electron for Tc required promotion to $d^5ps$ to allow all seven electrons to be used in bonding and thus a change from a configuration with less than 0.5 p electrons per atom (bcc) to one with one p electron per atom (hcp). Engel pointed out that the Hume-Rothery correlation could be extended to the transition metals if only the p/s electrons were counted. The Engel correlation is based on the concept that the inner-shell d or f orbitals are somewhat localized and interact primarily with nearest neighbors and do not influence long-range order. The outer-shell p/s electrons range far out and are decisive in controlling long-range order which determines the crystal structure. The discussion of II,E on the degree of promotion for Ni to Pt and Cu to Au is in agreement with the Engel structure correlation. To account for the high cohesion of these metals, corresponding to five or four bonding electrons, it was necessary to promote a sufficient number of d electrons to p orbitals to achieve 2.5 or more p/s electrons per atom. All of these metals have the ccp structure which changes to hcp structure upon going from Cu to Zn with a ps configuration.

## IV.   APPLICATION OF A SEMIEMPIRICAL CORRELATION

A successful model should have a sound theoretical foundation so that it does much more than reproduce the original data that were used to fix the parameters—in the case of the Engel model, the variations of bonding enthalpy per electron across the Periodic Table. In the several decades since the presentation of the Engel model and its predictions of phase behavior for multicomponent transition-metal systems, a large amount of data has accumulated to check these predictions. The reliability of the predictions has been excellent. The percent of observations in agreement with the predictions has been in the high nineties. A few striking examples can be given here to illustrate the power of such semiempirical correlations. The key feature of the Engel model is the distinction between outer-shell p/s electrons and the inner-shell d/f electrons. As mentioned in the introduction, it is desirable to limit the range of application of a model to keep it from becoming too complicated. The noble gases have some of the same crystal structures as the transition metals, but one would not expect the Engel model to say anything about the structures of van der Waals bonded materials. Nor would it be applied to ionic structures. Although the model is consistent with the behavior of nontransition metals and nonmetals, it does not have anything new to add to the models that exist for those materials. So the region of application properly should be transition-metal systems for which there is a clear distinction between the inner-shell d electrons and the outer-shell p/s electrons. This would be true for the transition metals starting at the right with Cu, Ag, and Au and going to the left until the d orbitals have become so expanded that they are not distinct from the p,s orbitals. The d orbitals are still distinct for Ti, Zr, and Hf. The d orbitals have become expanded enough for the third group metals that predictions based on localization of the d orbitals would be given a large uncertainty. For the alkaline earths, the distinction would be very small.

Using spectroscopic data and the bonding enthalpy values for the different orbitals, the stabilities of the bcc and hcp phases have been calculated by Brewer (1967) for Ca to Cu, Sr to Ag, and Ba to Au, and for the lanthanides— 100 calculations in total. In addition, where spectroscopic data were available or could be estimated for the fcc phases, similar calculations were made. Within the uncertainty of the calculations, the only discrepancies were the occurrence of fcc phases of Ca, Sr, and Yb. Brewer (1968) has noted that the occurence of the $\alpha$- and $\beta$-Mn phases is consistent with the Engel correlation; thus the Engel correlation scores 98% concurrence even including the divalent metals. Excellent agreement is also found by Brewer (1965) for multicomponent systems where the limits of the bcc, hcp, and ccp phases are found to be consistent with the Engel correlation.

The effects of variables such as pressure and solute addition upon phase transformations such as the bcc–hcp transitions of Ti, Zr, and Hf provide a direct check of the electronic factors. For small solute additions, contributions other than electronic largely cancel out. The partial localization of the d orbitals is the decisive factor. Anything that pushes the nuclei closer together will substantially improve the overlap and therefore the bonding of d electrons. Thus pressure should stabilize the structure with the most bonding d electrons. The following summary from Brewer (1965, 1974b, 1977) presents the general predictions of the Engel model in regard to stabilization of the various transition metal structures.

   a.   Small additions of interstitial electron-donors (B, C, N, O, H) stabi-
        lize structures with the most p electrons: ccp > hcp > bcc. Interstitial
        electron acceptors (F, higher H conc.) favor structures with the fewest
        p electrons: bcc > hcp > ccp.
   b.   Small additions of nontransition elements stabilize the structure with
        the fewest d bonds.
        For Group VI and to the left: ccp > hcp > bcc.
        For Group VIII and to the right: bcc > hcp > ccp.
        For Group VII: bcc, ccp > hcp or > $\beta$-Mn.
   b'.  Since pressure stabilizes the structures with the most d bonds, the
        following corollary results: The effect of pressure will be just the re-
        verse of the summary under (b).
   c.   For transition metals of Group V and to the left, the addition of any
        transition metal to the right of the solvent metal will stabilize the struc-
        ture with the most d bonds (most d electrons for these groups): bcc >
        hcp > ccp.
            For the same metals, the addition of any transition metal to the left
        of the solvent metal will stabilize the structure with the fewest d elec-
        trons: ccp > hcp > bcc.
        These results also apply to the metals of Group VI, except that the
        order of stabilization upon adding transition metals to the right re-
        verses for addition of metals of Group VIII and beyond. For transition
        metals of Group VII and to the right, the addition of any transition
        metal (4d or 5d) to the left of the solvent metal will stabilize the struc-
        ture with the fewest p and the most d electrons: bcc > hcp > ccp.

Reviews by Brewer (1965, 1974a,b) of the available data show agreement
with the predictions from the Engel model of better than 98% for (a) and (b)
and better than 99% for (b') and (c).

The most striking aspect of the Engel model is the prediction of intermetallic compounds of extremely high stability for mixtures of transition metals of
the left-hand side with vacant d orbitals with platinum groups metals that
have nonbonding electron pairs. This prediction has been confirmed most
strikingly by Brewer and Wengert (1973). With more space, many additional
examples can be given of the power of models such as the Engel model which
have firm theoretical bases. As one last example of the power of this model,
complete thermodynamic data have been calculated by Brewer and
Lamoreaux (1980) for 100 of the binary systems of molybdenum from hydrogen to lawrencium, and these data have been used to calculate all of the phase
diagrams. It is this ability to make reliable predictions for unstudied systems
which makes such models so useful.

## ACKNOWLEDGMENT

This work was supported by the Division of Materials Sciences, Office of Basic Energy Sciences, U.S. Department of Energy under contract no. W-7405-End-48.

## REFERENCES

Brewer, L. (1965). *In* "High-Strength Materials" (V. F. Zackay, ed.), pp. 12–103. Wiley, New York.
Brewer, L. (1967). *In* "Phase Stability in Metals and Alloys" (P. Rudman, J. Stringer, and R. I. Jaffee, eds.), pp. 39–61, 241–249, 344–346, 560–568. McGraw-Hill, New York.
Brewer, L. (1968). *Science* **161**, 115–122.
Brewer, L. (1971). *J. Opt. Soc. Am.* **61**, 1101–11.
Brewer, L. (1974a). *Rev. Chim. Miner.* **11**, 616–623.
Brewer, L. (1974b). *J. Nucl. Mater.* **51**, 2–11.
Brewer, L. (1977). "The Cohesive Energies of the Elements," Lawrence Berkeley Lab. Rep. LBL-3720 Rev., with revision of the Se and Te values based on Drowart and Smoes (1977).
Brewer, L., and Lamoreaux, R. (1980). "Molybdenum, Physicochemical Properties of Its Compounds and Alloys," At. Energy Rev., Spec. Issue 7, Parts I and II. IAEA, Vienna.
Brewer, L., and Wengert, P. R. (1973). *Metall. Trans.* **4**, 83–104, 2674.
Brewer, L., and Winn, J. S. (1980). *Symp. Faraday Soc.* **14**, 126–135.
Cotton, F. A. and Wilkinson, G. (1980). "Advanced Inorganic Chemistry," 4th ed. Wiley (Interscience), New York.
Drowart, J., and Smoes, S. (1977). *J. Chem. Soc., Faraday Trans. 2* **73**, 1755–1767.
Engel, N. (1939). *Ingenioeren* **N101**.
Engel, N. (1940). *Ingenioeren* **M1**.
Engel, N. (1945). "Haandogi Metallare." Selskabet for Metal-forskning, Copenhagen.
Engel, N. (1949). *Kem. Maanedsbl.* **30**(5), 53; (6), 75; (8), 97; (9), 105; (10), 114.
Engel, N. (1954). *Powder Met. Bull* **7**, 8.
Engel, N. (1964). *ASM Trans. Q.* **57**, 610.
Engel, N. (1967). *Acta Metall.* **15**, 557.
Hume-Rothery, W. (1931). "The Metallic State." Oxford Univ. Press, London and New York.
Hume-Rothery, W. (1936). "Structures of Metals and Alloys." Institute of Metals, London.
Knight, L. B., Jr., Brittain, R. D., Duncan, M., and Joyner, C. H. (1975). *J. Phys. Chem.* **79**, 1183.
Martin, W. C., Zalubas, R., and Hagan, L. (1978). "Atomic Energy Levels—The Rare Earth Elements." US Govt. Printing Office, Washington, D.C.
Moore, C. E. (1949). "Atomic Energy Levels," Vol. I. US Govt. Printing Office, Washington, D.C.
Moore, C. E. (1952). "Atomic Energy Levels," Vol. II. US Govt. Printing Office, Washington D.C.
Moore, C. E. (1958). "Atomic Energy Levels," Vol. III. US Govt. Printing Office, Washington D.C.
Pauling, L. (1947). *J. Am. Chem. Soc.* **69**, 542.
Pauling, L. (1949). *Proc. R. Soc. London, Ser. A* **196**, 343.
Pauling, L. (1950). *Proc. Natl. Acad. Sci. U.S.A.* **36**, 533.
Pauling, L. (1960). "The Nature of the Chemical Bond," 3rd ed. Cornell Univ. Press, Ithaca, New York.

# 8

# Theoretical Probes of Bonding in the Disiloxy Group

## MARSHALL D. NEWTON

|       |                                        |     |
|-------|----------------------------------------|-----|
| I.    | Introduction . . . . . . . . . . . . . . . . . . . . . . . . . . . . . . . . . . | 175 |
| II.   | Theoretical Considerations . . . . . . . . . . . . . . . . . . . . | 176 |
|       | A.   Many-Electron Wavefunctions . . . . . . . . . . . . . . | 176 |
|       | B.   Covalency and Ionicity . . . . . . . . . . . . . . . . . | 177 |
|       | C.   Hybrid Atomic Orbitals . . . . . . . . . . . . . . . . . | 179 |
| III.  | Computational Details . . . . . . . . . . . . . . . . . . . . . | 184 |
| IV.   | Results and Discussion . . . . . . . . . . . . . . . . . . . . . | 185 |
|       | A.   Atomic Populations . . . . . . . . . . . . . . . . . . | 185 |
|       | B.   SiO Overlap Populations . . . . . . . . . . . . . . . . | 187 |
|       | C.   Hybridization of the Oxygen AOs . . . . . . . . . . . | 187 |
| V.    | Conclusions . . . . . . . . . . . . . . . . . . . . . . . . . . | 191 |
|       | References . . . . . . . . . . . . . . . . . . . . . . . . . | 192 |

## I. INTRODUCTION

The techniques of *ab initio* quantum chemistry provide powerful and versatile tools for elucidating the microscopic details of chemical bonding (Schaefer, 1977). The present day capacity of high-speed digital computers is such that these techniques are no longer confined primarily to the first row of the periodic table. In particular, a variety of recent *ab initio* and related semiempirical studies have shed considerable light on the nature of bonding in the disiloxy group (—Si—O—Si—), which plays a fundamental role in the chemistry of silica polymorphs and glasses, silicates, and siloxanes (Gilbert *et al.*, 1973; Pantelides and Harrison, 1976; Chelikowsky and Schluter, 1977; Ciraci and Batra, 1977; Calabrese and Fowler, 1978; Schneider and Fowler, 1978; Sauer and Zurawski, 1979; Newton and Gibbs, 1980; O'Keeffe *et al.*, 1980; Gibbs *et al.*, this volume, Chapter 9). An especially encouraging result from these studies is the apparent ability of simple discrete molecular clusters

**175**

Structure and Bonding in Crystals, Vol. I

[e.g., $(SiH_3)_2O$, $(Si(OH)_3)_2O$, or $Si_2O$] to give a good quantitative account of the properties of related crystalline materials. These properties include structural features such as the correlation of bond lengths and angles in the $Si_2O$ unit and spectroscopic data such as x-ray fluorescence and photo-electron spectra. Of course, the ability to calculate with reasonable accuracy a certain property does not *ipso facto* "explain" the property; i.e., in this sense, an *ab initio* calculation can be considered simply as an alternative to other more conventional "experimental" probes of chemical structure. However, a major advantage of such a computational determination lies in the fact that one can subsequently attempt a meaningful "explanation" of the property in terms of the detailed wavefunction which yielded the property of interest.

In this chapter we shall attempt to analyze the electronic charge distribution and the bond-angle and bond-length variations exhibited by the SiOSi linkage in terms of simple orbital concepts which can be quantitatively tested by relatively simple *ab initio* calculations. In the next two sections we outline some pertinent orbital models of chemical bonding and computational schemes for implementing them. In Section IV, specific calculated results for the disiloxy group are presented and discussed.

## II. THEORETICAL CONSIDERATIONS

### A. Many-Electron Wavefunctions

Most discussions of the electronic structure of discrete molecular species are based on many-electron wavefunctions constructed from appropriate one-electron orbitals. The basic building block is the electronic configuration, which is a product of orbitals antisymmetrized so as to conform with the Pauli exclusion principle. For closed-shell species, to which we shall confine our attention in the present study, a single configuration often serves as a useful approximation to the true eigenfunction of the electronic Schrodinger equation. For an atom, the one-electron orbitals are the familiar atomic orbitals $(\chi_j)$. In the case of a molecule, we generalize from atomic orbitals to molecular orbitals $(\phi_i)$, and the latter are generally expanded in terms of appropriate atomic orbitals (AOs) located on the various atomic centers. Hence we can write

$$\psi_{\substack{\text{many} \\ \text{electron}}}^{\text{MO}} = \prod_{i=1}^{n} \phi_i \tag{1}$$

where

$$\phi_i = \sum_{j}^{m} \chi_j C_{ji} \tag{2}$$

and where $n$ is the total number of electrons, and $m$ is the total number of AOs. The atomic orbital basis functions $\chi_j$ have nonlinear parameters which are determined along with the linear coefficients, $C_{ji}$, by minimizing the total energy,

$$\bar{E} = \int \psi^* H\psi \, d\tau \tag{3}$$

for a given molecular geometry. The equilibrium geometry is then determined by minimizing the total energy with respect to the geometrical parameters.

An alternative to the molecular orbital wavefunction is the so-called valence-bond many-electron wavefunction, $\psi^{VB}$, which is a linear combination of various atomic orbital configurations (i.e., antisymmetrized AO products) constructed so as to correspond to a particular type of chemical bonding such as a covalent or ionic valence structure. By taking linear combinations of appropriate valence structures, one obtains additional energy lowering which is generally labeled resonance stabilization. Resonance between valence bond structures is a particular example of what is called configuration interaction (CI); CI can also be carried out among molecular orbital configurations [Eq. (1)], although in this work we consider only single-configuration MO wavefunctions.

One of the persistent ironies of modern-day quantum chemistry is that while molecular orbital techniques are generally much more convenient for computational implementation, most chemists still feel more at home with the concepts of valence bond theory. Thus one of the major goals of this study is to exemplify the procedures whereby one translates the information contained in molecular orbital wavefunctions into valence-bond-like language.

## B. Covalency and Ionicity

### 1. Valence Bond Definition

The degree of ionicity in chemical bonds is a topic of great interest in the context of mineral substances. A straightforward definition which arises from valence bond theory is illustrated here for a diatomic molecule AB.[†] If the atoms A and B have atomic orbitals $\chi_A$ and $\chi_B$, respectively, one may construct three electron-paired valence-bond wavefunctions which have singlet spin symmetry. The wavefunction,

$$\psi^{AB}_{\text{covalent}} = [\chi_A(1)\chi_B(2) + \chi_B(1)\chi_A(2)][\alpha(1)\beta(2) - \beta(1)\alpha(2)] \tag{4}$$

[†] Shull (1962) has emphasized need for more general characterizations of ionic character.

corresponds to a covalent electron-pair bond between A and B, whereas the two wavefunctions,

$$\psi_{\text{ionic}}^{A^-B^+} = [\chi_A(1)\chi_A(2)][\alpha(1)\beta(2) - \beta(1)\alpha(2)] \tag{5}$$

and

$$\psi^{A^+B^-} = [\chi_B(1)\chi_B(2)][\alpha(1)\beta(2) - \beta(1)\alpha(2)] \tag{6}$$

represent electrostatic bonding characteristic of ion pairs. In Eqs. (4)–(6), $\alpha$ and $\beta$ refer to spin functions, and the total wavefunction is given as the product of a spatial and a spin factor. Resonance among the three simple valence structures yields

$$\psi = (C_{AB})\psi^{AB} + (C_{A^-B^+})\psi^{A^-B^+} + (C_{A^+B^-})\psi^{A^+B^-} \tag{7}$$

where the first term is the covalent contribution and the next two terms give the ionic contribution. Pauling (1960) has shown how the relative weighting of covalent and ionic terms can be related to the difference in effective electronegativities of atoms A and B. However, it is important to note that Pauling's criterion refers only to a particular type of ionicity—namely, that related to the net polarity arising from unequal weighting of $\psi^{A^-B^+}$ and $\psi^{A^+B^-}$. In the prototype case of the $H_2$ molecule, calculations have indicated an appreciable degree of ionic character (i.e., $C_{A^-B^+} = C_{A^+B^-} = 0.26\, C_{AB}$) (Weinbaum, 1933), even though symmetry clearly precludes any net polarity. Of course in bonds between electropositive atoms A (e.g., Al or Si) and electronegative atoms B (e.g., O or F), one can plausibly neglect the contribution from $\psi^{A^-B^+}$ relative to that of $\psi^{A^+B^-}$, and Pauling's electronegativity criterion can be employed directly in estimating $C_{AB}/C_{A^+B^-}$. Strictly speaking, of course, this ratio is not expected to be a simple intrinsic property of atoms A and B and can be expected to depend on the environment: e.g., when a molecule is placed in a crystal, the crystal potential will often support a greater degree of ionicity than that characteristic of the isolated molecule (e.g., see Newton and Gibbs, 1980). Nevertheless, small molecular clusters often provide good models for electronic structure in solids.

## 2. General Population Analysis Techniques

Turning now to molecular orbital or more general wavefunctions, we must seek alternative definitions of bonding character. By integrating the product $\psi^*\psi$ over all but one of its electronic coordinates one obtains the ordinary electron density, $\rho(r)$, which is generally normalized so as to yield the total number of electrons, $n$:

$$\int \rho(\mathbf{r})\, d\tau = n \tag{8}$$

If the many-electron wavefunction can be expanded as a sum of products of AOs (as is generally the case for MO, VB, or CI wavefunctions), then the density $\rho$ can be partitioned into AO product contributions:

$$\rho(\mathbf{r}) = \sum_{i,j}^{m} \rho_{ij}\chi^*_i(\mathbf{r})\chi_j(\mathbf{r}) \tag{9}$$

where $\rho_{ij}$ is called the density matrix. Integration then leads to individual AO overlap populations which by construction add up to the total electron population, $n$, such that

$$\sum_{i,j} \rho_{ij}\int \chi_i^*(\mathbf{r})\chi_j(\mathbf{r})\, d\tau = \sum_{i,j} \rho_{ij}S_{ij} = n \tag{10}$$

where $S_{ij}$ is an AO overlap integral. A diagonal term, $\rho_{ii}$, is called a net orbital population, $q_i'$ (Mulliken, 1955), which when summed over all AOs on a given atom A yields a net atomic population, $q'(A)$. On the other hand, summing $\rho_{ij}S_{ij}$ over all AOs $\chi_{i_A}$ on center A and AOs $\chi_{j_B}$ on center B yields the net overlap or bond population, $n(AB)$. Large positive values of $n(AB)$ give a rough indication of strong covalent bonding (e.g., Newton and Gibbs, 1980), whereas a pure ionic bond would yield essentially zero. Strong exchange repulsion between nonbonded atoms often leads to negative values of $n(AB)$ (O'Keeffe $et\ al.$, 1980). Although overlap populations offer at best only a semiquantitative measure of bonding, in certain selected situations they provide useful quantitative correlations as discussed below (see also Newton and Gibbs, 1980).

Since the net atomic populations $q'(A)$ do not add up to $n$, it is desirable to modify them so as to yield populations which do have this desired property. There is clearly no unique procedure for doing this, and we simply employ Mulliken's original suggestion—namely, to assign half of $n(AB)$ to $q'(A)$, and half to $q'(B)$; this leads to the gross atomic populations $q(A)$:

$$q(A) = \sum_{i_A} \left[ \sum_{j}^{m} \rho_{ij}S_{ij} \right] \tag{11}$$

where the inner summation over all AOs yields the gross orbital population $q_{i_A}$, in orbital $\chi_{i_A}$ centered on atom A. Effective atomic charges are obtained by subtracting the atomic nuclear charges ($Z$) from the atomic populations.

## C.  Hybrid Atomic Orbitals

### 1.  General Definition

While most AO basis sets used in computations are pure angular momentum types (i.e., s, p, d, etc.), much insight is given by analyzing bonding in terms of hybrid AOs. In this paper we shall be interested in s–p hybridization, and we define a normalized $sp^\lambda$ hybrid AO as

$$\chi_{sp(\lambda)} = [1/(1 + \lambda)^{\frac{1}{2}}](\chi_s + \sqrt{\lambda}\, \chi_p) \tag{12}$$

where $\chi_s$ and $\chi_p$ are valence s and p AOs. A population analysis analogous to that given above reveals that the fractional s character in such a hybrid AO can be expressed as

$$f_s = 1/(1 + \lambda) \tag{13}$$

## 2. Definition in Terms of Molecular Orbital Wavefunctions

Since molecular orbitals are expanded in terms of AOs, they implicity contain hybrid AOs. The problem is that each MO in general represents a different degree of AO hybridization. Thus what is needed is a procedure for determining MOs such that a given set of AOs contributes significantly only to one MO. The particular linear combination of AOs in this MO then defines the corresponding hybrid AO. To achieve this goal we exploit the fact than an MO wavefunction [Eq. (1)] is invariant when the occupied MOs are subjected to a unitary transformation (Roothaan, 1951). Molecular orbitals are generally obtained in so-called canonical form, in which they are delocalized and transform as irreducible representations of the molecular point group. In this situation the AOs on a given atom usually contribute appreciably to several different MOs. However, unitary transformations can be found which optimally localize the MOs according to a suitable criterion. We employ the criterion of Lennard-Jones and Pople (1950) and Edmiston and Ruedenberg (1963), according to which the MOs are rotated until electronic repulsion between different MOs is minimized. The localized MOs (LMOs) then correspond generally to two-center two-electron bonds or one-center lone pairs (Newton et al., 1970), and each bond or lone pair automatically defines a set of hybrid AOs—i.e., a bond LMO can be written as

$$\phi^b_{AB} = C_A h^b_A + C_B h^b_B + \text{small terms} \tag{14}$$

and a lone pair LMO as

$$\phi^{lp}_A = h^{lp}_A + \text{small terms} \tag{15}$$

where $h_A$ and $h_B$ are the hybrid AOs on the dominant atomic centers in a given LMO, and contributions from other centers are generally quite small (Newton et al., 1970).

It is important to emphasize that atomic orbital hybridization, which is clearly revealed in terms of LMOs, may be completely obscured in the more common delocalized MOs because of symmetry; e.g., in the linear

SiOSi group, mixing of s and p AOs on oxygen is forbidden by symmetry, while no such restriction pertains to the LMOs which represent the individual SiO bonds. As emphasized above, the localized and delocalized representations correspond to the same many-electron MO wavefunction, $\psi^{MO}$, and all observable properties are the same for both. It is simply a question of choosing the most convenient MO representation for analyzing a particular property: i.e., LMOs and their hybrid AOs are very useful for studying local bonding properties, even though "hybridization" is not a true quantum-mechanical observable. On the other hand, spectroscopic properties such as photoemission and fluorescence, which are not considered here, are usually more easily discussed in terms of delocalized MOs.

### 3.   Relationship to Chemical Bonding

We review briefly the role of hybrid AOs in models of chemical bonding. As discussed by Coulson in his book (1961), the degree of bonding between two atoms is intimately associated with the degree of hybridization of the relevant AOs. In general, the strength of a bond depends on the degree of atomic orbital overlap and the matching of effective orbital electronegativities. The overlap associated with sp hybrid AOs [i.e., $\lambda = 1$, Eq. (12)] is generally much greater than that for pure s or p AOs (the overlap of $sp^2$ and $sp^3$ hybrid AOs lies between these extremes). For a given atom, the orbital electronegativity increases with the s character of the $sp^\lambda$ hybrid since the mean distance of an s electron from the atomic nucleus is less than that for the corresponding p orbital. In a homopolar bond, orbital electronegativities are of course exactly matched by symmetry, irrespective of hybridization, but in heteropolar bonding, the flexibility associated with hybridization can be important in yielding an optimal matching. Equilibrium geometry depends on hybridization in two important aspects: bond angles are strongly influenced by the directionality of hybrid AOs, and bond distances associated with a given atom tend to decrease with increasing fractional s character as a result of the differences in effective radii of s and p AOs noted above. This dependence of bond length on s character is well documented for carbon–carbon bonds where, for example, a good linear correlation has been established between average hybrid AO s character [$f_s$, Eq. (13)] and the length of carbon single bonds (Dewar and Schmeising, 1960). The latter cover a range of $\sim 1.35–1.55$ Å, in spite of the fact that they all have a formal bond order of unity.

Although specific details of calculations will be given in Sec. IV, we will specialize the discussion at this point in anticipation of our primary interest in the SiOSi linkage. In particular, we focus on the divalent oxygen atom, which in a zeroth order picture can be viewed as being in a valence state containing two unpaired valence electrons in equivalent bonding hybrid AOs

$(h^b)$ and four valence electrons in equivalent lone pair hybrid AOs $(h^{lp})$. The directionality of two equivalent and orthogonal $sp^\lambda$ hybrids can be expressed as (Coulson, 1961),

$$\lambda = -\sec\theta \tag{16}$$

where $\theta$ is the interhybrid angle. For real s–p hybrids, $\theta$ is seen to be restricted to the range 90–180°. In this section we consider only orthogonal hybrids and assume that they point directly at the atoms they are bonded to; i.e., in the case of the SiOSi group it it assumed that the oxygen bonding hybrids perfectly follow the silicon atoms as the SiOSi angle varies. It is because of the great variation observed in this angle for silica, silicates and siloxanes, that we concentrate on hybridization at the oxygen atom (Hill and Gibbs, 1979; O'Keeffe and Hyde, 1978; Gibbs et al., this volume, Chapter 9). The silicon AOs are also expected to be hybridized $(\sim sp^3)$, but there is no evidence of any strong *variations* in the silicon hybridization.

The valence state of divalent oxygen can be summarized as

$$\psi^{VS} = (sp^{\lambda_b})_1^1(sp^{\lambda_b})_2^1(sp^{\lambda_{lp}})_1^2(sp^{\lambda_{lp}})_2^2 \tag{17}$$

where the bonding hybrids form an angle, $\theta_b = \sec^{-1}(-\lambda_b)$, and the lone pair hybrids form an angle $\theta_{lp} = \sec^{-1}(-\lambda_{lp})$.[†] Both pairs of equivalent hybrids are symetrically disposed with respect to the local $C_{2v}$ axis. Furthermore, each bonding hybrid $(h^b)$ is orthogonal to each lone pair hybrid $(h^{lp})$, and the $h^b$–$h^{lp}$ pairs form angles

$$\theta_{b,lp} = \cos^{-1}[-\cos(\theta_b/2)\cos(\theta_{lp}/2)]$$

In Table I, pertinent information is summarized for a variety of bonding angles $(\theta_b)$, illustrating how the fractional s and p character of the six valence electrons is coupled to the interhybrid angles.

We emphasize the mutual relationships among the hybrids and the bond angles which they are involved in (Walsh, 1947; Bent, 1961). Thus perturbations which are viewed as causing a change of bond angle would, according to the simple model entertained here, lead to a rehybridization of the AOs, while electronic perturbations which are viewed as changing the degree of hybridization would lead to a change in the bond angle (and also bond strength). We note some of the more common perturbations, which can be separated into external and intraatomic factors. Among the external factors are local effects such as nonbonded repulsions between the two atoms bonded to oxygen—i.e., Si $\cdots$ Si repulsion in the present context, as emphasized by O'Keeffe and Hyde (1978)—which tend to increase $\theta_b$, and a variety of packing factors which are operative in solid materials. If the possibility of pi

---

[†] In eq (17), the occupation numbers (1 or 2) are indicated by superscripts, while members of a pair of equivalent hybrids are distinguished by subscripts (1 or 2).

TABLE I

**Properties of Orthogonal Hybrid AOs on Divalent Oxygen**[a]

| Interhybrid angles | | | $\lambda$ | | Orbital population | |
|---|---|---|---|---|---|---|
| $h^b h^b(\theta_b)$ | $h^{lp} h^{lp}(\theta_{lp})$ | $h^b h^{lp}(\theta_{b,lp})$ | $h^b$ | $h^{lp}$ | 2s | 2p |
| 90.0 | 180.0 | 180.0 | $\infty$ | 1 | 2.0 | 4.0 |
| 120.0 | 101.5 | 108.4 | 2.0 | 5.0 | 1.3 | 4.7 |
| 140.0 | 94.1 | 103.5 | 1.3 | 14.1 | 1.1 | 4.9 |
| 160.0 | 90.9 | 97.0 | 1.1 | 63.0 | 1.0 | 5.0 |
| 180.0 | 90.0 | 90.0 | 1.0 | $\infty$ | 1.0 | 5.0 |

[a] $\lambda$ is defined by Eq. (12); a pure 2p hybrid corresponds to the limit, $\lambda \to \infty$; the orbital populations are based on the valence-state configurations given by Eq. (17) and the definition of fractional s character [Eq. (13)].

bonding exists, then additional angle widening is possible. For the disiloxy-group this would involve back donation from oxygen 2p $\pi$ AOs (i.e., those components of the oxygen 2p AOs perpendicular to the SiO bond directions) into empty $\pi$ AOs associated with the silicon 3p or 3d AOs. We also note that the electronegativity of the atoms bonded to oxygen can affect the relative degree of s character in the bonding and lone pair hybrids and, hence, the bond angles and length. Thus replacing the hydrogen atoms in $H_2O$ by the more electropositive Si atoms (as in the disiloxy group) is expected to increase (decrease) the s character of the bonding (lone pair) hybrids on the oxygen atom (Bent, 1961).

Among the intraatomic factors controlling hybridization we emphasize:

1.   Promotion energy associated with changing the s-orbital population. Since it requires $\sim 15$ eV to promote an oxygen atom from its ground state configuration ($2s^2\ 2p^4$) to the excited configuration ($2s^1\ 2p^5$), compared to $\sim 4$ and $\sim 6$ eV for the analogous processes in C and Si (Moore, 1949), it is clear that such promotion in oxygen will be relatively unfavorable.

2.   Exchange repulsion between electrons in different oxygen hybrid AOs (Pople, 1950; Gillespie, 1972). These are generally observed to increase in importance as follows: bond–bond < bond–lone pair < lone pair–lone pair.

3.   The increase in SiO bond strength associated with increasing s character.

Thus we see that the first and second intraatomic effects tend to favor small bond angles ($\theta_b$), while the third factor as well as Si $\cdots$ Si repulsions for the case of the disiloxy group favor larger angles. Clearly a delicate balance is involved in achieving the optimal compromise, and it is perhaps not surprising that when the additional degrees of freedom provided by the full range of packing factors are included, a large range of bond angles is observed in solid materials containing the disiloxy group. Of course, the conflicting demands associated with the above factors may require a relaxation of the assumption that bonding hybrids are always perfectly directed towards the bonded atoms. Furthermore, there is no intrinsic reason (except convenience) for constraining hybrid AOs on a given atom to be mutually orthogonal. In fact the hybrids defined by the localized MOs (LMOs) in Sec. II,C,2 are in general neither perfectly directed nor orthogonal, except as constrained by molecular symmetry. Nevertheless, simple hybridization arguments of the type given above have successfully accounted for many bond-length–bond-angle correlations in organic systems (Walsh, 1947; Dewar and Schmeising, 1960; Bent, 1961; Newton *et al.*, 1970) and appear to offer a natural basis for analyzing some of the analogous correlations observed for the disiloxy group by Gibbs and coworkers in recent years (see literature cited by Hill and Gibbs, 1979). Thus as the SiOSi angle varies in response to the numerous influences outlined above, the hybridization at oxygen, and hence also the bond length and strength, is expected to adjust continuously. It remains to demonstrate to what extent these expectations are borne out by detailed calculations. The results of Newton and Gibbs (1980) have been shown to be *consistent* with a hybridization model, and in Sec. IV more direct quantitative evidence is offered.

### III. COMPUTATIONAL DETAILS

The MO calculations reported below employed a minimal basis of Slater-type atomic orbitals (STOs), supplemented in some cases with 3d AOs on Si. The STOs were expanded in terms of Gaussian-type orbitals (GTOs), with three GTOs assigned to each s or p STO (STO-3G) as described by Hehre *et al.* (1969, 1970) and with each STO 3d AO represented by a single GTO (STO-3G*), as described by Collins *et al.* (1976). The molecular orbital calculations were carried out using the Gaussian 76 computer program (Binkley *et al.*, 1978).

Disiloxane (**Ia**) and pyrosilicic acid (**Ib**) were employed as models for the —SiOSi— linkage. Newton and Gibbs (1980), O'Keeffe *et al.*, (1980), and Gibbs *et al.* (this volume, Chapter 9) have documented the great utility of these molecules in analyzing properties of silica, silicates and siloxanes. Both molecules were constrained to have $C_{2v}$ symmetry and fully staggered conformational geometries (i.e., HSiOSi, HOSiO, and OSiOSi bonded sequences were all assigned dihedral angles of 60 or 180°). Aside from the SiOSi angle, which

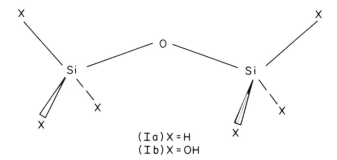

(Ia) X = H
(Ib) X = OH

was varied, tehrahedral angles were assumed. The SiH and OH bond lengths were taken as 1.49 and 0.96 Å, respectively (Newton and Gibbs, 1980), and for the most part the SiO distance was kept fixed at 1.62 Å, the average value observed for silicates (Louisnathan and Gibbs, 1972).

The LMO calculations were carried out for disiloxane as described by Newton et al. (1970), but not for pyrosilicic acid due to the computational effort required. However, the bridging oxygen atom hybrid AOs are expected to be very similar for the two species. Results based on LMOs are presented below for SiOSi angles between 120 and 170°. For angles close to 180°, the iterative minimization procedure used to determine the LMOs (Sec. II,C,2) converges very slowly, suggesting that the defining equations may be approaching a situation where no unique solution exists (Newton et al., 1970). Hence no hybrid AO analysis of the LMOs was attempted for the SiOSi angle of 180°. A similar situation was noted by Chipman et al. (1980) in a hybrid orbital analysis of $H_2O$.

## IV.   RESULTS AND DISCUSSION

### A.   Atomic Populations

Information regarding the effective charge on the bridging oxygen atom of the disiloxy group is presented in Tables II and III. Table II gives results for disiloxane and pyrosilicic using a simple STO-3G basis and also one supplemented with 3d AOs on Si (STO-3G*). It is seen that the calculated bridging oxygen populations and equilibrium SiOSi angles are similar for both molecules. The 3d AOs have a small effect on equilibrium angle and lead to a significant reduction of the atomic charges, as noted by Newton and Gibbs (1980).

The reduction of positive charge on the Si atoms due to the 3d AOs would be expected to yield smaller SiOSi angles through reduced electrostatic

TABLE II

Dependence of Atomic Population and Equilibrium SiOSi Angle on Basis Set

|  | Basis set | SiOSi angle (deg) | $q$(Si) | $q$(O)[a] |
|---|---|---|---|---|
| $(SiH_3)_2O$[b] | STO-3G | 129 | 13.16 | 8.64 |
|  | STO-3G* | 131 | 13.40 | 8.43 |
|  | STO | 137 | 12.59 | 8.66 |
| $\{Si(OH)_3\}_2O$[c] | STO-3G* | 133 | 13.17 | 8.42 |

[a] Bridging oxygen atom.
[b] Disiloxane (**Ia**).
[c] Pyrosilicic acid (**Ib**).

repulsion, and this effect appears to be dominant in the case of pyrosilicic acid. The 3d AOs have less effect on the Si charges in disiloxane, and a slight *increase* in SiOSi is observed. At any rate, for purposes of studying angle-dependence, Table II suggests that 3d AOs on Si are not of primary importance, and we thus restrict our attention to the minimal basis (STO-3G) results. Table III indicates an appreciable increase in ionic character with increasing SiOSi angle. This is expected from Table I, which shows how the s character (and hence electronegativity) of the oxygen bonding hybrids increases with SiOSi angle. The resulting sigma-bonding charge transfer (Si → O) is somewhat offset by $\pi$-electron back donation (O → Si), which

TABLE III

Angle Dependence of Properties Calculated for $\{Si(OH)_3\}_2O$ (**Ib**)[a]

| SiOSi angle (deg) | At fixed SiO distance (1.62 Å) | | Optimized SiO distance[b] (Å) |
|---|---|---|---|
|  | $n$(SiO)[b] | $q$(O)[c] |  |
| 120 | 0.520 | 8.63 | 1.63 |
| 140 | 0.550 | 8.67 | 1.60 |
| 160 | 0.572 | 8.69 | 1.57 |
| 180 | 0.579 | 8.70 | 1.56 |

[a] STO-3G basis.
[b] Refers to bridging SiO bonds. The $n$(SiO) values and optimal SiO distances for 140°, 160°, and 180° were taken from Newton and Gibbs (1980); the equilibrium value for the terminal SiO bonds is 1.66 Å.
[c] $q$(O) is the electron population on the bridging oxygen atom.

leads, for example, to a population of $\sim 1.88$ electrons in the oxygen $2p\pi$ AO perpendicular to the SiOSi plane in disiloxane, irrespective of SiOSi angle. The analogous $2p\pi$ AO in the reference species, $H_2O$, must, of course, have a population of 2.00, since the hydrogen atoms have no $\pi$ AOs. The STO-3G oxygen population for $H_2O$ (with bond angle of HOH $= 140°$) is only $\sim 8.5$ electrons, compared with $\sim 8.6$–$8.7$ for the siloxyl oxygen, thus confirming our expectation that the $SiX_3$ group (X = H or OH) is somewhat more electropositive than a hydrogen atom.

The calculated Si and O atom populations displayed in Tables II and III are in rough accord with Pauling's (1960) estimate of $\sim +1.0$ and $-0.5$, respectively, for the effective atomic charges, as discussed by Newton and Gibbs (1980). Pauling deduced this result from the SiO bond length in silicates and the electronegativities of Si and O, which lead to a bond order of 1.55 and 51% ionic character for the SiO bond. Thus relative to four-coordinate $Si^{4+}$, $4 \times 1.55 \times (1 - 0.51)$ electrons must be transferred, yielding a net charge of $+0.96$. Experimental support for a value in the vicinity of 1.0 has been provided by Gibbs *et al.* (1978) and Stewart *et al.* (1980) from data for coesite and low quartz, respectively.

## B.  SiO Overlap Populations

Table III shows that as the SiOSi angle in pyrosilicic acid increases, not only do the effective atomic charges (a measure of ionic character) increase, but also the SiO overlap population (related to covalent bond strength). Newton and Gibbs (1980) and Gibbs *et al.* (this volume, Chapter 9) have demonstrated that $n(SiO)$ variations at fixed Si—O distance can be used to predict accurately the actual change in SiO bond length with angle through the use of a simple linear relation obtained from a regression analysis. They also emphasized the increasingly important contribution (both relative and absolute) of the oxygen 2s AOs to $n(SiO)$ with increasing angle, consistent with the model of rehybridization at oxygen inferred from bond-length–bond-angle variations (see Table I). The optimal SiO distances are noted in the last column of Table III.

## C.  Hybridization of the Oxygen AOs

Before turning to hybrid AOs in the context of LMOs, we ask a more basic question: is the oxygen 2s AO expected to mix appreciably with any other valence AOs, and to what extent is it appropriate to consider it as part of the oxygen atom core? Table IV displays the 2s populations for the bridging oxygen atom in disiloxane, distributed among the various totally symmetric canonical (or delocalized) valence MOs. In "solid state" terminology, the 2s

TABLE IV

**Distribution of Bridging Oxygen 2s Population among Valence
MOs of Disiloxane (Ia)**

| SiOSi angle (deg) | | | | | | | |
|---|---|---|---|---|---|---|---|
| 120 | | 140 | | 160 | | 180 | |
| $\varepsilon$(a.u.)[a] | $q_{2s}$ | $\varepsilon$(a.u.)[a] | $q_{2s}$ | $\varepsilon$(a.u.)[a] | $q_{2s}$ | $\varepsilon$(a.u.)[a] | $q_{2s}$ |
| $-1.234$ | 1.63 | $-1.209$ | 1.65 | $-1.190$ | 1.66 | $-1.183$ | 1.66 |
| $-0.649$ | 0.11 | $-0.634$ | 0.09 | $-0.627$ | 0.07 | $-0.624$ | 0.07 |
| $-0.483$ | 0.02 | $-0.477$ | 0.01 | $-0.474$ | 0.00 | | |
| $-0.338$ | 0.07 | $-0.317$ | 0.04 | $-0.302$ | 0.01 | | |
| $(-0.326$ | 0.00) | $(-0.311$ | 0.00) | $(-0.301$ | 0.00) | $(-0.300$ | 0.00) |
| Total: | 1.83 | | 1.79 | | 1.75 | | 1.73 |

[a] Molecular orbital one-electron energy. One atomic unit (a.u.) = 27.21 eV or 2625.47 KJ/mole. All totally symmetric valence MOs (and, in parentheses, the highest filled MO) are included.

AO is clearly confined primarily to the bottom of the valence band. However, there is some 2s character in higher lying valence MOs, and the detailed distribution is seen to depend on SiOSi angle. The 2s population changes by $\sim 0.1$ electron over the range, $120 \rightarrow 180°$. Further evidence suggesting that the 2s AO should not be relegated to the oxygen core is given in Table V, which lists the matrix elements of the Hartree–Fock effective one-electron Hamiltonian (and the analogous overlap elements) which couple the oxygen 2s AO with other valence AOs. These matrix elements, including the one-center 2s–2p element (when the bond angle of SiOSi $\neq 180°$), are clearly large and are of the order of the width of the valence band (see Table IV). Hence it is not surprising that MO calculations in which these matrix elements (and the corresponding overlap elements) are artificially set to zero yield electronic structure properties substantially different from the proper MO results.

Table IV reveals that the 2s population of oxygen *decreases* with increasing SiOSi angle, even though the total atomic population *increases* (Table III). This, of course, is exactly what is expected on the basis of the idealized hybrid AO arguments summarized in Table I, although the decrease in 2s population due to rehybridization is partially offset as a result of increasing bond polarity. The 2s populations in the range 1.73–1.83 correspond to a modest degree of s $\rightarrow$ p promotion, as expected (see Sec. II,C,3).

The hybrid AOs inferred from localized molecular orbitals are displayed in Table VI for SiOSi angles between 120 and 170°. At first glance, these

**TABLE V**

**Off-Diagonal Matrix Elements $F_{ij}$ of the Effective One-electron Hamiltonian Involving the Bridging Oxygen 2s Atomic Orbital[a,b]**

| | Absolute value (a.u.) | |
|---|---|---|
| Matrix element | SiOSi angle = 145° | SiOSi angle = 180° |
| $F_{2s_O2p_O}$ | 0.07 (0.00)[c] | 0.00 (0.00) |
| $F_{2s_O3s_{Si}}$ | 0.57 (0.26) | 0.56 (0.26) |
| $F_{2s_O3p_{Si}}$ | $\begin{cases} 0.75\ (0.36)^d \\ 0.23\ (0.11)^e \end{cases}$ | 0.78 (0.37) |

[a] Based on the disiloxane MO wavefunction (STO-3G); the $SiO_2$ framework is in the $xz$ plane, with the two-fold axis coincident with the $z$ axis.

[b] One a.u. = 27.21 eV = 2625.47 kJ/mole; all nonzero off-diagonal elements of the type $F_{(2s)(j)}$, where j ≡ an O or Si valence AO, are included; quantities in parentheses are the corresponding overlap elements.

[c] Refers to the oxygen $2p_z$ AO.

[d] Refers to the silicon $3p_x$ AO.

[e] Refers to the silicon $3p_z$ AO.

**TABLE VI**

**Bridging Oxygen Hybrid AOs Obtained from Localized Moelcular Orbitals for Disiloxane (Ia)**

| | Oxygen atom hybrid AO $(sp^\lambda)^a$ | | | |
|---|---|---|---|---|
| | Bonding | | Lone pair | |
| SiOSi angle | $\lambda$ | $f_s$ | $\lambda$ | $f_s$ |
| 120 | 3.81 | 0.208 | 1.95 | 0.340 |
| 130 | 3.52 | 0.221 | 2.14 | 0.319 |
| 140 | 3.26[b] | 0.235[b] | 2.36 | 0.298 |
| 150 | 3.01 | 0.249 | 2.61 | 0.277 |
| 160 | 2.79 | 0.264 | 2.90 | 0.256 |
| 170 | 2.58 | 0.279 | 3.18 | 0.239 |

[a] These hybrid AOs differ from those in Table I because they are not constrained to be orthogonal, nor is the SiOSi angle constrained to coincide with $\theta_b$. Also in contrast to Table I, it is to be emphasized that the total oxygen 2s populations (see Table IV) are not simple additive functions of the $f_s$ values.

[b] The corresponding values for $H_2O$ at the same bond angle are 3.72 and 0.212. The slightly larger $f_s$ value for disiloxane is consistent with the fact that $H$ is a slightly more electronegative substituent than the $SiH_3$ group (see discussion in Secs. II,C,3 and IV,A).

latter (nonorthogonal) hybrids appear quite different from those of Table I. However, both sets of hybrid AOs exhibit the same qualitative dependence on angle, and in one very important respect there is a quantitative relation between the two: namely, the $f_s$ values (Eq. 13) associated with the two sets of bonding hybrids are linearly related ($r^2 = 0.93$) over the range of angles included in Table VI:

$$f_s^{IDEAL} = -0.1102 + 2.242 f_s^{LMO} \qquad (18)$$

where $f_s^{IDEAL}$ refers to the orthogonal bond hybrids of Table I, which are assumed to point directly at the Si atoms. As a result, the correlation between SiO bond distance and angle, which Newton and Gibbs (1980) nominally analyzed in terms of the idealized hybrids (Table I), carries over equally well to the presumably more realistic hybrids obtained from the *ab initio* LMOs (Table VI), to the extent that the linear relation of Eq. (18) is obeyed; i.e., the expression

$$d(SiO) = 1.750 - 0.311 f_s^{IDEAL} \qquad (r^2 = 0.99) \qquad (19)$$

which correlates SiO bond lengths in silica polymorphs with SiOSi angle via Eq. (13) (see Fig. 9a of Newton and Gibbs, 1980), can be reexpressed in terms of $f_s^{LMO}$ through the use of Eq. (18). Thus the concept of linear dependence of a property on percent s character is seen to be justified in spite of the fact that no unique definition of hybrid AO exists. A similar situation arises in the case of nuclear spin–spin coupling constants (Newton, 1977).

We note in passing the work of O'Keeffe and Hyde (1978), who emphasized variations in $d(SiO)$ and suggested the possibility of a nearly constant Si $\cdots$ Si nonbonded contact distance. They have derived least-squares equations analogous to Eq. (19) based on silicate data, but with $f_s^{IDEAL}$ [which is equal to $1/(1 - \sec\theta)$] replaced by $\csc(\theta/2)$, where $\theta$ is the SiOSi angle. They found a very small slope for small values of $\csc(\theta/2)$ (i.e., $\theta$ between 145 and 180°), but an appreciable slope ($+0.81$ Å) for large $\csc(\theta/2)$ (i.e., $\theta$ between 123 and 145°). The ability of Eq. (19) to handle the entire range of angles is perhaps made more understandable by recognizing that $1/(1 - \sec\theta)$ varies linearly with the *square* of $\csc(\theta/2)$, since $1 - 1/(1 - \sec\theta) = [\csc(\theta/2)]^2/2$. An increasing slope in terms of the $\csc(\theta/2)$ variable corresponds to a constant slope when the variable is $f_s = 1/(1 - \sec\theta)$.

The LMO bonding (lone pair) hybrids have appreciably less (more) s character than the idealized ones, and correspondingly smaller (larger) interhybrid angles; e.g., the LMO bond hybrids are directed apart by only $\sim 95°$, irrespective of the SiOSi angle, a result which appears to underscore the dominant role of repulsions associated with the lone pair AOs as noted in Sec. II,C,3. Thus the directional requirements of the two lone pairs yield "bent" SiO bonds in the sense that the hybrids deviate from the SiO inter-atomic vectors by as much as $\sim 30°$. Such a phenomenon is quite common in organic molecules (Newton, 1977). See also the results of Chipman et al. (1980) for $H_2O$.

In contrast to the marked dependence of the calculated oxygen atom hybrid AOs on SiOSi angle, the Si atom hybrids involved in the bridging bonds remain nearly constant ($\lambda \sim 1.4$) as expected, although the $\lambda$ value is surprisingly small.

The reader is reminded (see Section III) that while the MO method can be easily applied in the limit when the SiOSi group is linear, and observables such as total energy, ionization energy, and atomic charge can be straight-forwardly evaluated, the analysis of the MO wavefunction in terms of a single simple valence-bond structure (i.e., two-electron two-center "bonds" or one-center lone pairs) seems to breakdown in this limit; i.e., the LMOs become quite complicated, suggesting in valence-bond terms that several valence structures may be necessary to account for this extreme situation. Physically, the problem arises from the fact that at 180°, a simple valence-bond model would be expected by symmetry to place the bonding electrons in AOs of sigma symmetry and the lone pair electrons in $2p\pi$ AOs, thus forcing the angle between lone pairs to be 90°, a situation which is ener-getically unfavorable (Pople, 1950; Gillespie, 1972).

## V.  CONCLUSIONS

Molecular orbital calculations are useful in analyzing structural data related to the disiloxy group in a variety of ways:

1.   Total energies reproduce the variation of equilibrium SiO bond length with SiOSi angle observed in silica polymorphs.

2.   Atomic and overlap populations obtained from population analysis of the MO wavefunctions show the increasing importance of the oxygen 2s AOs in bonding as the SiOSi angle increases.

3.   The AO hybridization model inferred on the basis of the above results is justified in terms of ab initio hybrid AOs obtained from localized MOs. The s character of these hybrids is shown to vary linearly with SiOSi angle, as implied by the analysis of Newton and Gibbs (1980) and Gibbs et al. (this volume, Chapter 9).

4.   The hybrid AO analysis demonstrates the expected dominance of lone pair repulsion; this leads to large interhybrid angles between lone pair hybrids and, correspondingly, small angles between bonding hybrids, with the result that the SiO bonds are appreciably bent.

5.   Direct inspection of Hamiltonian matrix elements demonstrates the importance of overlap of the oxygen 2s AO with other valence AOs in accounting for the electronic structure of the siloxyl group.

## ACKNOWLEDGMENT

Research was carried out at Brookhaven National Laboratory under contract with the U.S. Department of Energy and supported by its Office of Basic Energy Sciences.

I express my considerable debt to Professor G. V. Gibbs for introducing me to the structural problems considered herein, and for innumerable discussions on these topics. Many of the ideas presented in this chapter have arisen from work carried out in collaboration wih him.

## REFERENCES

Bent, H. A. (1961). *Chem. Rev.* **61**, 275–311.

Binkley, J. S., Whiteside, R., Haribaran, P. C., Seeger, R., Hehre, W. J., Lathan, W. A., Newton, M. D., Ditchfield, R., and Pople, J. A. (1978). "Gaussian 76—An *Ab Initio* Molecular Orbital Program." Quantum Program Chemistry Exchange, Bloomington, Indiana.

Calabrese, E., and Fowler, W. B. (1978). *Phys. Rev. B: Condens. Matter* [3] **18**, 2888–2896.

Chelikowsky, J. R., and Schluter, M. (1977). *Phys. Rev. B: Solid State* [3] **15**, 4020–4029.

Chipman, D. M., Palke, W. E., and Kirtman, B. (1980). *J. Am. Chem. Soc.* **102**, 3377–3383.

Ciraci, S., and Batra, I. P. (1977). *Phys. Rev. B: Solid State* [3] **15**, 4923–4934.

Collins, J. B., von R. Schleyer, P., Binkley, J. S., and Pople, J. A. (1976). *J. Chem. Phys.* **64**, 5142–5151.

Coulson, C. A. (1961). "Valence." Oxford Univ. Press, London and New York.

Dewar, M. J. S., and Schmeising, H. N. (1960). *Tetrahedron* **11**, 96–120.

Edmiston, C., and Ruedenberg, K. (1963). *Rev. Mod. Phys.* **35**, 457–465.

Gibbs, G. V., Hill, R. J., Ross, F. A., and Coppens, P. (1978). *Abstr. Prog.* **3**, 407.

Gibbs, G. V., Meagher, E. P., Newton, M. D., Swanson, D. K., this volume.

Gilbert, T. L., Stevens, W. J., Schrenk, H., Yoshimine, M., and Bagus, P. S. (1973). *Phys. Rev. B: Solid State* [3] **8**, 5977–5998.

Gillespie, R. J. (1972). "Molecular Geometry." Van Nostrand-Reinhold, Princeton, New Jersey.

Hehre, W. J., Stewart, R. F., and Pople, J. A. (1969). *J. Chem. Phys.* **51**, 2657–2664.

Hehre, W. J., Ditchfield, R., Stewart, R. F., and Pople, J. A. (1970). *J. Chem. Phys.* **52**, 2769–2773.

Hill, R. J., and Gibbs, G. V. (1979). *Acta Crystallogr., Sect. B* **B35**, 25–30.

Lennard-Jones, J. E., and Pople, J. A. (1950). *Proc. R. Soc. London, Ser. A* **202**, 166–180.

Louisnathan, S. J., and Gibbs, G. V. (1972). *Am. Mineral.* **57**, 1647–1663.

Moore, C. E. (1949). *Natl. Bur. Stand. (U.S.),* Circ. **467**.

Mulliken, R. S. (1955). *J. Chem. Phys.* **23**, 1833–1840.

Newton, M. D. (1977). *In* "Applications of Electronic Structure Theory" (H. F. Schaefer, III, ed.), pp. 223–275. Plenum, New York.

Newton, M. D., and Gibbs, G. V. (1980). *Phys. Chem. Miner.* **6**, 221–246.

Newton, M. D., Switkes, E., and Lipscomb, W. N. (1970). *J. Chem. Phys.* **53**, 2645–2657.

Newton, M. D., O'Keeffe, M., and Gibbs, G. V. (1981). *Phys. Chem. Miner.* **6**, 305–312.

O'Keeffe, M., and Hyde, B. G. (1978). *Acta Crystallogr., Sect. B* **B34**, 27–32.

Pantelides, S. T., and Harrison, W. A. (1976). *Phys. Rev. B: Solid State* [3] **13**, 2667–2691.

Pauling, L. (1960). "The Nature of the Chemical Bond," 3rd ed. Cornell Univ. Press, Ithaca, New York.

Pople, J. A. (1950). *Proc. R. Soc. London, Ser. A* **202**, 323–336.

Roothaan, C. C. J. (1951). *Rev. Mod. Phys.* **23**, 69–89.

Sauer, J., and Zurawski, B. (1979). *Chem. Phys. Lett.* **65**, 587–591.

Schaefer, H. F., ed. (1977). "Modern Theoretical Chemistry," Vol. 4. Plenum, New York.

Schneider, P. M., and Fowler, W. B. (1978). *Phys. Rev. B: Solid State* [3] **12**, 7122–7133.

Shull, H. (1962). *J. Appl. Phys.* **33**, 290–292.

Stewart, R. F., Whitehead, M. A., and Donnay, G. (1980). *Am. Mineral.* **65**, 324–326.

Walsh, A. D. (1947). *Discuss. Faraday Soc.* **2**, 18–23.

Weinbaum, S. (1933). *J. Chem. Phys.* **1**, 593–596.

# 9

# A Comparison of Experimental and Theoretical Bond Length and Angle Variations for Minerals, Inorganic Solids, and Molecules

## G. V. GIBBS, E. P. MEAGHER,

## M. D. NEWTON, and D. K. SWANSON

|  |  |  |
|---|---|---|
| I. | Introduction | 195 |
| II. | Molecular Orbital Method | 197 |
| III. | The Molecular Structure of Orthosilicic Acid, $Si(OH)_4$: A Comparison with the Shapes of $SiO_3(OH)^{3-}$ and $SiO_2(OH)_2^{2-}$ Anions in Hydrated Silicates | 199 |
| IV. | Force Constants and Optimized Geometry for the Disiloxy Unit of the Pyrosilicic Acid Molecule, $H_6Si_2O_7$: A Comparison with Experimental Geometries and the Bulk Modulus of the Silica Polymorphs | 202 |
|  | A. Si—O Bond Length and SiOSi Angle Variations | 202 |
|  | B. Stretching and Bending Force Constants of the Disiloxy Unit | 206 |
|  | C. SiOSi Angle Distribution | 207 |
|  | D. Bonding in Tetrahedral AlOSi Linkages | 210 |
| V. | Si—O Bridging Bond Length–Bond Strength Sum and Angle Variations | 211 |
|  | A. The Electrostatic Valence Rule | 211 |
|  | B. Correlation of Si—O Bond Length with $p(O)$ and SiOSi Angle | 212 |
|  | C. Theoretical Correlation between Tetrahedral Si—O Bond Length, $p(O)$, and TOT angle (T = Si, Al) | 213 |
| VI. | Geometries of Molecules and Related Groups in Solids | 216 |
| VII. | Conclusions | 221 |
|  | References | 222 |

## I. INTRODUCTION

In recent years, molecular orbital (MO) theory has yielded powerful and reliable techniques for calculating equilibrium geometries, electronic charge distributions, bond energies, and heats of reactions for small molecules and has provided valuable and important insights for understanding these

**195**

Structure and Bonding in Crystals, Vol. I

properties (Hermann *et al.*, 1973; Schaefer, 1977). In the realm of crystal chemistry, it has substantiated a connection between bond length and angle variations for numerous silicates, phosphates, sulfates, germanates, and other inorganic solids and has provided new insights into (1) the origins of shared polyhedral edge distortions in close-packed solids, (2) the stereochemical requirements of pentacoordinated aluminum, (3) the tendency for dimers in most pyrosilicates to adopt a doubly eclipsed configuration, (4) the restricted range of two-repeat chain configurations in the monopolysilicates, and (5) the compressibilities of the silica polymorphs (Bartell *et al.*, 1970; Gibbs *et al.*, 1972, 1974, 1976, 1977a,b; Louisnathan *et al.*, 1972a,b, 1977; Lager and Gibbs, 1973; Chiari and Gibbs, 1974; Chelikowsky and Schluter, 1977; Tossell and Gibbs, 1977, 1978; Meagher *et al.*, 1979, 1980; Hill *et al.*, 1977, 1979; McLarnan *et al.*, 1979; Harrison, 1978; Peterson *et al.*, 1979; Sauer and Zurawski, 1979; Tossell, 1979; Newton and Gibbs, 1980; De Jong and Brown, 1980; Meier and Ha, 1980; Newton *et al.*, 1980; Newton, 1981). In addition, $X\alpha$, SCF-MO *ab initio*, and other moderately accurate calculations have been shown to be useful in constructing one-electron energy diagrams and interpreting emission, absorption, and photoelectron spectra of silica and silicate minerals (Collins *et al.*, 1972; Gilbert *et al.*, 1973; Tossell, 1973, 1975, 1979; Tossell *et al.*, 1973; Yip and Fowler, 1974; Pantelides and Harrison, 1976; Griscom, 1977; Tossell and Gibbs, 1977; Dikov *et al.*, 1977; Fowler *et al.*, 1978; Newton and Gibbs, 1979; Meier and Ha, 1980, De Jong and Brown, 1980).

Recently, Newton and Gibbs (1980), Meagher *et al.* (1980), and Swanson *et al.* (1980) completed *ab initio* calculations for a number of molecules containing tetrahedral $T—O$ (T = Al, Si) bonds and TOT angles and found that energy-optimized geometries compare well with local geometries in the silica polymorphs, the silicates and siloxanes. In addition, Swanson *et al.* (1980), Meagher *et al.* (1980), Gupta *et al.* (1981), and Geisinger *et al.* (in preparation) calculated energy-optimized geometries for a number of first- and second-row-atom oxide and sulfide molecules using *ab initio* methods and observed that the optimized bond lengths are in close agreement with those observed for solids. In this chapter, we will examine and review the model calculations presented in these papers and the close correspondence that obtains between bond lengths and angles calculated for the molecules and those in solids.

This chapter is divided into five parts. In the first part, the theoretical method is described and Mulliken's recipe for calculating bond overlap population is related to the total MO electron density. In the second part, the molecular structure of orthosilicic acid, $Si(OH)_4$, as determined by *ab initio* theory, will be compared with the geometries of $SiO_3(OH)^{3-}$ and $SiO_2(OH)_2^{2-}$ anions in solids, and the fundamental vibration mode of the molecule will be compared with the experimental frequency of the silicate ion in aqueous solution. In the third part, a potential energy surface produced for the pyrosilicic acid molecule, $H_6Si_2O_7$, and a potential energy curve for $H_6SiAlO_7^{1-}$ will be examined in terms of bond length and angle variations in the silica polymorphs, siloxanes and silicates. Force constants obtained from the surface will be compared with experimental trends involving the SiOSi angle. Then, a Boltzmann weighted SiOSi angle distribution generated for $H_6Si_2O_7$ will be compared with an experimental distribution of the angle for the siloxanes. In the fourth part, a multiple linear regression analysis completed for Si—O bond lengths and angles generated with STO-3G calculations for several molecules will be examined in terms of the observed tetrahedral bond lengths of a number of framework silicates. In the fifth and final part of the chapter, the geometries for a number of first- and second-row-atom oxide and sulfide molecules and solids will be examined. The close geometric correspondence between these molecules and their counterparts in solids will be presented as evidence that local bonding forces in molecules and related solids are similar.

## II.  MOLECULAR ORBITAL METHOD

Thanks to the power of present generation computers, the molecular orbital method is serving as an increasingly important probe of molecular structure, as reviewed, for example, in the recent work edited by Schaefer (1977). In addition to computational feasibility, the method lends itself easily to simple interpretative models, thus allowing the full amount of information contained in the total wavefunction to be distilled into a relatively small number of chemically informative quantities such as electronic populations on the atoms and in the bonds of a molecule.

The central concept in molecular orbital theory is the idea that a complicated many-electron wavefunction for a molecule can be usefully approximated as a antisymmetrical product of one-electron functions, $\phi$, called molecular orbitals (MOs):

$$\psi_{\text{many-electron}} \approx \prod_{i=1}^{n} \phi_i \tag{1}$$

where $n$ is the total number of electrons in the molecule. The optimal wavefunction of this type (referred to as a Hartree–Fock wavefunction) for a molecule in its ground electronic state is the one which minimizes the total molecular energy. The latter quantity can be expressed as

$$E_{\text{mol}} = \int \psi^* \mathscr{H} \psi \, d\tau \tag{2}$$

where $\psi$ is the many-electron wavefunction defined in Eq. (1) and $\mathscr{H}$ is the many-electron Hamiltonian operator which contains the kinetic energy and the various Coulombic contributions to the potential energy. The criterion of the energy minimization leads to the familiar effective one-electron Schrödinger equation

$$F\phi_i = \varepsilon_i\phi_i, \tag{3}$$

where the operator $F$ is the Hartree–Fock or effective one-electron Hamiltonian, and $\varepsilon_i$ is the one-electron energy, which often serves as a useful estimate of the energy to ionize an electron out of the $i$th MO. In spite of the practice employed in some approximate methods, the total energy cannot be expressed solely as a function of $\varepsilon_i$.

In practice, the above equation is generally solved in an approximate manner by expanding the MO, $\phi_i$, in terms of a convenient basis set of $N$ atomic orbitals, $\chi_j$, centered on the various atoms of the molecule,

$$\phi_i \approx \sum_{j=1}^{N} \chi_j C_{ji} \tag{4}$$

In the present application, we adopt a minimal basis set, $\chi_j$, in which each occupied atomic orbital of the constituent atoms is represented by a single Slater-type atomic orbital (STO) basis function. Thus, for H, O, and Si atoms, one is dealing with one, five, and nine STO basis functions, respectively. For ease of computation, the latter functions are in turn expanded in terms of so-called Gaussian-type orbitals (GTOs) (Hehre *et al.*, 1969, 1970). In the minimum basis set calculations used in this study (referred to as STO-3G basis set calculations) each STO is represented by a linear combination of three Gaussian functions. Calculations with STO-3G basis sets yield structural results very similar to those obtained with STO bases (Pitzer and Merrifield, 1970; Stevens, 1970).

Once the basis set has been selected, the above one-electron Schrödinger equation is recast as a finite matrix equation

$$\mathbf{FC} = \mathbf{SC}\varepsilon_i \tag{5}$$

where $\mathbf{C}$ is a column vector of MO coefficients already defined, $\mathbf{F}$ is the matrix whose elements are defined as

$$F_{ij} = \int \chi_i^* F \chi_j \, d\tau \tag{6}$$

and $\mathbf{S}$ is the overlap matrix with elements

$$S_{ij} = \int \chi_i^* \chi_j \, d\tau \tag{7}$$

In this study, the Gaussian 76 computer program (Binkley *et al.*, 1978) was used to calculate the necessary matrix elements and solve the matrix equation, thereby yielding the desired MO coefficients $C_{ji}$. Once the latter are available, all molecular properties can be obtained. Two such properties are of particular interest here: the total molecular energy and the orbital population analysis. Minimization of the total energy with respect to geometrical parameters yields the equilibrium molecular geometry. Orbital population analysis is a convenient device for partitioning the total number of electrons in a molecule into various atomic and bond contributions. To achieve this result we first note that the total molecular orbital electron density

$$\rho(\mathbf{r}) = \sum_{i=1}^{n} \phi_i^*(\mathbf{r})\phi_i(\mathbf{r}) \tag{8}$$

can be expanded in terms of the AO basis as

$$\rho(\mathbf{r}) = \sum_{jk}^{N} \sum_{i=1}^{n} C_{ji}C_{ki}\chi_j^*(\mathbf{r})\chi_k(\mathbf{r}) \tag{9}$$

Integration over all space then yields the total number of electrons, $n$:

$$n = \sum_{jk}^{N} \sum_{i=1}^{n} C_{ji}C_{ki}S_{jk} \tag{10}$$

to which the $j$–$k$ pair of AOs contributes

$$n(j\text{--}k) = \sum_{i=1}^{n} C_{ji}C_{ki}S_{jk} \tag{11}$$

The latter quantity, when summed over all AOs $j$ on center $J$ and AOs $k$ on the center $K$, is defined as the bond or overlap population, $n(J\text{--}K)$, for the $J$–$K$ atom pair. As discussed by Mulliken (1932, 1935) in his original formulation, it can be either positive (bonding) or negative (antibonding).

## III.  THE MOLECULAR STRUCTURE OF ORTHOSILICIC ACID, Si(OH)$_4$: A COMPARISON WITH THE SHAPES OF SiO$_3$(OH)$^{3-}$ AND SiO$_2$(OH)$_2^{2-}$ ANIONS IN HYDRATED SILICATES

When concentrated in aqueous solutions in amounts greater than about 100 ppm silica, the orthosilicic acid molecule polymerizes rapidly into dimers and higher molecular weight species of silicic acid. Because of its strong tendency to self-polymerize, the molecule has never been concentrated in sufficient amounts and purity for a structural analysis by conventional methods (Ilers, 1979). Since a knowledge of the structure is important to our understanding of the properties of soluble silica, we have predicted the structure of the molecule at 0 K, using the STO-3G method. Another reason for undertaking the analysis was to determine how well the shape of $Si(OH)_4$ conforms with the shapes of $SiO_3(OH)^{3-}$ and $SiO_2(OH)_2^{2-}$ anions in solid silicate hydrates.

In previous studies of silicic acid (Ilers, 1955; Collins *et al.*, 1972), the topology of $Si(OH)_4$ was assumed to consist of a central silicon atom bonded to four oxygen atoms which in turn are each bonded to a hydrogen atom. Within the constraints of this topology, we undertook a complete optimization of the geometry of the molecule with the Pulay (1969) gradient method. To a very close approximation, the resulting shape (Fig. 1) possesses $S_4 (=\bar{4})$ point symmetry with four Si—OH bond lengths of 1.65 Å, four O—H bonds of 0.98 Å, four SiOH angles of 108.8°, and four OSiO angles of 107.1 and two of 114.2°. Moreover, each O—H bond vector is rotated about the Si—O bond vector 98° out of the OSiO plane of the molecule. (See Table I for the atomic coordinates and the nonequivalent bond lengths and angles.) An earlier optimization of the Si—OH bond length, within the constraints of $D_{2d} (=\bar{4}2m)$ point symmetry, yielded a value of 1.65 Å, which is identical with the one calculated here (Newton and Gibbs, 1980). The success of *ab initio* theory in reproducing the observed geometries of a number of first- and second-row molecules indicates that the theory may be used as an alternative to conventional methods of molecular structure analysis (Collins *et al.*, 1976; Schmiedekamp *et al.*, 1979; Wallmeier and Kutzelnigg, 1979; Newton, 1981).

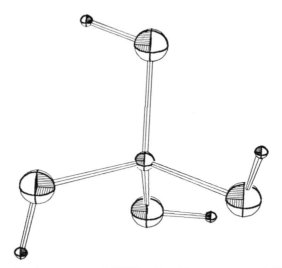

**Fig. 1.** The molecular structure of $Si(OH)_4$ ($S_4 = \bar{4}$ point symmetry). The large spheres represent oxygen, the small ones hydrogen, and the central sphere of intermediate size represents Si. No significance is attached to the sizes of the spheres. The bond lengths and angles of the molecule are given in Table I.

TABLE I

The Cartesian Coordinates of the $H_4SiO_4$ Molecule Optimized
with STO-3G Theory and a Comparison of the Optimized Bond
Lengths and Angles with Those Measured for a Number of Solid
Hydrated Silicates[a]

(a)  Atomic Cartesian coordinates (asymmetric unit, $S_4 = \overline{4}$ point symmetry) for $H_4SiO_4$ (Å)

| Atom | $x$ | $y$ | $z$ |
|------|-----|-----|-----|
| Si | 0.0 | 0.0 | 0.0 |
| O | 1.389 | 0.0 | 0.898 |
| H | 1.585 | −0.920 | 1.178 |

(b)  Comparison of bond lengths and angles

| | $H_4SiO_4$ | Hydrated silicates |
|---|---|---|
| $d(Si—OH)$, Å | 1.654 | 1.63–1.70 |
| $d(O—H)$, Å | 0.981 | 1.06 |
| $\angle OSiO$, deg. | 107.1[4][b], 114.2[2] | 106–116 |
| $\angle SiOH$, deg. | 108.8 | 108–125 |

[a] From Newton and Gibbs, 1980.
[b] Number inside bracket gives the multiplicity of the angle.

Thus, the *ab initio* optimized geometry of $Si(OH)_4$ is expected to be in close agreement with the actual structure. Moreover, the close correspondence between the shapes of the orthosilicic acid molecule and $Si_3(OH)^{3-}$ and $SiO_2(OH)_2^{2-}$ anions in solid silicate hydrates (Table II; cf. Glasser and Jamieson, 1976) attests to the ability of the molecule to give a good quantitative account of local geometries and bonding in solids.

In addition to predicting the structure of $Si(OH)_4$, the potential energy surfaces generated in the calculations were used to calculate the molecule's force constants and breathing frequency vibrations. The fundamental breathing frequency (8.14 m$^{-1}$) calculated from the Si—O stretching force constant (665 Nm$^{-1}$)* compares well with that observed (8.19 m$^{-1}$) for the silicate ion in aqueous solution (Nakamoto, 1970; Basile *et al.*, 1973). Calculation of the OSiO bending force constant (103 Nm$^{-1}$) indicates that the bond is considerably stiffer than the angle. The stiffness with respect to bond compression agrees with recent high-pressure studies which show that the bond is virtually unchanged (shortens by 0.02 Å) up to pressures of $\sim 50$ kbars (Levien *et al.*,

* Obtained by calculating the curvature of the potential energy surface at the minimum-energy bond length.

1980; Levien and Prewitt, 1981). In addition, the SiOSi angle, which is deformed by as much as $8°$, conforms with the small symmetric bending force constant ($\sim 10 \text{ Nm}^{-1}$) calculated for the SiOSi angle of $H_6Si_2O_7$ (Newton et al., 1980).

## IV. FORCE CONSTANTS AND OPTIMIZED GEOMETRY FOR THE DISILOXY UNIT OF THE PYROSILICIC ACID MOLECULE, $H_6Si_2O_7$: A COMPARISON WITH EXPERIMENTAL GEOMETRIES AND THE BULK MODULUS OF THE SILICA POLYMORPHS

In addition to determining the structure of orthosilicic acid, it was demonstrated in the last section that a close correspondence exists between the geometry of the isolated species and that of the silicate anions in solid hydrated silicates. In this section we will show that a close similarity also exists between the calculated geometry of the disiloxy unit in the pyrosilicic acid molecule and the unit in the silica polymorphs. Potential energy surfaces calculated for the molecule will be used to explain the broad distribution of SiOSi angles exhibited by silicates, siloxanes, and silica glasses and the correlations that obtain between the angle, the bridging bond length and the stretching and bending force constants of the disiloxy unit.

### A. Si—O Bond Length and SiOSi Angle Variations

The SiOSi angle in silicates and siloxanes exhibits a broad range of values from 120 to 180° (Tossell and Gibbs, 1978). As the angle widens, the bridging bond length is observed to shorten by a small but significant amount (Cannillo et al., 1968; Brown et al., 1969; Brown and Gibbs, 1970; Baur, 1971, 1977; Gibbs et al., 1972, 1977b; Hill and Gibbs, 1979; Newton and Gibbs, 1980). In addition, the stretching force constant of the bridging bond in the siloxanes has been observed to increase with increasing angle while the angle bending force constant decreases (Lazarev et al., 1967). In an effort to explain these properties with modern methods of quantum chemistry, Newton and Gibbs (1980) first calculated a potential energy curve for the pyrosilicic acid molecule as a function of the SiOSi angle, using the STO-3G method. In the calculations, the bridging and terminal bonds of the molecule were all fixed at 1.62 Å (the average bond length recorded for silicate tetrahedra in solids; Liebau, 1972; Griffen and Ribbe, 1979), all the O—H bonds were fixed at 0.96 Å, all the OSiO and SiOH angles were fixed at 109.5° and the SiOSi angle was varied from 110 to 180°. The shape of the resulting curve (the middle one in Fig. 5) is similar to those generated with the CNDO/2 method for disiloxane and various silicic acid molecules (Tossell and Gibbs, 1978; Meagher et al., 1979; De Jong and Brown, 1980). The curve has a relatively

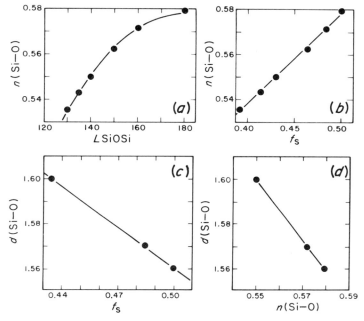

**Fig. 2.** (a) Mulliken bond overlap population, $n(\text{Si—O})$, calculated for the bridging bond of $H_6Si_2O_7$ ($C_{2v}$) plotted as a function of the SiOSi angle (see Fig. 4, Newton and Gibbs, 1980, for a drawing of the molecule); (b) $n(\text{Si—O})$ plotted against $f_s = 1/(1 + \lambda^2)$, the fraction s character of the hybrid AOs of the bridging oxygen of $H_6Si_2O_7$ where $\lambda^2 = -\sec \angle \text{SiOSi}$; (c) minimum-energy Si—O bridging bond length, $d(\text{Si—O})$, calculated SiOSi angles of 140, 160, and 180° for $H_6Si_2O_7$ vs. $f_s$; (d) $d(\text{Si—O})$ vs. $n(\text{Si—O})$. The $n(\text{Si—O})$ values used to prepare Figs. 2a, 2b, and 2d were calculated with all Si—O bonds frozen at 1.62 Å.

flat-lying segment from 150 to 180° with a shallow minimum at 137° and a steeply rising segment at $\sim 120°$. The minimum energy angle of 137° is slightly less than that (144°) recorded, on the average, for solid silicates (Tossell and Gibbs, 1978).

The Mulliken bond overlap population, $n(\text{Si—O})$, calculated for the bridging bonds is plotted as a function of the SiOSi angle in Fig. 2a. As discovered earlier with the extended Hückel method, $n(\text{Si—O})$ increases nonlinearly as the angle widens (Gibbs et al., 1972). However, when the assumption was made that the hybrid AOs on the bridging oxygen are equivalent, orthogonal, and directed along the Si—O bond vectors, Newton and Gibbs (1980) found that $n(\text{Si—O})$ is linearly correlated with $f_s$, the fractional s character of the hybrid orbitals (Fig. 2b). In their treatment, $f_s$ was taken to equal $1/(1 + \lambda^2)$ where $\lambda^2 = -\sec \angle \text{SiOSi}$ (Coulson, 1961).* In a

---

\* The notation $\lambda^2$ corresponds to the notation of $\lambda$ used by Newton (1981).

more sophisticated treatment, Newton (1981) has since undertaken a study of the hybrid AOs by analyzing the appropriate localized molecular orbitals (LMOs) of the disiloxy unit. Despite the discovery that the LMO hybrid AOs are appreciably different from those employed by Newton and Gibbs (1980), he did find that the $f_s$ values calculated in the two treatments are linearly dependent ($r = 0.96$). This is reassuring; it shows that the LMO $f_s$ value must also vary linearly with the overlap population and, as we will see below, with the optimized bridging bond lengths of the pyrosilicic acid molecule.

To explore how the Si—O bond lengths vary with angle, Newton and Gibbs (1980) optimized the bridging and terminal bond lengths for the molecule at angles of 140, 160, and 180°. The bond lengths calculated for the terminal bonds were found to be independent of the angle and to equal the value (1.66 Å) obtained for the orthosilicic acid molecule. On the other hand, the bridging bond lengths were found to be linearly and inversely* correlated with the fractional s character of each hybrid (Fig. 2c). Since the overlap population of the bridging bond is positively correlated with $f_s$ (Fig. 2b), then $d(\text{Si}—\text{O})$ must be inversely correlated with $n(\text{Si}—\text{O})$ as well as with $f_s$ (Fig. 2d). This serves to indicate as observed previously in terms of extended Hückel and CNDO/2 calculations for a large number of silicates that Si—O bonds with larger overlap populations should be shorter than those with smaller populations (Gibbs et al., 1972, 1974, 1977a,b; Louisnathan and Gibbs, 1972a,b; Tossell and Gibbs, 1977; Meagher et al., 1979). It is important to note that the $n(\text{Si}—\text{O})$ values used to prepare Fig. 4 were calculated with all bond lengths fixed at the mean value of 1.62 Å. The fact that these populations (based on the fixed *unrelaxed* $d(\text{Si}—\text{O})$ value) correlate with *relaxed* "equilibrium bond lengths" simply reflects the ability of changes in $n(\text{Si}—\text{O})$ caused by angle variations *alone* to mimic the tendency of $d(\text{Si}—\text{O})$ values to *relax* (relative to 1.62 Å) as the SiOSi angle is varied. Of course, $n(\text{Si}—\text{O})$ values based on *relaxed* Si—O bond lengths (i.e., optimized as a function of the SiOSi angle) also vary inversely with $d(\text{Si}—\text{O})$ and, in fact, with significantly greater slopes because shorter bonds usually involve larger overlap integrals and vice versa (Gibbs et al., 1972).

When the Si—O bond lengths measured for the high-pressure silica polymorph coesite (Gibbs et al., 1977b) and the optimized bridging bond lengths for $H_6Si_2O_7$ are both plotted against SiOSi angle, nonlinear trends obtain (Fig. 3) similar to that calculated between $n(\text{Si}—\text{O})$ and the SiOSi angle. On the other hand, linear trends obtain when both sets of bond lengths are plotted against $f_s$. Since an s electron is closer on the average to the nucleus of an atom than a p electron, it follows that the effective radius of the bridging oxygen should decrease as the angle widens and that the length of the bridging bond should shorten linearly (Fig. 4) with increasing s character, all other things being equal (Dewar and Schmeising, 1960).

---

* "Inversely" is used here to denote a negative slope.

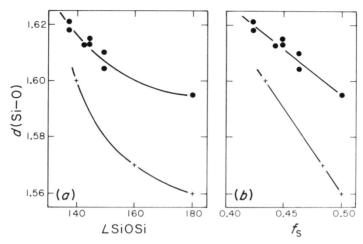

**Fig. 3.**   Experimental Si—O bond lengths, $d$(Si—O), for the silica polymorph coesite (upper curve, solid points) and minimum-energy Si—O bond lengths for $H_6Si_2O_7$ (lower curve, crosses) plotted against (a) SiOSi angle and (b) $f_s = 1/(1 + \lambda^2)$ where $\lambda^2 = -\sec \angle$ SiOSi.

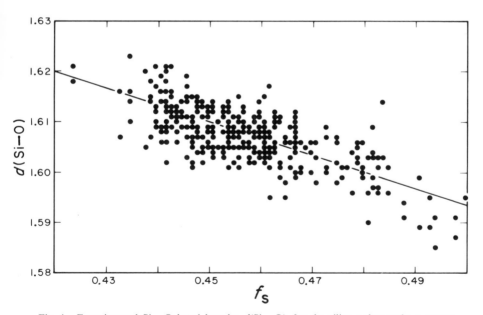

**Fig. 4.**   Experimental Si—O bond lengths, $d$(Si—O), for the silica polymorphs α-quartz (Zachariasen and Plettinger, 1965), coesite (Gibbs *et al.*, 1977b), low cristobalite (Dollase, 1965), and low tridymite (Konnert and Appleman, 1978), plotted against $f_s = 1/(1 + \lambda^2)$ where the values of $1/(1 + \lambda^2)$ were estimated from the observed SiOSi angles.

### B.  Stretching and Bending Force Constants
###     of the Disiloxy Unit

Three potential energy vs. SiOSi angle curves calculated for $H_6Si_2O_7$ by Newton and Gibbs (1980) are drawn in Fig. 5. In the calculations of these curves, the bridging bonds of the molecule were set in succession at 1.59, 1.62, and 1.65 Å, whereas the terminal bonds were set at 1.63, 1.62, and 1.61 Å, respectively, to maintain the mean bond length of the molecule at 1.62 Å. By fitting a parabola to each of these curves, the symmetric bending force constants of the molecule were found to increase from 8.21 to 13.1 to 17.9 Nm$^{-1}$ as the minimum energy angle narrows from 145 to 137 to 130°, respectively (Newton et al., 1980). In other words, the curves indicate as observed that the SiOSi bending force constant and the Si—O bridging bond length should both increase as the angle narrows.

It is pertinent that the bending force constants generated for $H_6Si_2O_7$ are in rough agreement with those derived for $\alpha$-quartz in lattice dynamical studies (Newton et al., 1980). Moreover, assuming that the dominant contribution to the bulk modulus of quartz is due to the SiOSi bending force constant, Newton et al. (1980) have calculated the bulk modulus of the mineral ($3.97 \times 10^{10}$ Pa, compared with $3.93 \times 10^{10}$ Pa observed) using the bending force constant of the molecule (without appealing to experimental data other than the Si—O bond length and several fundamental constants). The success of this calculation suggests that the nature of the

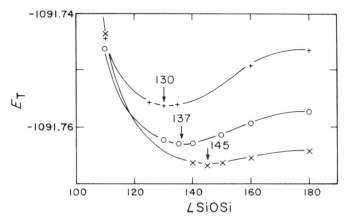

**Fig. 5.**   The total energy (atomic units) for $H_6Si_2O_7$ calculated as a function of the SiOSi angle for bridging bond lengths of 1.65 Å (upper curve), 1.62 Å (middle curve), and 1.59 Å (lower curve). The minimum energy SiOSi angle for each curve is indicated by a value located at the end of an arrow pointing toward the minimum. (1 atomic unit = 27.21 eV = 2.625 × 10$^3$ kJ mole$^{-1}$).

bonding forces in the disiloxy units of the two systems is nearly the same and that the force constants of the unit in the molecule are transferable to the corresponding group in a solid like $\alpha$-quartz.

The variation of the symmetric stretching force constant of the bridging bond has been calculated by Meagher et al. (1980) by evaluating the potential energy curve of $H_6Si_2O_7$ as a function of the bond length for bridging angles of 140, 160, and 180°. In the calculation, all the terminal bonds were fixed at the optimized value of 1.66 Å (Newton and Gibbs, 1980), the O—H separations were fixed at 0.96 Å, and each OSiH angle was fixed at 109.5°. The force constants obtained by fitting a parabola to the total energy as a function of the bridging bond length were found to increase from 732 Nm$^{-1}$ at 140° to 833 Nm$^{-1}$ at 160° to 878 Nm$^{-1}$ at 180°. Hence, as observed for the siloxanes, the stretching force constant of the disiloxy unit for $H_6Si_2O_7$ increases as the bridging angle widens and bridging bond shortens. Also, a comparison of the stretching force constant calculated for the Si—O bonds of the orthosilicic and pyrosilicic acid molecules indicates that the force constant of the Si—O bond of a disiloxy unit is similar to that of a silanol (Si—OH) group. This concurs with previous studies of the vibrational spectra of siloxanes and silanols (Voronkov et al., 1975).

Finally, linear regression analyses of the stretching and bending force constants for the disiloxy unit as functions of $f_s$ yielded 0.99 for the coefficient of determination, which indicates that 99 percent of the variation of each force constant can be explained in terms of a linear dependence on the fractional s character of the hybrid AOs of the bridging oxygen.

## C.  SiOSi Angle Distribution

In order to explore the factors that account for the distribution of the angle in the siloxanes and silica, we employed the potential energy surface for $H_6Si_2O_7$ calculated by Meagher et al. (1980) as a function of the SiOSi angle and the bridging bond length. The surface depicted in Fig. 6 shows a long, narrow valley surrounded on three sides by steep energy barriers with a shallow minimum at a Si—O bond length of 1.60 Å and a SiOSi angle of 140°. It is noteworthy that the minimum energy bond length and angle provided by this surface are in close correspondence with those (SiOSi angle = 144°; $d$(Si—O) = 1.61 Å) measured for $\alpha$-quartz (Levien et al., 1980). In addition, the valley shows a slight curvature which conforms with the curvilinear Si—O bond length–SiOSi angle trend observed in Fig. 3 for coesite. A small energy barrier ($\sim 3\ kT$ at room temperature) between the minimum-energy value at 140° and the minimum-energy value at 180° indicates that a relatively small amount of energy is required to deform the SiOSi angle of $H_6Si_2O_7$ from its minimum energy value to linearity. If the

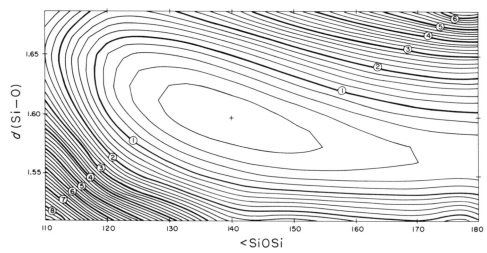

**Fig. 6.** Potential energy surface for $H_6Si_2O_7$ plotted as a function of the bridging bond length, $d(Si{-}O)$, and the SiOSi angle, $\angle SiOSi$. The energy is represented by contours which correspond to increments of 0.001 a.u. (0.6275 kcal/mole) relative to the minimum energy point ($-1091.76678$ a.u.) denoted by the cross. For convenience, the 0.005 a.u. increments (3.1 kcal/mole) are indicated by heavier lines labeled from 1 to 8 in order of increasing energy. Distances and angles are given in Å and degrees, respectively.

bonding forces of the disiloxy units in silica and the siloxanes are similar to those in the molecule, then the angles in these materials may be readily altered to satisfy the local bonding and packing requirements of a continuous structure without excessively destabilizing the resulting structure (Tossell and Gibbs, 1978). In the process, a broad continuum of angles should result, in conformity with the broad range of SiOSi angles observed for compounds and glasses with disiloxy units. Also, because of the steep energy barrier encountered at angles of about 120°, it is doubtful whether structures with SiOSi angles less than $\sim 120°$ will form. Thus, the potential energy surface calculated for $H_6Si_2O_7$ conforms with the broad range of SiOSi angles in solids, the absence of angles narrower than $\sim 120°$ and the curvilinear relationship observed between bond length and angle for coesite. In addition, the width of the valley over the expected range of angles is relatively narrow ($\sim 0.1$ Å), in agreement with the relatively small range of Si—O bond lengths (1.55–1.69 Å) recorded for silicate minerals. The minimum energy Si—O bond length of 1.60 Å and SiOSi angle of 140° requires a Si $\cdots$ Si nonbonded separation for $H_6Si_2O_7$ of 3.00 Å, in close agreement with the lower limit (3.0–3.1 Å) proposed by Glidewell (1977) and O'Keeffe and Hyde (1978) from considerations of nonbonded radii (Newton et al., 1980). Also, the

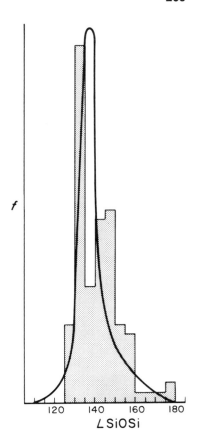

**Fig. 7.**   A comparison of a Boltzmann weighted SiOSi angle distribution function calculated for $H_6Si_2O_7$ with an experimental frequency distribution of the SiOSi angles for gas-phase and molecular crystal siloxanes. The theoretical distribution was calculated with the expression $K \sin(\angle SiOSi)$ $\exp(-\Delta E/kT)$ where $\Delta E = E_{op}(140°) - E_{op}(\angle SiOSi)$, $k$ is Boltzmann's constant, and $K$ is an arbitrary constant used to match the theoretical distribution with that observed for the siloxanes; $E_{op}(140°)$ denotes the optimized energy of $H_6Si_2O_7$ at $140°$ and $E_{op}(\angle SiOSi)$ denotes the energy of the molecule optimized at the remaining SiOSi angles used to prepare Fig. 6.

observation that the terminal bonds of the molecule are about 0.06 Å longer than the bridging bonds conforms with the tendency of Si—OH bonds in solids to be about 0.05 Å longer, on the average, than Si—O bonds.

To learn whether the STO-3G method is capable of generating a distribution of SiOSi angles that conforms with the experimental distribution of the angle in gas phase and molecular crystal siloxanes, a theoretical distribution was calculated with the data used to prepare Fig. 6 and a procedure employed by Newton *et al.* (1979) in their study of the room temperature distribution of the O—H $\cdots$ O angle. In this approach the probability that a given SiOSi angle is adopted is assumed to be proportional to the product of a room temperature Boltzmann factor, $\exp(-\Delta E/kT)$, and a geometrical factor, $\sin(\angle SiOSi)$, where $\Delta E$ is the total energy as a function of $\angle SiOSi$. The resulting theoretical distribution, based on calculated $\Delta E(\angle SiOSi)$ values, is drawn in Fig. 7 and shows a close correspondence with the experimental distribution. Both exhibit peaks at about 135°, and both are asymmetric and skewed to wide angles. On the other hand, while the theoretical curve shows a well-defined single peak, the experimental one shows at least two. As will be discussed in a forthcoming paper (Chakoumakos and Gibbs, 1980), these peaks are due to a relatively large population of angles from siloxanes with six- and eight-membered rings whose angles are fairly restricted in value and range between 125 and 135° and 140 and 150°, respectively (Voronkov *et al.*, 1975). The small peak at 180° arises from structures with disiloxane groups and may be ascribed to a combination of steric effects and an easily deformable SiOSi angle.

### D.  Bonding in Tetrahedral AlOSi Linkages

In addition to SiOSi linkages, a large number of silicates like the feldspars, the feldspathoids, and the zeolites contain tetrahedral SiOAl linkages. In these minerals, both Al and Si play similar roles and have little effect on the structures as a whole except that the Al—O bond (1.75 Å) is somewhat longer on the average than the Si—O bond (1.61 Å) (Smith and Bailey, 1963) and that the AlOSi angle is slightly narrower (137°) on the average than the SiOSi angle (144°) (Tossell and Gibbs, 1978). To determine whether the STO-3G method is capable of reproducing the geometries of the SiOAl bonds in these minerals, Meagher *et al.* (1980) undertook an optimization of the Si—O and Al—O bond lengths and the AlOSi angle for $H_6SiAlO_7^{-1}$. In these calculations, each O—H bond was fixed at 0.96 Å, and each OTO (T = Al, Si) and TOH angle was fixed at 109.5°. The minimum-energy Si—O (1.59 Å) and Al—O (1.70 Å) bond lengths calculated for the SiOAl linkage of the molecule compare well with averaged values (Si—O 1.58 Å, Al—O 1.71 Å) measured for two-coordinated SiOAl linkages in feldspar (Phillips *et al.*, 1973). In addition, the minimum-energy SiOAl angle (131°) is within 1° of the CNDO/2 value (130°) obtained by De Jong and Brown (1980) and within 6° of the average value (137°) found for a larger number of silicates with AlOSi bonds (Liebau, 1961; Brown and Gibbs, 1970; Tossell and Gibbs, 1978).

Since, as noted above, Si and Al play similar roles in the tetrahedral framework of a silicate, a potential energy curve for $H_6SiAlO_7^{-1}$ was calculated as a function of the AlOSi angle to see whether it is similar to the one calculated for $H_6Si_2O_7$. As expected, the shape of the resulting curve shown in Fig. 8b is nearly identical with the one calculated for pyrosilicic acid (Fig. 8a), except that the minimum is slightly deeper (barrier to linearity $\sim 6\,kT$ at room temperature) and shifted to a narrower angle (131°). Nonetheless, the overall shape of the curve is consistent with the broad range of AlOSi angles recorded for silicates with tetrahedral Al (115 to 170°) (Tossell and Gibbs,

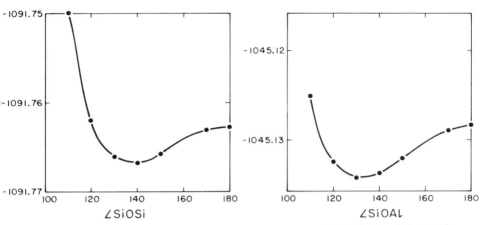

**Fig. 8.**  A comparison of the potential energy curves for $H_6Si_2O_7$ (left) and $H_6SiAlO_7^{1-}$ (right) plotted as a function of the bridging angle.

1978). The close similarity of the curves calculated for $H_6Si_2O_7$ and $H_6SiAlO_7^{-1}$ provides an appreciation of why SiOSi and SiOAl tetrahedral linkages exhibit similar configurations and bonding properties in a silicate. In addition, the curves account for the highly flexible nature of these angles and provide insight into the glass-forming tendencies and the wide range of topologies exhibited by many framework silicates (Tossell and Gibbs, 1978; De Jong and Brown, 1980; Newton and Gibbs, 1980).

## V.  Si—O BRIDGING BOND LENGTH–BOND STRENGTH SUM AND ANGLE VARIATIONS

### A.  The Electrostatic Valence Rule

In 1929 Pauling defined the strength of an electrostatic bond between a bonded cation and anion in a crystal as the ionic valence of the cation divided by its coordination number. With this definition, he then proposed his famous electrostatic valence rule which states that the bond strength sum of the bonds to each anion in a stable ionic crystal should exactly or nearly equal the ionic valence of the anion. Although this rule is satisfied by a number of crystals, Baur (1970) has since demonstrated that bond strength sums to the oxide ions in numerous silicates, borates, phosphates, and sulfates may deviate by as much as 40 percent from the valence of the ion. In addition, he showed by preparing scatter diagrams of the observed bond length vs. the bond strength sum, $p(O)$, for each oxide ion, that a positive correlation obtains between $p(O)$ and lengths of the bonds to the oxide ion. (J. V. Smith first discovered this correlation for the Si—O bond in 1954.)

## B.  Correlation of Si—O Bond Length with $p(O)$ and SiOSi Angle

In a study of the pyrosilicates, Baur (1971) found that the bridging Si—O bond lengths correlate with both $p(O)$ and SiOSi angle. Yet, because a correlation between bond length and angle could not be established for several silica polymorphs and four pyrosilicates [in which $p(O) = 2.0$ for all oxygen atoms], he concluded that the Si—O bond length and angle correlation found for the pyrosilicates follows as a corollary of a cause and effect relation between bond length and $p(O)$. According to Baur's arguments, when a bridging oxygen is involved in a narrow angle, it is more likely to be bonded to other metal atoms in addition to the two Si atoms, thereby increasing its bond strength sum to a value greater than 2.0. On the other hand, when the oxygen atom is involved in a wide angle, he argued that the oxygen, for steric reasons, cannot bond to more than two Si atoms, so that it must receive a bond strength sum of exactly 2.0. In other words, it is the increased number of bonds formed at narrow angles that produces the lengthening of the Si—O bond, and *not* the intrinsic bonding effects associated with a narrow angle. If these arguments are valid, then $p(O)$ should be highly correlated with SiOSi angle. However, a recent linear regression analysis by Swanson *et al.* (1980) has shown that less than 15 percent of the $p(O)$ variation in the pyrosilicates studied by Baur can be explained in terms of a linear dependence on the SiOSi angle. Further, since the Si—O bond lengths of the silica polymorphs [where $p(O) = 2.0$] correlate with SiOSi angle* (Hill and Gibbs, 1979; Baur, 1977), it appears, in contradiction to Baur's (1971, 1977, 1978) conclusions, that the Si—O bond lengths in silicates may depend on SiOSi angle as well as $p(O)$. As noted by Tossell and Gibbs (1978), an empirical approach like Baur's can establish a correlation between two variables, but it cannot establish a cause and effect relationship inasmuch as other variables besides the ones considered can change from one system to another. For example, we cannot dis-

---

* Baur's (1971) failure to find this correlation is due to his inclusion of several bond lengths and angle data in the data set from structures other then the silica polymorphs (Gibbs *et al.*, 1977b; Baur, 1977).

criminate by empirical methods whether the accompanying decrease in the SiOAl angle is due to the substitution of Si by Al, or to a reduction of $p(O)$, or to both factors, or to some other factors. However, by completing *ab initio* calculations for a model system in which $p(O)$ varies over a range of values, we may distinguish between the effects logically and objectively by optimizing the bridging bond lengths and angles for the model systems and analyzing the results as done in the next section.

### C.   Theoretical Correlation between Tetrahedral Si—O Bond Length, $p(O)$, and TOT Angle (T = Si, Al)

As indicated above, Baur (1971) has shown that the Si—O bonds for a large variety of silicates can be ranked with $p(O)$ with longer bonds involving oxygen atoms with larger bond strength sums (Fig. 9). A linear regression analysis of these data yields the equation $d(Si—O) = 0.091\ p(O) + 1.440$ and $r^2$ value of 0.48, indicating that slightly less than half the variation in the bond length can be explained in terms of a linear dependence on $p(O)$. To determine how well the STO-3G method is capable of reproducing this correlation, Meagher *et al.* (1980) and Swanson *et al.* (1980) optimized the bridging bond lengths and angles for the nine molecules listed in Table II. The resulting Si—O bond lengths plotted against $p(O)$ follow the general trend established for the silicates quite well (Fig. 10a). In fact, a linear regression analysis of the data returned an equation $d(Si—O) = 0.08\ p(O) + 1.46$ with $r^2 = 0.68$, which is statistically identical with the one obtained for the silicates (Baur, 1971).

An examination of the data in Table II shows that the Si—O bonds in $[Si(OH)_3]_2O$ and $(SiH_3)_2O$ range in length, for example, from 1.60 to 1.66 Å,

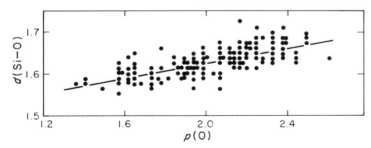

**Fig. 9.**   A scatter diagram of the experimental Si—O bond lengths, $d(Si—O)$, for 293 bonds vs. $p(O)$, the bond strength sum of each bond formed with the oxygen (modified after Baur, 1970).

TABLE II

Optimized Tetrahedral T—O (T = Si, Al) Bridging Bond Lengths, TOT Angles, $p(O)$ Values, and $f_s = 1.0/(1 + \lambda^2)$ where $\lambda^2 = -\sec \angle SiOSi$

| Molecule | $d(Si—O)$ | $d(Al—O)$ | $\angle TOT$ | $p(O)$ | $f_s$ |
|---|---|---|---|---|---|
| $H_6Si_2O^{a,b,c}$ | 1.655 | — | 126° | 2.0 | 0.370 |
| $H_6SiAlO^{1-d}$ | 1.64 | 1.74 | 124 | 1.75 | 0.359 |
| $H_6SiAl(OH)^d$ | 1.70 | 1.84 | 131 | 2.75 | 0.396 |
| $H_6Si_2ONaH_3^{2-e}$ | 1.640 | — | 136 | 2.25 | 0.418 |
| $H_6Si_2OMgH_3^{1-e}$ | 1.670 | — | 127 | 2.50 | 0.376 |
| $H_6Si_2OAlH_3^e$ | 1.693 | — | 123 | 2.75 | 0.353 |
| $H_6Si_2O_7^{e,f}$ | 1.60 | — | 140 | 2.00 | 0.434 |
| $H_6SiAlO_7^{1-e}$ | 1.59 | 1.70 | 130 | 1.75 | 0.391 |
| $H_6SiAlO_6(OH)^e$ | 1.67 | 1.795 | 133 | 2.75 | 0.405 |

[a] Sauer and Zurawski (1979).
[b] Newton and Gibbs (1979).
[c] Meier and Ha (1980).
[d] Swanson et al. (1980).
[e] Meagher et al. (1980).
[f] Newton and Gibbs (1980).

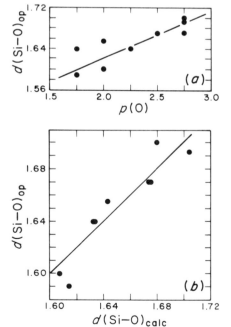

Fig. 10. (a) Optimized Si—O bridging bond lengths, $d(Si—O)_{op}$, calculated for the molecules compiled in Table II, plotted against $p(O)$. The solid line represents the experimental line of $d(Si—O)$ vs. $p(O)$ drawn in Fig. 9. (b) A comparison of $d(Si—O)_{op}$ vs. $d(Si—O)_{calc} = 1.713 + 0.069\ p(O) - 0.563f_s$. The $p(O)$ and $f_s$ values used to calculate $d(Si—0)_{calc}$ are given in Table II.

despite a constant bond strength sum of 2.0 for the bridging oxide ion. On the other hand, if the bridging angles of the molecules are taken into consideration, then an inverse correlation emerges between the bond length and the angle with the bond lengthening as the angle narrows. As observed for the silica polymorphs it appears that part of the bond length variations recorded for the molecules in Table II may be ascribed to variations in the bridging angle. To test this assertion, a multiple linear regression analysis was completed for the Si—O bond length data as a function of $p(O)$ and $f_s$ (Table II). The analysis yielded the equation $d(Si—O) = 1.713 + 0.069p(O) - 0.563f_s$ and indicates that 89 percent of the variation of the Si—O bond length can be explained in terms of a multiple linear dependence on $p(O)$ and $f_s$. In other words, $f_s$ is indicated to make a significant contribution to the regression sum of squares and to explain more than 65 percent of the variation in $d(Si—O)$ left unexplained by $p(O)$. For sake of comparison, the optimized bond lengths are plotted against those calculated with the above equation in Fig. 10b.

Since the $d(Si—O)$ vs. $p(O)$ and $f_s$ variations for the molecules in Table II are similar to those for the silicates, the Si—O bond lengths for several framework silicates were calculated to see how well the above equation serves to rank the experimental bond lengths (Meagher et al., 1980). Using the equation $d(Si—O) = 1.713 + 0.069p(O) - 0.563f_s$ and the experimental values of $f_s$ and $p(O)$, Meagher et al. (1980) calculated the lengths of 123 Si—O bonds for anorthite, $CaAl_2Si_2O_8$ (Wainwright and Starkey, 1971); low microcline, $KAlSi_3O_8$ (Brown and Bailey, 1964); low albite, $NaAlSi_3O_8$ (J. E. Wainwright and J. Starkey, personal communication); low cordierite, $Mg_2Al_4Si_5O_{18}$ (Cohen et al., 1977); and the silica polymorphs $\alpha$-quartz (Zachariasen and Plettinger, 1965), low cristobalite (Dollase, 1965), coesite (Gibbs et al., 1977b), and synthetic low tridymite (Baur, 1977). In the calculation, each structure was assumed to be completely ordered. Notwithstanding the fact that some of the calculated values depart from the experiment by as much as 0.05 Å, the equation serves to rank the observed bond lengths fairly well (Fig. 11), particularly since it was derived with the bond lengths and angles of the molecules in Table II. When a multiple linear regression analysis was undertaken for the 123 experimental bond lengths as a function of $p(O)$ and $f_s$, both variables were found to make a highly significant contribution to the regression sum of squares. However, only 46 percent of the variation in Si—O can be explained in terms of a linear dependence on $p(O)$, while 70 percent of the variation can be explained with the addition of $f_s$ to the regression model. Hence, as indicated by the STO-3G method, it appears that the Si—O bond length in a silicate depends on the SiOSi angle as well as $p(O)$. Finally, an equation based on the theoretical geometries of molecules that serves to rank the experimental bond lengths of a solid is further supporting evidence for the belief that local bonding forces in solids are not significantly different

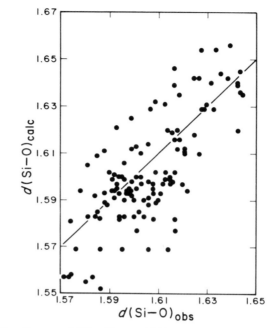

**Fig. 11.**   Scatter diagram of $d(\text{Si}-\text{O})_{\text{calc}} = 1.713 + 0.69p(\text{O}) - 0.563f_s$ vs. $d(\text{Si}-\text{O})_{\text{obs}}$, the observed Si—O bond lengths in anorthite, low albite, low cordierite, $\alpha$-quartz, low cristobalite, coesite, and low tridymite. The solid curve defines the line of perfect agreement.

from those in molecules containing the same atoms with identical coordination numbers.

## VI.   GEOMETRIES OF MOLECULES AND RELATED GROUPS IN SOLIDS

It has been recognized for some time that the shapes of $SiO_4$ and disiloxy units in molecules and solids are similar (Gibbs *et al.*, 1972, 1974, 1977a,b; Tossell and Gibbs, 1977, 1978; Meagher *et al.*, 1979; Glidewell, 1977; O'Keeffe and Hyde, 1979; Newton and Gibbs, 1980; De Jong and Brown, 1980; Newton, 1981). To illustrate this fact, the shape of the disiloxy unit of the gas-phase molecule disilyl ether is compared in Table III with that of the molecule in the three-dimensional molecular crystal. Despite differences between the forces acting on the disiloxy unit of the free molecule and that of the molecular crystal, the shapes of two are practically identical. Indeed, as noted by Barrow

**TABLE III**

**Experimental Bridging Bond Lengths and Angles in Gas-Phase Molecules and Solids**[a]

| Molecules | $d(T-X)$, Å | $\angle TXT$, deg. | Solids | $\langle d(T-X) \rangle$, Å | $\langle \angle TXT \rangle$, deg. |
|---|---|---|---|---|---|
| $(H_3Si)_2O$ (ED)[b] | 1.63 | 144 | $\{(H_3Si)_2O$ (X-ray)[f] | 1.63 | 142 |
|  |  |  | $\{$ Silicates[g] | 1.62 | 144 |
| $(H_3Si)_2NH$ (ED)[c] | 1.73 | 128 | Silicon nitrides[h] | 1.74 | 122 |
| $(H_3Si)_2S$ (ED)[d] | 2.14 | 97 | Silicon sulfides[i] | 2.13 | 109 |
| $(H_3Ge)_2O$ (ED)[e] | 1.77 | 126 | Germanates[j] | 1.76 | 129 |

[a] Modified after Tossell and Gibbs (1978).
[b] Almenningen et al. (1963b)
[c] Rankin et al. (1969)
[d] Almenningen et al. (1963a)
[e] Glidewell et al. (1970)
[f] Barrow et al. (1979)
[g] Tossell and Gibbs (1979).
[h] Nakajima et al. (1980)
[i] Geisinger et al. (1981)
[j] Hill et al. (1977)

*et al.*, (1979), the packing of the molecule in the crystal appears to conform with the geometry of the free molecule rather than vice versa. Moreover, when the bond lengths and angles of the disiloxy unit of the molecule are compared with typical values for a silicate, close agreement obtains notwithstanding the continuous and periodic nature of the structure of the silicate (Table III). As a matter of fact, just as the packing of the molecules in the molecular crystal seems to be governed by the inherent shape of the free molecule, it appears that the configuration adopted by a disiloxy unit in a silicate is prescribed by the same kinds of forces that determine the shape of the group in the molecule. In addition, when the shapes of disiloxy units in siloxanes and silicates are compared, they are found to be statistically identical, with shorter bonds tending to involve the wider angles (Newton and Gibbs, 1980).

The close similarity between the shape of a gas-phase molecule and that of a corresponding unit in a solid is not, however, restricted to siloxanes and silicates. For example, it also holds for the gas-phase molecule disilazane, $(SiH_3)_2NH$, and solid silicon nitrides (Nakajima *et al.*, 1980); the gas-phase molecule disilyl sulphide, $(SiH_3)_2S$, and solid silicon sulfides (Geisinger *et al.*, in preparation); and even the gas-phase molecule digermyl ether $(GeH_3)_2O$, and solid germanium oxides (Hill *et al.*, 1977) (Table III). Because the bond lengths and angles of these molecules and solids are similar, we believe that the local chemical interactions in the two systems are essentially the same and that carefully chosen molecules can therefore be employed as useful models to provide a quantitative account of local geometries in solids (Tossell and Gibbs, 1978).

In view of the close correspondence that obtains between molecules and solids and the ability of STO-3G calculations for molecules like $Si(OH)_4$ and $(Si(OH)_3)_2O$ to give a good account of the shapes of silicate and disiloxy units in silicates, bond lengths have been calculated for 25 first- and second-row-atom oxide and sulfide triangular, tetrahedral, and octahedral molecules. The resulting bond lengths are given in Table III (Newton and Gibbs, 1980; Meagher *et al.*, 1980; Swanson *et al.*, 1980; Gupta *et al.*, 1981; Geisinger *et al.*, in preparation; Gupta and Tossell, 1981). The configurations assumed for the three molecular geometries are shown in Fig. 12. Their point symmetries, neglecting the hydrogen atoms, are $D_{3h}$ ($= 62m$) for the triangular configuration, $T_d$ ($= \overline{4}3m$) for the tetrahedral configuration, and $O_h$ ($= m3m$) for the octahedral configuration. The negative charge on each molecule was partly or completely neutralized by attaching hydrogen atoms to each of its anions. The hydrogen atoms were fixed at 0.96 Å and 1.40 Å from each oxide and sulfide anion, respectfully, and the angle at each anion was fixed at 109.5°. The equilibrium bond length for each molecule was found by fitting a parabola to three points near the minimum in its potential energy curve. The resulting bond lengths are compared with experimental values obtained for

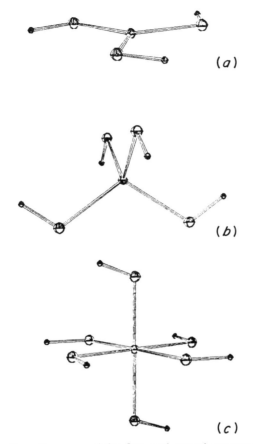

**Fig. 12.** The configuration assumed for first- and second-row-atom oxide and sulfide molecules: (a) triangular molecule ($C_{3h} = \bar{6}$ point symmetry); (b) tetrahedral molecule ($D_{2d} = \bar{4}2m$ point symmetry); and (c) octahedral molecule ($C_i = \bar{1}$ point symmetry). The large spheres represent the anion (oxygen or sulfur), the small one represents hydrogen, and the intermediate-size sphere at the center of each molecule represents the first- and second-row atom. No significance is attached to the relative sizes of the spheres.

the corresponding polyhedra in solids in Table IV and in Fig. 13. The absolute mean error in the agreement between the experimental and theoretical values for the 25 molecules is 0.05 Å. On the one hand, the bond lengths calculated for the neutral molecules show the best agreement (absolute mean error of 0.02 Å). On the other hand, the bonds calculated for the positively charged molecules or for ones with a negative charge of two to five tend to show the largest errors (mean square error of 0.07 Å). It is not clear whether these

TABLE IV

**A Comparison of Bond Lengths (Å) Optimized for Triangular, Tetrahedral, and Octahedral Molecules with Corresponding Bond Lengths in Solids**

| | $d(M\text{—}O)_{op}$ | $d(M\text{—}O)_{exp}$ | | $d(M\text{—}O)_{op}$ | $d(M\text{—}O)_{exp}$ |
|---|---|---|---|---|---|
| **Triangular $M(OH)_3^{n-}$ molecules ($C_{3h}$)** | | | | | |
| $Be(OH)_3^{1-}$ | $1.52^a$ | $1.52^b$ | $C(OH)_3^{1+}$ | $1.33^a$ | $1.27^b$ |
| $B(OH)_3$ | $1.37^a$ | $1.37^b$ | $N(OH)_3^{2+}$ | $1.35^a$ | $1.23^b$ |
| **Tetrahedral $M(OH)_4^{n-}$ molecules ($D_{2d}$)** | | | | | |
| $Li(OH)_4^{3-}$ | $2.06^c$ | $2.01^b$ | $Na(OH)_4^{3-}$ | $2.12^a$ | $2.34^b$ |
| $Be(OH)_4^{2-}$ | $1.63^c$ | $1.64^d$ | $Mg(OH)_4^{2-}$ | $1.85^a$ | $1.84^b$ |
| $B(OH)_4^{1-}$ | $1.48^c$ | $1.48^d$ | $Al(OH)_4^{1-}$ | $1.72^a$ | $1.75^d$ |
| $C(OH)_4$ | $1.43^c$ | $1.40^e$ | $Si(OH)_4$ | $1.65^g$ | $1.62^d$ |
| $N(OH)_4^+$ | $1.44^c$ | $1.39^f$ | | | |
| **Octahedral $M(OH)_6^{n-}$ molecules ($C_2$)** | | | | | |
| $Na(OH)_6^{5-}$ | $2.36^a$ | $2.37^b$ | $Al(OH)_6^{3-}$ | $1.88^a$ | $1.88^b$ |
| $Mg(OH)_6^{4-}$ | $2.02^a$ | $2.07^b$ | $Si(OH)_6^{2-}$ | $1.77^a$ | $1.75^b$ |

| | $d(M\text{—}S)_{op}$ | $d(M\text{—}S)_{exp}$ | | $d(M\text{—}S)_{op}$ | $d(M\text{—}S)_{exp}$ |
|---|---|---|---|---|---|
| **Tetrahedral $M(SH)_4^{n-}$ molecules ($D_{2d}$)** | | | | | |
| $Li(SH)_4^{3-}$ | $2.51^h$ | $2.44^i$ | $Na(SH)_4^{3-}$ | $2.57^h$ | $2.81^i$ |
| $Be(SH)_4^{2-}$ | $2.09^h$ | $2.09^i$ | $Mg(SH)_4^{2-}$ | $2.29^h$ | $2.42^i$ |
| $B(SH)_4^{1-}$ | $1.91^h$ | $1.94^i$ | $Al(SH)_4^{1-}$ | $2.17^h$ | $2.26^i$ |
| $C(SH)_4$ | $1.84^h$ | | $Si(SH)_4$ | $2.13^h$ | $2.13^i$ |

[a] Swanson *et al.* (1980).
[b] Shannon and Prewitt (1969).
[c] Gupta *et al.* (1981).
[d] Griffen and Ribbe (1979).
[e] Mijlhoff *et al.* (1973).
[f] Jansen (1979).
[g] Newton and Gibbs (1980).
[h] Geisinger *et al.* (in preparation).
[i] Shannon (1981).

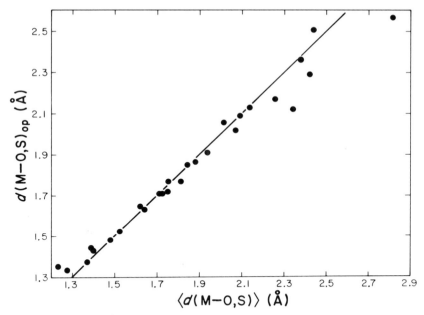

**Fig. 13.** A comparison of the optimized bond lengths, $d(M\!-\!O, S)_{op}$, for the first and second row atom (M) oxide and sulfide molecules given in Table IV vs. $\langle d(M\!-\!O, S)\rangle$, the average experimental bond length taken from oxide and sulfide solids. The configurations assumed for the calculations are given in Fig. 12.

errors are due to an inherent defect of the model or to extraneous electron–electron repulsions as suggested by Gupta and Tossell (1981). In all likelihood, the errors probably indicate that bond lengths calculated for charged molecules do not correspond with values in solids where the overall system is neutral. Nevertheless, the correlation between the theoretical and experimental bond lengths is well developed $(r = 0.98)$ and indicates once again that the local bonding forces in a molecule and a chemically similar solid are quite similar.

## VII.   CONCLUSIONS

A STO-3G molecular orbital structure determination of $Si(OH)_4$ indicates that the shape of the molecule is very similar to that of $SiO_3(OH)^{3-}$ and $SiO_2(OH)_2^{2-}$ polyhedra in solid silicate hydrates. Good agreement obtains between calculated and observed breathing frequencies for the molecule and a silicate anion in aqueous solution. In addition, STO-3G-generated potential energy curves have been used to determine the minimum-energy bridging bond lengths and angle for the pyrosilicic acid molecule. Although obtained for a molecule, the curves conform with (1) bond length and angle trends observed for the silica polymorphs, (2) the force constant–angle trends observed for the siloxanes, (3) the broad range of bridging angles observed for solid and vitreous silica, silicates, and siloxanes, and (4) the nonbonded Si $\cdots$ Si separations observed for the silica polymorphs. The observed distribution of the SiOSi angles for molecular crystal siloxanes is shown to conform reasonably well with a theoretical distribution calculated from the potential energy function of $H_6Si_2O_7$ assuming a Boltzmann distribution.

The close correspondence of the potential energy vs. bridging angle curves for $H_6SiAlO_7^{-}$ [1] and $H_6Si_2O_7$ provides insight into why Si and Al play similar roles in the tetrahedral polymer of a silicate. The glass-forming properties and the wide range of topologies exhibited by these silicates are ascribed to the highly flexible nature of the bridging angle. A multiple linear regression analysis of bond lengths and angles obtained with the *ab initio* method for a number of molecules indicates that the resulting bond-length variations depend on both the bridging angle and bond strength sum variations of the bridging oxygen. A regression equation calculated in the analysis was found to order the experimental bond lengths for the framework silicates fairly well. Finally, a close geometric similarity observed for a number of oxide and sulfide molecules and comparable units in solids indicates that the local bonding forces in these molecules and solids are nearly the same.

### ACKNOWLEDGMENTS

We are pleased to acknowledge the National Science Foundation and the National Science and Engineering Research Council for funding this study with grants EAR77-23114 (with P. H. Ribbe) and NSERC 67-7061, respectively. This study was also supported by the Division of Basic Energy Sciences of the U.S. Department of Energy. Bryan Chakoumakos, Jim Downs, and Karen Geisinger read the manuscript and made several helpful suggestions for its improvement. We thank Ramonda Haycocks for typing the manuscript and Sharon Chiang for drafting the figures.

### REFERENCES

Almenningen, A. K., Hedberg, K., and Seip, R. (1963a). *Acta Chem. Scand.* **17**, 2264–2270.
Almenningen, A. K., Bastiansen, O., Ewing, V., Hedberg, K., and Traetteberg, M. (1963b). *Acta Chem. Scand.* **17**, 2455–2460.
Barrow, M. J., Ebsworth, E. A. V., and Harding, M. M. (1979). *Acta Crystallogr., Sect. B* **B35**, 2093–2099.
Bartell, L. S., Su, L. S., and Yow, H. (1970). *Inorg. Chem.* **9**, 1903–1912.

Basile, L. J., Ferraro, J. R., LaBonville, P., and Wall, M. C. (1973). *Coord. Chem. Rev.* **11**, 21–69.

Baur, W. H. (1970). *Trans. Am. Crystallogr. Assoc.* **6**, 129–154.

Baur, W. H. (1971). *Am. Mineral.* **56**, 1573–1599.

Baur, W. H. (1977). *Acta Crystallogr., Sect. B* **B33**, 2615–2619.

Baur, W. H. (1978). *Acta Crystallogr., Sect. B* **B34**, 1751–1756.

Binkley, J. S., Whiteside, R., Haribaran, P. C., Seeger, R., Hehre, W. J., Lathan, W. A., Newton, M. D., Ditchfield, R., and Pople, J. A. (1978). "Gaissian 76—An *Ab Initio* Molecular Orbital Program." Quantum Program Chemistry Exchange, Bloomington, Indiana.

Brown, B. E., and Bailey, S. W. (1964). *Acta Crystallogr.* **17**, 1391–1400.

Brown, G. E., and Gibbs, G. V. (1970). *Am. Mineral.* **55**, 1587–1607.

Brown, G. E., Gibbs, G. V., and Ribbe, P. H. (1969). *Am. Mineral.* **54**, 1044–1061.

Cannillo, E., Rossi, G., and Ungaretti, L. (1968). *Accad. Naz. Dei Lincei* **45**, 399–414.

Chakoumakos, B. C., and Gibbs, G. V. (1980). *Geol. Soc. Am. Abstr.* **12**, 400.

Chelikowsky, J. R., and Schluter, M. (1977). *Phys. Rev. B: Solid State* [3] **15**, 4020–4029.

Chiari, G., and Gibbs, G. V. (1974). *Atti Accad. Sci. Torino* **108**, 879–901.

Cohen, J. P., Ross, F. K., and Gibbs, G. V. (1977). *Am. Mineral.* **62**, 67–68.

Collins, G. A. D., Cruickshank, D. W. J., and Breeze, A. (1972). *J. Chem. Soc., Farday Trans. 2* **68**, 1189–1195.

Collins, J. B., Schleyer, P., Binkley, J. S., and Pople, J. A. (1976). *J. Chem. Phys.* **64**, 5141–5151.

Coulson, C. A. (1961). "Valence." Oxford Univ. Press, London and New York.

De Jong, B. H. W. S., and Brown, G. E. (1980). *Geochim. Cosmochim. Acta* **44**, 491–511.

Dewar, M. J. S., and Schmeising, H. N. (1960). *Tetrahedron* **11**, 96–120.

Dikov, Yu. P., Debolsky, E. I., Romashenko, Yu. N., Dolin, S. P., and Levin, A. A. (1977). *Phys. Chem. Miner.* **1**, 27–41.

Dollase, W. A. (1965). *Z. Kristallogr.* **121**, 369–377.

Fowler, W. B., Schneider, P. M., and Calabrese, E. (1978). *In* "The Physics of SiO$_2$ and its Interfaces" (S. T. Pantelides, ed.), pp. 70–74. Pergamon, Oxford.

Gibbs, G. V., Hamil, M. M., Louisnathan, S. J., Bartell, L. S., and Yow, H. (1972). *Am. Mineral.* **57**, 1578–1613.

Gibbs, G. V., Louisnathan, S. J., Ribbe, P. H., and Phillips, M. W. (1974). *In* "The Feldspars" (W. S. MacKenzie and J. Zussman, eds.), pp. 49–67. Manchester Univ. Press, Manchester.

Gibbs, G. V., Chiari, G., Louisnathan, S. J., and Cruickshank, D. W. J. (1976). *Z. Kristallogr., Kristallgeom., Kristallphys., Kristallchem.* **143**, 166–176.

Gibbs, G. V., Meagher, E. P., Smith, J. V., and Pluth, J. J. (1977a). *Acs Symp. Ser.* **40**, 19–29.

Gibbs, G. V., Prewitt, C. T., and Baldwin, K. J. (1977b). *Z. Kristallogr.* **145**, 102–123.

Gilbert, T. L., Stevens, W. J., Schrenk, H., Yoshimine, M., and Bagus, P. S. (1973). *Phys. Rev. B: Solid State* [3] **8**, 5977–5998.

Glasser, L. S. Dent., and Jamieson, R. B. (1976). *Acta Crystallogr., Sect. B* **B32**, 705–710.

Glidewell, C. (1977). *Inorg. Nucl. Chem. Lett.* **1** , 65–68.

Glidewell, C., Rankin, D. W. H., Robiette, A. G., and Sheldrick, G. M. (1970). *J. Chem. Soc. A* pp. 315–317.

Griffen, D. T., and Ribbe, P. H. (1979). *Neues Jahrb. Mineral., Abh.* **137**, 54–73.

Griscom, D. L. (1977). *J. Non-Cryst. Solids* **24**, 155–234.

Gupta, A., and Tossell, J. A. (1981) *Phys. Chem. Miner.* (submitted for publication).

Gupta, A., Swanson, D. K., Tossell, J. A., and Gibbs, G. V. (1981). *Am. Mineral.*

Harrison, W. A. (1978). *In* "The Physics of SiO$_2$ and its Interfaces" (S. T. Pantellides, ed.), pp. 105–110. Pergamon, Oxford.

Hehre, W. J., Steward, R. F., and Pople, J. A. (1969). *J. Chem. Phys.* **51**, 2657–2664.

Hehre, W. J., Ditchfield, R., and Pople, J. A. (1970). *J. Chem. Phys.* **53**, 932–935.

Hermann, F., McLean, A. D., and Nesbet, R. K., eds. (1973). "Computational Methods for Large Molecules and Localized States in Solids." Plenum, New York.

Hill, R. J., and Gibbs, G. V. (1979). *Acta Crystallogr. Sect. B* **B35**, 25–30.

Hill, R. J., Louisnathan, S. J., and Gibbs, G. V. (1977). *Aust. J. Chem.* **30**, 1673–1684.

Hill, R. J., Gibbs, G. V., and Peterson, R. C. (1979). *Aust. J. Chem.* **32**, 231–241.

Ilers, R. K. (1955). "The Colloid Chemistry of Silica and Silicates." Cornell Univ. Press, Ithaca, New York.

Ilers, R. K. (1979). "The Chemistry of Silica." Wiley, New York.

Jansen, M. (1979). *Angew. Chem., Int. Ed. Engl.* **18**, 698–699.

Konnert, J. H., and Appleman, D. E. (1978). *Acta Crystallogr. Sect. B* **B34**, 391–403.

Lager, G. A., and Gibbs, G. V. (1973). *Am. Mineral.* **58**, 756–764.

Lazarev, A. M., Poiker, K. R., and Tenisheva, R. F. (1967). *Dokl. Akad. Nauk SSSR* **175**, 1322–1324.

Levien, L., and Prewitt, C. T. (1981). *Am. Mineral.* **66**, 324–333.

Levien, L., Prewitt, C. T., and Weidner, D. J. (1980). *Am. Mineral.* **65**, 920–930.

Liebau, F. (1961). *Acta Crystallogr.* **14**, 1103–1109.

Liebau, F. (1972). *In* "Handbook of Geochemistry" (K. H. Wedepohl ed.), Vol. II/3, 14-A-1–14-A-32. Springer-Verlag, Berlin and New York.

Louisnathan, S. J., and Gibbs, G. V. (1972a). *Mater. Res. Bull.* **7**, 1281–1292.

Louisnathan, S. J., and Gibbs, G. V. (1972b). *Am. Mineral.* **57**, 1614–1642.

Louisnathan, S. J., Hill, R. J., and Gibbs, G. V. (1977). *Phys. Chem. Miner.* **1**, 53–69.

McLarnan, T. J., Hill, R. K., and Gibbs, G. V. (1979). *Aust. J. Chem.* **32**, 949–959.

Meagher, E. P., Tossell, J. A., and Gibbs, G. V. (1979). *Phys. Chem. Miner.* **4**, 11–21.

Meagher, E. P., Swanson, D. K., and Gibbs, G. V. (1980). *Trans. Am. Geophys. Union, EOS* **61**, 408.

Meier, R., and Ha, T. K. (1980). *Phys. Chem. Miner.* **6**, 37–46.

Mijlhoff, F. C., Giese, H. J., and Van Schaick, E. J. M. (1973). *J. Mol. Struct.* **20**, 393–401.

Mulliken, R. S. (1932). *Phys. Rev.* [2] **41**, 49–71.

Mulliken, R. S. (1935). *J. Chem. Phys.* **3**, 573–585.

Nakajima, Y., Swanson, D. K., and Gibbs, G. V. (1980). *Trans. Am. Geophys. Union, EOS* **61**, 408.

Nakamoto, K. (1970). "Infrared Spectra of Inorganic and Coordination Compounds," 2nd ed. Wiley, New York.

Newton, M. D. (1981). *In* "Structure and Bonding in Crystals" (M. O'Keeffe and A. Navrosky, eds.). Academic Press, New York.

Newton, M. D., and Gibbs, G. V. (1979). *Trans. Am. Geophys. Union, EOS* **60**, 415.

Newton, M. D., and Gibbs, G. V. (1980). *Phys. Chem. Miner.* **6**, 221–246.

Newton, M. D., Jeffrey, G. A., and Takagi, S. (1979). *J. Am. Chem. Soc.* **101**, 1997–2002.

Newton, M. D., O'Keeffe, M., and Gibbs, G. V. (1980). *Phys. Chem. Miner.* **6**, 305–312.

O'Keeffe, M., and Hyde, B. G. (1978). *Acta Crystallogr., Sect. B* **B34**, 27–32.

O'Keeffe, M., and Hyde, B. G. (1979). *Trans. Am. Crystallogr. Assoc.* **15**, 65–75.

Pantelides, S. T., and Harrison, W. A. (1976). *Phys. Rev. B: Solid State* [3] **13**, 2667–2691.

Pauling, L. (1929). *J. Am. Chem. Soc.* **51**, 1010–1026.

Peterson, R. C., Hill, R. J., and Gibbs, G. V. (1979). *Can. Mineral.* **17**, 703–711.

Phillips, M. W., Ribbe, P. H., and Gibbs, G. V. (1973). *Am. Mineral.* **58**, 495–499.

Pitzer, R. M., and Merrifield, D. P. (1970). *J. Chem. Phys.* **52**, 4782–4787.

Pulay, P. (1969). *Mol. Phys.* **17**, 197–204.

Rankin, D. W. H., Robiette, A. G., Sheldrick, G. M., and Sheldrick, W. S. (1969). *J. Chem. Soc. A* pp. 1224–1227.

Sauer, J., and Zurawski, B. (1979). *Chem. Phys. Lett.* **65**, 587–591.

Schaefer, N. F., ed. (1977). "Modern Theoretical Chemistry," Vol. 4. Plenum, New York.

Schmiedekamp, A., Cruickshank, D. W. J., Skaarup, S., Pulay, P., Hargittai, I., and Boggs, J. E. (1979). *J. Am. Chem. Soc.* **101**, 2002–2010.

Shannon, R. D. (1981). In "Structure and Bonding in Crystals" (M. O'Keeffe and A. Navrosky, eds.). Academic Press, New York.

Shannon, R. D., and Prewitt, C. T. (1969). *Acta Crystallogr., Sect. B* **B25**, 925–946.

Smith, J. V. (1954). *Am. Mineral.* **38**, 643–661.

Smith, J. V., and Bailey, S. W. (1963). *Acta Crystallogr.* **16**, 801–811.

Stevens, R. M. (1970). *J. Chem. Phys.* **52**, 1397–1402.

Swanson, D. K., Meagher, E. P., and Gibbs, G. V. (1980). *Trans. Am. Geophys. Union, EOS* **61**, 409.

Tossell, J. A. (1973). *J. Phys. Chem. Solids* **34**, 307–319.

Tossell, J. A. (1975). *J. Am. Chem. Soc.* **97**, 4840–4844.

Tossell, J. A. (1979). *Trans. Am. Crystallogr. Assoc.* **15**, 47–63.

Tossell, J. A., and Gibbs, G. V. (1977). *Phys. Chem. Miner.* **2**, 21–57.

Tossell, J. A., and Gibbs, G. V. (1978). *Acta Crystallogr., Sect. A* **A34**, 463–472.

Tossell, J. A., Vaughan, D. J., and Johnson, K. H. (1973). *Chem. Phys. Lett.* **20**, 329–334.

Voronkov, M. G., Yuzhelevskii, Yu. A., and Milehkevich, V. P. (1975). *Russ. Chem. Rev.* (*Engl. Transl.*) **44**(4), 355–372.

Wainwright, J. E., and Starkey, J. (1971). *Z. Kristallogr.* **133**, 75–84.

Wallmeier, H., and Kutzelnigg, W. (1979). *J. Am. Chem. Soc.* **101**, 2804–2814.

Yip, K. L., and Fowler, W. B. (1974). *Phys. Rev. B: Solid State* [3] **10**, 1400–1408.

Zachariasen, W. H., and Plettinger, H. A. (1965). *Acta Crystallogr.* **18**, 710–714.

# 10

## The Role of Nonbonded Forces in Crystals

### M. O'KEEFFE and B. G. HYDE

|      |                                                          |     |
|------|----------------------------------------------------------|-----|
| I.   | Introduction                                             | 227 |
| II.  | Structures Derived from That of Cristobalite            | 229 |
| III. | Nonbonded or "One-Angle" Atomic Radii                   | 237 |
| IV.  | Applications of Nonbonded Radii in Crystal Chemistry    | 239 |
| V.   | Nonbonded Interaction Potentials                        | 244 |
|      | A.   The Interaction Energy                             | 244 |
|      | B.   Force Constants                                    | 246 |
|      | C.   Force Constants and Nonbonded Replusions in Diamond | 247 |
|      | D.   Force Constants in Silicates                       | 250 |
| VI.  | What Is the Size of an Atom?                            | 251 |
|      | References                                              | 253 |

## I. INTRODUCTION

In this chapter we are concerned with the "size" of an atom (or ion) in a crystal. In particular we are concerned with how atom $\cdots$ atom interactions affect structural detail such as bond lengths, bond angles, and coordination numbers. The emphasis is on other than nearest neighbor (i.e. on "nonbonded") interactions in crystals (such as silicates) that until recently (cf. Chapters 8, 9) were usually discussed within the framework of the ionic model.

There are many facts that do not fit into the generally accepted ionic model—with its basis of ions of fixed radius, cation coordination number based on radius ratio rules, and so on. Some of these have been detailed by O'Keeffe (1977). Such inconvenient facts are often ignored in standard texts which discuss structures based on "close-packed" anion arrays. We follow Wells (1975) in also pointing out that (generally) emphasis on the coordination of cations has led to a tendency to underestimate the importance of the

Structure and Bonding in Crystals, Vol. I
Copyright © 1981 by Academic Press, Inc.
All rights of reproduction in any form reserved.
ISBN 0-12-525101-7

coordination of anions. (The two coordination numbers are not, of course, independent: in a binary compound they are related via the stoichiometric ratio anions/cations.) It should also be added that, in what follows, the use of the terms anion and cation are without prejudice: they in no way imply an acceptance of the electrostatic concept of charged ions, implicit in the "ionic model." The terms, now conventional, simply imply less and more electropositive atom species. Treating atoms simply as atoms (and not worrying about charges) makes it possible to slip smoothly from the geometry of nonmolecular crystals to that of small gas molecules (Chapters 8, 9).

Our aim here is to examine the known geometry of many nonmolecular crystal structures to see whether a reconsideration of already available facts can lead to a model that succeeds in at least some places where the ionic model fails—perhaps particularly in attempting to understand coordination numbers in topologically simple structures. It is not to develop a comprehensive and rigorous theory, but rather to develop notions, as simple as those in the ionic model, that a crystal chemist can use as simple rules of thumb to cope with the vast variety of structure types, and puzzling details of their structure. (Examples are: "Why is Si usually four-coordinate in oxides—but not always?"; "Why is the structure of cristobalite not $C9$ type—as the ionic model would lead us to expect?") Sophisticated explanations may be available (and indeed are in a bewildering profusion!), but something simple is desirable, if possible.

The early history of the ionic model is intriguing: Bragg (and Wyckoff and Niggli) first ($\sim 1920$) gave tables of radii with large cations and small anions (e.g., Bragg, 1920). The basis of the tables was virtually constant distance (in various structures) between the centers of two neighboring atoms of given kinds. Its justification was to use this fact as an aid to solving unknown structures: it was a constraint on the number of possible structure models that needed to be considered. (And so Bragg warned, more than once, that "This additive law is only intended to be regarded as a working approximation, an aid to the analysis of complex structures"—advice that everyone subsequently ignored.) It is worth recalling that the observation is that of constant bond length, not of constant ion size. The former we accept; the latter is now known not to be true (cf. Chapters 13, 16).

Bragg's radii gave a good approximation to bond lengths in all types of crystal (covalent, ionic, metallic). The scale was fixed by the bond lengths in elemental crystals such as diamond, and the sizes of atoms reflected Lothar Meyer's curve of atomic volumes. Using the greatly enlarged data base available to him, Slater (1965) refined Bragg's radii, developing a set which we will refer to as Bragg–Slater radii. These radii are in general quite different from the ionic radii very much in use by crystal chemists.

The later, now accepted, ionic radii are based on the notion of anion–anion contact. The possible alternative, cation–cation contact, has not been considered. Furthermore, the spherical ion was clearly assumed to present the same radius to its nearest neighbor atom as to its more distant neighbors. This is in stark contrast to the approach to organic structures where nearest-neighbor (covalent) radii are markedly less than minimum next-nearest-neighbor distances, which are at least only slightly less than the sum of *much larger* van der Waals radii. In our earlier work (O'Keeffe and Hyde, 1978a,b), we did not emphasize this vital point, which now seems to be the clue to, and the justification for, our approach to structures. It is now clear that, under the term *configurational analysis*, the organic chemist has been using a similar approach to molecular geometry for a long time. And, only slightly removed, Kitaigorodsky (1973) and others have long used similar ideas to account for the structure and thermodynamics of (organic) molecular crystals. It is a familiar fact in drawings of molecular structures that van der Waals spheres overlap: to a very considerable extent for first-nearest-neighbor atoms, and to a lesser (but still significant) extent for second- and even third-nearest-neighbor atoms (see, for example, Kitaigorodsky, 1973; Pauling, 1960).

In this chapter we review the development and applications of the idea of nonbonded radii in crystal chemistry.

## II.  STRUCTURES DERIVED FROM THAT OF CRISTOBALITE

The structure of the high cristobalite form of $SiO_2$ and some "stuffed" cristobalites such as $KAlO_2$ and $KFeO_2$ were originally reported to consist of a framework of corner-connected $BX_4$ tetrahedra, stoichiometry $BX_2$, of the C9 type; cubic, space group $Fd3m$. Within this framework there are larger interstices ($X_{12}$ truncated tetrahedra) which can accommodate the larger "stuffing" cation A. This is the parent of very many derived structures which play a key role in the development of ideas concerning nonbonded interactions.

A great deal of subsequent work on high cristobalite suggested that the bond angle $B\hat{X}B$ in high cristobalite was less than the 180° obtaining in C9—that $Fd3m$ was the space group of the *average* rather than the real structure. Leadbetter *et al.* (1973), with Wright and Leadbetter (1975), eventually deduced the correct structure. It transpires that the 16 oxygen atoms in the pseudo-$Fd3m$ cell are in a special subset of its 96(h) sites [Si in 8(a)] so that the symmetry is in fact tetragonal, space group $F\bar{4}d2$. Choosing the smaller, more conventional cell of the same symmetry, $I\bar{4}2d$, Si is in 4(a) (0, 0, 0 etc.) and O in 8(d) $(x, \frac{1}{4}, \frac{1}{8},$ etc.) with $x = 0.09$. Figure 1 shows a projection of the

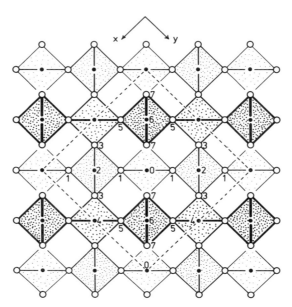

**Fig. 1.** Projection of the $C_9$ structure ($BX_2$) on (001). Small filled circles are B atoms, open circles are $X$ atoms, heights in multiples of $c/8$.

$C9$ structure (proposed for high cristobalite) on (001) of the $Fd3m$ cell. Figure 2 shows two corresponding projections of the real high cristobalite structure, on (001) and (110) of the $I\bar{4}2d$ unit cell [equivalent to (001) and (010) of the $F\bar{4}d2$ and $Fd3m$ supercells].

The difference between the two structures—most clearly seen by comparing Figs. 1 and 2a—is systematic rotation of the $BX_4$ tetrahedra about their $\bar{4}$ axes (while retaining their corner-connectedness) parallel to [001] of both the $Fd3m$ and $I\bar{4}2d$ unit cells. These rotations (or "tilts") break no bonds, but they change the $B\hat{X}B$ bond angles (180° in $C9$) to lower values.

We have examined all such ways of tilting within the limits of the same sized $Fd3m$ parent unit cell. (The variety arises because each tetrahedron has three equivalent $\bar{4}$ axes—six if we take sense into account.) Only three combinations are possible. If the tetrahedron at the origin (Fig. 1) is rotated (clockwise) about [001], then the tetrahedra connected to its four corners must be rotated about

$$
\begin{array}{llll}
\text{(I)} & [00\bar{1}], & [001], & [00\bar{1}]; \\
\text{(II)} & [0\bar{1}0], & [001], & [010]; \\
\text{(III)} & [00\bar{1}], & [001], & [010];
\end{array}
$$

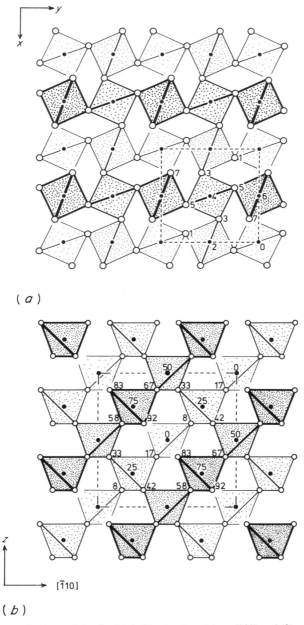

**Fig. 2.**    Two projections of the $\beta$-cristobalite structure ($a$) on (001) and ($b$) on (110) of the $I\bar{4}2d$ unit cell. Heights are ($a$) in multiples of $c/8$ or ($b$) $a[110]/100$.

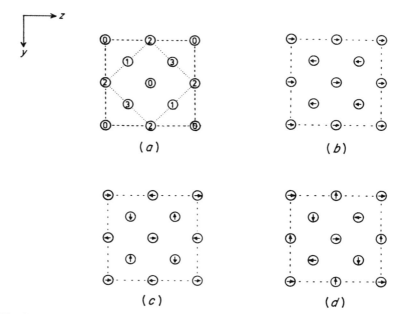

**Fig. 3.** The possible patterns of rotation axes within the $Fd3m$ cell of the $C_9$ structure [projected on (100)] in (a) the centers of the tetrahedra (heights in multiples of $a/4$) are shown; (b), (c), and (d) show the sense and directions of rotations that produce the $I\bar{4}2d$, the $P4_12_12$ and the $Pna2_1$ structures, respectively.

as in Fig. 3. (The three axes in each set are, in order, for tetrahedra at $x/a = -\frac{1}{4}$, 0, and then $+\frac{1}{4}$.)

Sequence I produces the true high cristobalite structure at a rotation angle of $\phi \simeq 19.8°$, when $B\hat{X}B = \theta \simeq 147°$. Sequence II produces the low cristobalite structure (Fig. 4) at $\phi \simeq 23.5°$, $\theta \simeq 146°$ (at R.T.). (This structure is also tetragonal, space group $P4_12_12$ or $P4_32_12$.) In both these cases, "stuffed" derivatives $ABX_2$ are also known. Sequence III produces an orthorhombic unit cell, space group $Pna2_1$. Only "stuffed" derivatives are known in this case (Fig. 5). Full details have been given elsewhere (O'Keeffe and Hyde, 1976) and will not be repeated here.

A large number of known compounds have structures derivable from $C9$ by these tilt sequences. The data for a sequence I selection are shown in Fig. 6, in which the axial ratio of the $I\bar{4}2d$ unit cell is plotted against the anion parameter $x$ (equivalent to tetrahedron tilt angle $\phi$). Examples include "cristobalites" $BX_2$, substituted cristobalites with superstructures $B'B''X_4$ (e.g. $BPO_4$), space group $I\bar{4}$, and "stuffed" cristobalites $ABX_2$. Two especially simple cases may be conceived: that in which the tetrahedra remain perfectly

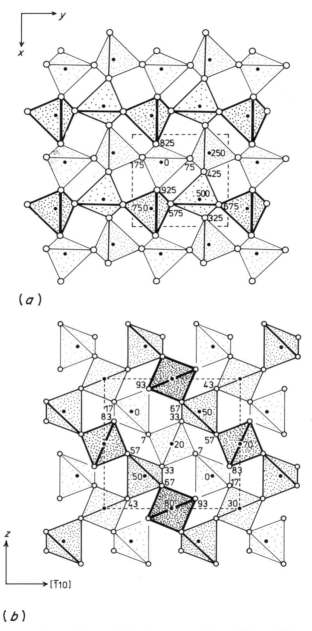

**Fig. 4.**   Two projections of the α-cristobalite structure (*a*) on (001) and (*b*) on (110) on the $P4_12_12$ unit cell.

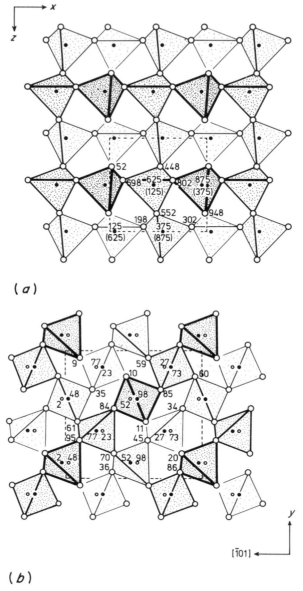

**Fig. 5.**   Two projections of the $Pna2_1$ "collapsed $C_9$ structure" with $\phi = 22\frac{1}{2}°$ (a) on (010) (cf. Figs. 4a and 2b) and (b) on (101) (cf. Figs. 4b and 2a).

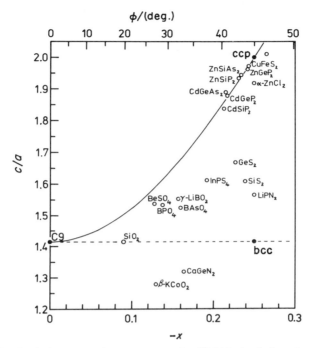

**Fig. 6.**   Graph of $c/a$ versus anion parameter $x$ for $\bar{I}42d$ ($\beta$-cristobalite-related structures). For regular tetrahedra (full line), $x = -(\tan \phi)/4$. The broken line is for constant unit cell shape (distorted tetrahedra).

regular, the cell edge $a$ decreasing ($c$ constant) as $\phi$ increases; and that in which $c$ and $a$ are both constant so that the tetrahedron is tetragonally distorted [by elongating the edges in (001)] as $\phi$ increases. These are shown in Fig. 6 by the full and broken lines respectively. The first ends, at $\phi = 45°$, with a ccp anion array; the second, also at $\phi = 45°$, with a bcc anion array; so that tetrahedra usually lie between the two extreme shapes well known in these arrays. Near the former (cf. Fig. 6), there is a cluster of $ABX_2$ compounds, the prototype of which is chalcopyrite, $CuFeS_2$, normally described as an approximate ccp S array with Cu and Fe ordered into half the tetrahedral interstices, i.e., a superstructure of zinc blende (or sphalerite). The description given here is more precise: a $C9$ $FeS_2$ array collapsed to $\phi \simeq 44.0°$, as in Fig. 7, at which point the truncated tetrahedron of $X_{12}$ around the A site (containing Cu) has collapsed to give a $CuS_4$ tetrahedron (not much larger than that around Fe) plus eight much more distant X atoms. (Its greater precision is perhaps emphasised by the detail that the $FeS_4$ tetrahedron is perfectly regular, while the $CuS_4$ one is slightly squashed along [001]—or extended in

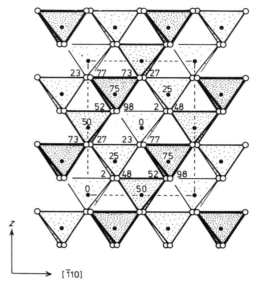

**Fig. 7.** The almost completely "collapsed" $I\bar{4}2d$ structure with regular tetrahedra, projected on (110) (compare with Fig. 2b) in the $ABX_2$ chalcopyrite structure compounds; the A atoms are superimposed on the B atoms in this projection, but with heights differing by 50.

(001)—exactly as expected for a C9 array of perfectly regular $FeS_4$ tetrahedra collapsed almost to $\phi = 45°$.)

In such $ABX_2$ cases it is clear that, for a C9 structure of $BX_2$ scaled by the length of the B—X bond, $l(B—X)$, the degree of collapse (the size of the rotation angle $\phi$) is that required to make the A—X distance exactly that of the appropriate bond length, $l(A—X)$. [If we assume perfectly regular $BX_4$ tetrahedra, $l(A—X)$ depends only on $\phi$. There is only one $\phi$ which will give the correct $l(A—X)$.] This immediately raises the question of what determines the angle $\phi$ (or anion parameter $x$) when there are no A cations in the structure, i.e. in $BX_2$ structures such as high cristobalite, $SiO_2$? This tilt angle, of course, is correlated with the bond angle $B\hat{X}B$. It will be recalled that $\theta \simeq 147°$ in high cristobalite, $\sim 146°$ in low cristobalite. That this is significant is supported by the fact that $\theta = 144°$ in quartz, $\sim 147°$ in silica glass, and is almost always $145°$ to $150°$ in silicate structures. Many explanations have previously been given—for example, $\pi$-bonding between empty d orbitals on the cation and filled p orbitals on the anion (Cruickshank, 1961; Grimm and Dorner, 1975)—but facts such as, for example, that $CsBeF_3$ (Steinfink and Brunton, 1968) has $\theta = 145°$ (with, probably, a similar $Be\hat{F}Be$ angle in the cristobalite form of $BeF_2$), persuaded us that such an explanation was, in general, unlikely. (Being a first row cation, Be, of course, would not have accessible d orbitals.) Our conclusion will be that the large bond angle is largely a consequence of nonbonded repulsions between nearest neighbor Si atoms.

### III.  NONBONDED OR "ONE-ANGLE" ATOMIC RADII

The notion of nonbonded contact between next-nearest-neighbor atoms (in the present context) derives from an original idea of Bartell (1960; see also Bartell, 1968), who used it to explain, quite accurately, the geometry of many (gas phase) organic molecules of the type

$$\begin{array}{ccc} X \diagdown & \diagup X'' & X \diagdown \\ & C{=}C & & C{=}X'' \\ X' \diagup & \diagdown X''' & \text{and} \quad X' \diagup \end{array}$$

(where the X's are H, C, N, O, F, Cl, Br).

No account need be taken of hybridization, hyperconjugation etc.; the only restriction is that there be only *one* bridging atom between the next-nearest neighbors, i.e. only

$$X \overset{\diagup Z \diagdown}{\cdots\cdots} X', \text{ and not } X \overset{\diagup Z \diagdown}{\underset{\diagdown Y \diagup}{\cdots\cdots}} X' \text{ etc.}$$

(Hence the term "one-angle" radius.) The hypothesis is that the $X \cdots X'$ atoms are "in contact"—that they may each be assigned a contact radius $R$ ("nonbonded" or "one-angle" radius), and that the distance between them is equal to the sum of these radii, $d(X \cdots X') = R(X) + R(X')$. These radii $R$ are the same from molecule to molecule.

In the intervening years this concept has been extensively developed for gas molecules by Bartell (e.g., Bartell *et al.*, 1979) and by Glidewell (1975). To come closer to our subject, the latter has shown that it works very well for gas molecules of the type

$$\begin{array}{ccccc} & \diagup O \diagdown & & \diagup NH \diagdown & & \diagup CH_2 \diagdown \\ H_3Si & & SiH_3, & H_3Si & SiH_3, & \text{and} \quad H_3Si & SiH_3 \cdots \end{array}$$

That is, the distance $d(Si \cdots Si) = 2R(Si)$ is virtually independent of the bridging atom. We found that it worked equally well for Si—O—Si in nonmolecular crystals—silicas and silicates—as well as in silica glasses (O'Keeffe and Hyde, 1978a). The distribution of distances $d(Si \cdots Si)$ between two silicon atoms bridged by a single oxygen atom (Fig. 8) was narrow (see also O'Keeffe and Hyde, 1979), and centered around 3.06 Å, giving $R(Si) = 1.53$ Å. It is interesting that Bragg had already noticed this: comparing the

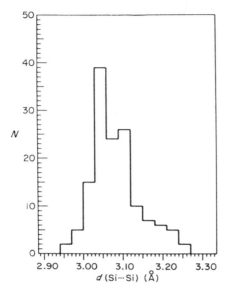

**Fig. 8.** Histogram showing the distribution of 141 nearest-neighbor $Si \cdots Si$ distances in various silicates and silicas.

five different conformations of $SiO_3^{2-}$ chains occurring in three different mineral structures [clinoenstatite, $MgSiO_3$, pigeonite, $(Ca,Mg,Fe)SiO_3$, and diopside, $CaMg(SiO_3)_2$; Bragg and Claringbull, 1965] he pointed out that "The Si—Si distance in a single chain, 3.05 Å, is the same in all three structures." But the significance of this observation was not pursued.

The argument can be made much more compelling. It transpires that similarly constant (though different) contact distances are observed in crystal structures containing

$$P\diagdown^{O}\diagup B, \quad B\diagdown^{O}\diagup B, \quad Ge\diagdown^{O}\diagup Ge \text{ etc.}$$

one-angle situations. Furthermore, in the mixed cases such as B—O—Si, $d(B \cdots Si) = R(B) + R(Si)$, and so on. That is, the one-angle radii are not only pretty constant for a given atom (and independent of the bridging atom), but they are also additive. The constancy is about as good as the constancy of bond lengths. Contact distances $d = 2R$ therefore do vary a little (usually $\ngtr \sim \pm 3\%$).

There are other observations that serve to substantiate these ideas. Thus McDonald and Cruickshank (1967), reporting a careful determination of the crystal structure of $Na_4P_2O_7 \cdot 10H_2O$, commented that in the pyrophosphate group ($P_2O_7^{4-}$, composed of two corner-connected $PO_4$ tetrahedra) the thermal parameters led to the following conclusion:

> The values of $U_{ij}$ are not consistent with appreciable rigid body vibrations of the whole $P_2O_7$ group, but suggest that the two $PO_4$ tetrahedra vibrate individually as rigid bodies, for the phosphorus atoms show small and approximately isotropic vibrations, while the oxygen atoms are markedly anisotropic. The bridge oxygen, O(1), has the largest individual $U_{ij}$. This is normal to the P—O—P plane, suggesting a vibration in which the $[P_2O_7]$ group "folds" along the symmetry axis while the phosphorus atoms remain stationary. The $U_{ij}$ of the terminal oxygen atoms are consistent with a vibration of this kind.

This clearly indicates that while the tetrahedra are "folding," the large phosphorus atoms are "rolling" on each other, $[d(P \cdots P) = 2.924$ Å, corresponding exactly to $R(P) = 1.46$ Å$]$.

To summarize up to this point:

1.   The simple hypothesis of nonbonded radii appears to work rather well in accounting for the geometry of structures in which there is only a single bridging anion between nearest neighbor cations. We have restricted considerations to mainly "tetrahedral" structures involving only atoms in the first few rows of the periodic table. (Heavier atoms often have bonds so long compared with their nonbonded radii that cation–cation contact does not occur at normal bond angles.) It works equally well for gas molecules and nonmolecular crystals.

2.   Nonbonded radii do not vary with the bridging atom, though this controls both the bond lengths and the bond angle.

3.   They are additive, i.e., $d(A \cdots B) = \frac{1}{2}[d(A \cdots A) + d(B \cdots B)]$.

4.   The question as to whether a structure (molecular or crystal) is ionic or covalent has not arisen.


### IV.   APPLICATIONS OF NONBONDED RADII IN CRYSTAL CHEMISTRY

Figure 9 shows some conformations of rings of three or four $SiO_4$ tetrahedra joined by sharing corners. For a "standard" bond length $l(Si—O) = 1.60$ Å and for regular tetrahedra, the $Si \cdots Si$ distances are as indicated in the figure. The value of $2R(Si) = 3.06$ Å implies that "four-rings" are appropriate, but that "three-rings" will be strained; and that consequently the former will be common and the latter relatively uncommon. This is in accord with observation in silicate structures; note that when three-rings do occur (as in benitoite) they are associated with anomalously long (strained) Si—O bonds. In germanates, by contrast, three-rings are much more common—this reflects the smaller value of $R(Ge)/l(Ge—O)$. The frequently occurring $Si_8O_{20}$ group of corner-sharing $SiO_4$ tetrahedra is derived from two joined

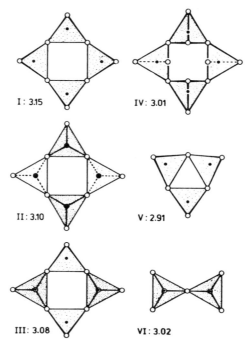

**Fig. 9.** Various symmetrical configurations of corner-connected $SiO_4$ tetrahedra (small filled circles are silicon). The numbers are Si $\cdots$ Si distances in Å for regular tetrahedra and a Si—O bond length of 1.60 Å.

four-rings and has all nearest Si $\cdots$ Si close to the optimum value. It is instructive to compare a drawing in which Si is represented by spheres of the nonbonded radius (Fig. 10b) to the conventional polyhedral drawing (Fig. 10a).

We may reconsider why Si is so frequently four-coordinate in oxides (silicas and silicates). It is not, as the radius ratio rule in the classical "ionic model" would have it, because Si is too small to crowd more than four oxygens around it. (Indeed, according to that rule and conventional radii it is too small for even four oxygens.) After all, $SiP_2O_7$ (and several other structures) contains $SiO_6$ octahedra (and they are *not* high pressure structures). Furthermore, $O^{2-}$ and $F^-$ are assigned approximately the same ion radii, but $SiF_6$ octahedra are very common indeed in crystal structures. It is simply that, in $SiO_2$ say, if the coordination number of Si is greater than 4, that of O must be greater than 2; i.e., there must be at least three Si atoms around an oxygen atom. But $OSi_3$ is impossible: it demands SiÔSi bond angles $\theta \not> 120°$, whereas the normal Si—O—Si bond angle is $\sim 146°$. [With a standard bond length

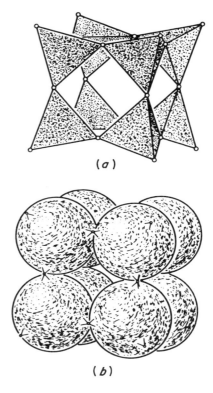

**Fig. 10.** Two representations of a $Si_8O_{20}$ unit: (*a*) the "conventional" one with $SiO_4$ tetrahedra and (*b*) with spheres of radius equal to the nonbonded radius of Si (1.53 Å) centered at the Si positions. Note the barely emergent vertices of the tetrahedra.

$l(Si-O) = 1.60$ Å, $\theta \leqslant 120°$ requires $d(Si \cdots Si) \leqslant 2.77$ Å, compared with the norm of 3.06 Å: much too short.] But, if the stoichiometric ratio anions/cations is large enough, then $SiX_6$ is possible while still maintaining $XSi_2$ coordination of the anions. This is achieved in silicofluorides because the anion valence is only half that of oxygen, and in $SiP_2O_7$ because the presence of higher-valence phosphorus raises the anion/cation stoichiometric ratio. In this last, Si is six-coordinated by oxygen, P is four-coordinated, and oxygen is two-coordinated only: $Si(VI)P_2(IV)O_7(II)$. $P_2O_7$ groups (cf. above) share their "outer" oxygens with six different $SiO_6$ octahedra.

Hence, in cases such as this the structure is constrained by cation–cation repulsion, not, as is normally supposed, anion–anion repulsion. This is consistent with the relative sizes of $R$ for cations and anions in the first few rows (see Table I), but disagrees with conventional wisdom. It means, for instance, that the compression of solids by applying high pressure involves overcoming cation–cation replusions: i.e., it is anion-centered polyhedra rather than cation-centered polyhedra that have domains of stability in $p$–$T$ space. To be

TABLE I

Nonbonded Radius $R$, Bond Length
to Oxygen in Tetrahedral Compounds $l$ (both in Å), and Their
Ratio for Some Atoms

| Atom | $R$ | $l$ | $R/l$ |
|------|-----|-----|-------|
| Li | 1.5 | 1.97 | 0.76 |
| Be | 1.35 | 1.65 | 0.82 |
| B | 1.26 | 1.49 | 0.85 |
| C | 1.25 | — | — |
| N | 1.14 | — | — |
| O | 1.12 | — | — |
| Na | 1.68 | 2.37 | 0.71 |
| Mg | 1.66 | 1.95 | 0.85 |
| Al | 1.62 | 1.77 | 0.92 |
| Si | 1.53 | 1.64 | 0.93 |
| P | 1.46 | 1.55 | 0.94 |
| S | 1.45 | 1.50 | 0.97 |
| Zn | 1.65 | 1.98 | 0.83 |
| Ga | 1.63 | 1.85 | 0.88 |
| Ge | 1.58 | 1.77 | 0.89 |
| As | 1.54 | 1.71 | 0.90 |

specific, the various low-pressure forms of silica transform to stishovite at
high pressure, not because high pressure is necessary to convert $SiO_4$ to $SiO_6$
but because it is necessary to convert $OSi_2$ to $OSi_3$. The former is a consequence of the latter, and not vice versa.

Additional credence for this attitude is provided by data available from
structural studies at high pressures. When forsterite ($Mg_2SiO_4$) is compressed
from 0 to 20 kbar (Hazen, 1978), the next-nearest-neighbor distances change
by: Si $\cdots$ Mg, $-0.2\%$; Mg $\cdots$ Mg, $-0.5\%$; and O $\cdots$ O, $-2.2\%$, suggesting
that the repulsive forces for O $\cdots$ O ($d \simeq 2.6$ Å) are less than for Mg $\cdots$ Mg
($d \simeq 3.0$ Å) and Si $\cdots$ Mg ($d \simeq 2.7$ Å). Similarly, when α-quartz is compressed
(Jorgensen, 1978; d'Amon et al., 1979) the $SiO_4$ tetrahedra remain undistorted:
both $l$(Si—O) and $\theta$(OSiO) remain constant within experimental uncertainty.
Second-nearest-neighbor oxygen distances (fourth-nearest-neighbors in the
structure:

$d$(O $\cdots$ O) decrease from 3.30 Å (at 1 bar) to 2.84 Å (at 68 kbar), a change
of 0.46 Å or $-13.9\%$. However, nearest-neighbor silicon atoms (second-

nearest-neighbor atoms in Si—O—Si links in the network of corner-connected $SiO_4$ tetrahedra) are almost equally spaced, but decrease their separation from $d(Si \cdots Si) = 3.07_3$ Å (at 1 bar) to $2.96_9$ Å (at 68 kbar), a change of only $0.10_4$ Å or $-3.4\%$. This result is not what one would expect for small Si cations and large O anions.

The important radius ratio is not $r_{cation}/r_{anion}$ but $R/l$ (ratio of nonbonded radius to bond length) for cations surrounding a central anion: for "one-angle" situations, C.N. $= 2$ demands $R/l \leqslant \sin 180°/2 = 1.000$; for C.N. $= 3$, $R/l \leqslant \sin 180°/3 = 0.866$; for C.N. $= 4$, $R/l \leqslant \sin 109°28'/2 = 0.816$; for C.N. $= 6$, $R/l \leqslant \sin 90°/2 = 0.707$; and for C.N. $= 8$, $R/l = 1/\sqrt{3} = 0.577$. And it would seem reasonable that $R/l$ be as high as possible—i.e., the number of bonds, or C.N., be as high as possible within the constraint that $d$ is not much less than $2R$, so that the nonbonded repulsion energy is not too high. This means that $R/l$ should, for preference, be rather close to the values just given. If, for a given structure, the last factor does not hold (i.e. $d$ is less than $2R$), then two possibilities are available: (1) the C.N. is reduced (e.g., diamond, $R/l = 0.812$, C.N. $= 4$, transforms to graphite, C.N. $= 3$); or (2) the "one-angle" situation B—X—B may be replaced by a "two-angle" situation (i.e. edge-sharing of cation-centered polyhedra in place of corner-sharing), which enables higher nonbonded repulsion energy to be balanced by higher bond energy because there are now twice as many bonds formed. Choice between these requires a knowledge of the relative energy changes from these two sources—a subject to which we will return later.

Very rough estimates of the relative energy changes in bond stretching and nonbonded compression yield a relation between the distortions of the tetrahedra in wurtzite structures, and the departure of $c/a$ (for the hexagonal unit cell) from the ideal value of $\sqrt{8/3} = 1.633$. Furthermore, together with $R$ values deduced elsewhere, they also indicate why some compounds can exist in both the sphalerite and wurtzite forms, but some only in the latter: these being the cases where $c/a < \sqrt{8/3}$ (O'Keeffe and Hyde, 1978b).

There is, also, a less well known tetrahedral structure, that of $\beta$-BeO (and also, almost certainly, that of $\delta$-AlN—the so called "expanded hexagonal" form). This structure is quite common among ternary analogs such as $LiAlO_2$, etc. Note that the $AlO_2$ framework is topologically that of cristobalite, and in fact belongs to the sequence II (low cristobalite) family mentioned above. The structure types are based on six-membered rings of "chair" and/or "boat" conformations (and four-membered rings in $\beta$-BeO). These structures are illustrated in Fig. 11. We (O'Keeffe and Hyde, 1979) have given reasons to believe that the order of stability sphalerite, wurtzite and $\beta$-BeO is related to increasing $R/l$.

Application of the $R/l$ size criterion also explains why $NaO_4$ tetrahedra are somewhat more common than $LiO_4$ tetrahedra in mixed oxides. Although

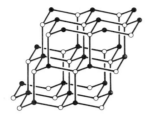  

**Fig. 11.** The bond networks in tetrahedral AX compounds: left with the sphalerite, center with the wurtzite, and right with the $\beta$-BeO structures.

Li is smaller than Na in terms of both $R$ and $l$ separately, it is considerably bigger in terms of $R/l$ (cf. Table I). Thus in $Li_2SiO_3$ the $OSi_2Li_2$ tetrahedron is very crowded—e.g., $d(Si \cdots Si) = 2.97$ Å is very short and $l(Si—O) = 1.68$ Å is rather long; $l(Li—O)$ is also long, 2.17 Å instead of the normal 1.97 Å. In the isostructural $Na_2SiO_3$ there is less strain: $d(Si \cdots Si) = 3.07$ Å [cf. $2R(Si)$ $= 3.06$ Å], $l(Si—O) = 1.67$ Å (still rather long), and $l(Na—O) = 2.40$ and 2.55 Å instead of the normal 2.37 Å. Note that the bond angles reflect the $R/l$ values: in $Li_2SiO_3$, $\theta(SiOSi) = 124°$, $\theta(LiOLi) = 97.5°$; and in $Na_2SiO_3$, $\theta(SiOSi) = 134°$, $\theta(NaONa) = 92.5°$ [cf. $R/l$ values which are $(Si) = 0.93$, $(Li) = 0.76$, $(Na) = 0.71$]. Such relative values are *not* to be expected in terms of ion radii. Note also that *shorter* $d$ values imply *longer* bonds, and not vice versa. (This also implies *smaller* bond angles.)

What we are concerned with here, of course, are the (Pauli) repulsions normally ascribed to the overlap of electron clouds when atom separations are less than the sum of the van der Waals radii—in the present instance, much less than the van der Waals radii. These are exactly the forces that the organic chemist considers when dealing with the configurational analysis of molecular geometry, or "steric strain."

The picture of nonmolecular ionic crystals that is emerging is one in which attractive bonding forces are balanced by repulsive nonbonded interactions. The latter are compressive, and bonds are in tension.

## V. NONBONDED INTERACTION POTENTIALS

### A. The Interaction Energy

The next level of sophistication beyond the use of hard-sphere radii is the development of nonbonded pair potential functions. This has been done, of course, in a number of cases for atomic and molecular gases, and used successfully to fit their thermodynamic and transport properties. However, we require functions $\phi(d)$ for the interaction energy for, e.g., Si atoms a distance $d$ apart. Clearly such functions cannot come directly from gas-phase data. For many atom pairs, notably $C \cdots C$, $N \cdots N$, $O \cdots O$, and $S \cdots S$, $\phi(d)$ has been derived from structural data such as details of molecular packing and thermodynamic data such as heats of sublimation of molecular crystals. The procedure has been described by Kitaigorodsky (1973).

Usually $\phi(d)$ is expressed in a "Lennard-Jones" form

$$\phi = Ad^{-12} - Bd^{-6} \tag{1}$$

or, more usually, the "Buckingham" form

$$\phi = A \exp(-\alpha d) - Bd^{-6} \tag{2}$$

although neither equation is entirely satisfactory over a wide range of $d$. Parson et al. (1972) and Farrar et al. (1973), for example, fit the $Ar \cdots Ar$ and $Ne \cdots Ne$ potentials piecewise and for small $d$ prefer the Morse potential

$$\phi = A[\exp(-2\alpha d) - 2 \exp(-\alpha d)] \tag{3}$$

It is remarkable in fact that the use of different data and different forms for $\phi(d)$ give quite closely agreeing results. (For a thorough analysis of $C \cdots C$ interactions, see Williams, 1967). A plot of such potentials has been presented before (O'Keeffe and Hyde, 1979) for $C \cdots C$, $N \cdots N$, $O \cdots O$, and $S \cdots S$. In Fig. 12, those data are shown again, supplemented with data for $Ar \cdots Ar$ (Parson et al., 1972) and $Ne \cdots Ne$ (Farrar et al., 1973). The inert-gas potentials come from a quite different kind of experiment but nevertheless fall exactly where expected.

A "theoretical" $F \cdots F$ potential has been quoted by Bartell et al. (1979). In deriving that potential, $F_2 \cdots F_2$ interaction energies were calculated using molecular orbital methods. In the range $d/2 = 1.0-1.3$ Å, the calculated $F \cdots F$ potential is almost indistinguishable from the $O \cdots O$ potential plotted in the figure. The MO calculation would not include a van der Waals term (an electron–electron correlation effect); if that were added, the $F \cdots F$ potential would fall between the $Ne \cdots Ne$ and the $O \cdots O$ curves very satisfactorily.

The curves in Fig. 12 allow us to make a fairly definite statement as to what a "one-angle" nonbonded radius is: the radius for X is half the $X \cdots X$ distance at which the $X \cdots X$ interaction energy is about $+20$ kJ mol$^{-1}$. This is the order of magnitude of the difference in energy between competing structures (the structural energy of Zunger, Chap. 5).

The van der Waals radius (Bondi, 1964) is close to the point where the interaction energy goes to zero.

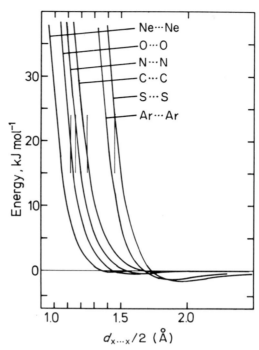

**Fig. 12.** Nonbonded interaction potentials for various atom pairs. The vertical lines are drawn at the nonbonded radii for O, N, C, and S.

## B. Force Constants

A related approach to the problem of determining parameters character-istic of nonbonded interactions has been taken by Bartell (Jacob *et al.*, 1967; Bartell, 1968; Bartell *et al.*, 1975; Fitzwalter and Bartell, 1976), following an earlier lead of Shimanouchi (1963). Shimanouchi analyzed the vibrational spectra of small molecules (such as $CCl_4$, $SiF_6$, etc.) using a "modified Urey–Bradley" (MUB) force field which included, as well as the usual valence stretch and bend force constants, specific nonbonded force constants of the form $\partial^2 \phi / \partial d^2$. The derived force constants for, e.g., F $\cdots$ F interactions are very close to those for the neighboring inert gas pair (Ne $\cdots$ Ne) which can be calculated from curves such as those in Fig. 11.

Fitzwalter and Bartell (1976) have made an extensive investigation into MUB force fields for hydrocarbons. These are designed to reproduce molec-ular conformations and vibration frequencies (the topics of "molecular mechanics"). Of particular interest in the present context is the derived C $\cdots$ C potential, and the *bonded* C—C potential. The nonbonded potential is expressed in the form of Eq. (2) with

$$A = 5.022 \times 10^{-16} \text{ J}$$
$$\alpha = 3.75 \times 10^{10} \text{ m}^{-1}$$
$$B = 2.799 \times 10^{-78} \text{ Jm}^6$$

These parameters give a value of $\phi(d)$ very similar to that given in Fig. 12. The corresponding bonded potential (for the single C—C bond) has a minimum at 1.28 Å, suggesting that the C—C bond in, e.g., ethane or diamond is very considerably stretched by nonbonded repulsions.

We have taken Bartell's $\phi(d)$ for C $\cdots$ C and applied it to some properties of diamond with results that are very instructive. Diamond was chosen because of its intrinsic interest, and because the shortest nonbonded $d(\text{C} \cdots \text{C})$ $(=2.52 \text{ Å})$ is close to $2R(\text{C})$ $(=2.50 \text{ Å})$.

## C.   Force Constants and Nonbonded Repulsions in Diamond

Force-constant models for the lattice dynamics and elastic constants of diamond have been the topic of innumerable papers over the years. They continue to appear with regularity (e.g., Gupta, 1979; Kushwaha and Kushwaha, 1979; Wall and Riter, 1980). Many are deficient in that they fit wrong data (incorrect elastic constants plagued the earlier literature) or, of more concern to us, even when nonbonded (nonnearest-neighbor) interactions are included, the models are not equilibrium ones (in the sense that the net force on the atoms is not zero). There have been rather sophisticated treatments in which the electronic energy is calculated for the distorted structure (Chadi and Martin, 1976). Valence force field treatments include those of Musgrave and Pople (1962) and of Keating (1966). Actually, the elastic constants of tetrahedral compounds are reasonably accounted for in a very simple model using only C—C stretch and C—C—C bending force constants (Harrison, 1980); however, we wish to include explicitly C $\cdots$ C nonbonded interactions.

Even the simplest model with only next-nearest-neighbor C $\cdots$ C interactions involves a number of force constants. We express the energy per bond as $E = \phi_1(l) + 3\phi_2(d) + 3\phi_3(\theta)$ where $l$ is the C—C bond length, $d$ the C $\cdots$ C nonbonded distance, and $\theta$ the C—C—C bond angle, and we exclude "cross" terms (bend–stretch interactions etc.); $\phi_3$ is assumed to be zero at the equilibrium value of $\theta$. Then there is an angle bending constant $K_\theta$ $(=\partial^2\phi_3/\partial\theta^2)$, a bond stretch constant $K_l$ $(=\partial^2\phi_1/\partial l^2)$, and a nonbonded force constant $K_d$ $(=\partial^2\phi_2/\partial d^2)$. There are also force constants $K'_l = (1/l)(\partial\phi_1/\partial l)$ and $K'_d = (1/d)(\partial\phi_2/\partial d)$. The equilibrium condition of zero net force on the atoms translates into $K'_l = -8K'_d$.

In order to get a feeling for the likely magnitudes of the force constants, we take Bartell's parameters for the nonbonded interactions, which give

$$K_d = 48.0 \text{ Jm}^{-2}$$
$$K_d' = -4.8 \text{ Jm}^{-2}$$

and hence
$$K_l' = 38.4 \text{ Jm}^{-2}$$

The negative sign of $K_d'$ indicates compression along the nonbonded directions. Conversely, a positive $K_l'$ indicates a stretched bond. Using this force field, the bulk modulus for the diamond structure is readily shown to be

$$K = (K_l + 8K_d)/3a$$

where $a$ is the lattice parameter. From the elastic constants (McSkimin et al., 1972), one has at $T = 0$, $K = 4.44 \times 10^{11}$ Pa and hence

$$K_l = 91.0 \text{ Jm}^{-2}$$

Note that in the absence of nonbonded repulsions ($K_d = 0$), one would calculate $K_l = 475 \text{ Jm}^{-2}$. This is close to the value usually quoted for C—C stretch (Herzberg, 1945). In our interpretation (and that of Bartell and his co-workers) it is mainly nonbonded interactions that contribute to "the stiffness of the C—C single bond."

One can readily derive that the Raman frequency $v_R$ is given by

$$3\pi^2 m v_R^2 = 2(K_l + 2K_l' + 8K_\theta/l^2)$$

whence

$$K_\theta/l^2 = 43.1 \text{ Jm}^{-2}$$

This set of force constants could be used to evaluate the full lattice dynamics of diamond. This has not been done, but it is sure that the result would be unsatisfactory. The important point is that all the force constants are physically reasonable. ($K_\theta$ is a little larger than, but comparable with, bending constants in hydrocarbon molecules.) Valence force fields that omit non-bonded repulsions do not work well. Thus when Musgrave and Pople (1962) fitted the Raman frequency they obtained an "embarrassingly small bulk constant." Conversely, Harrison's (1980) force constants, which fit the elastic constants, predict a Raman frequency that is too large by a factor of 2!

A simple way to compare the relative importance of first- and second-neighbor interactions is to consider the ratio of force constants $4\pi^2 m v_R^2/a(c_{11} - c_{12})$, where $c_{11} - c_{12}$ is the shear elastic constant. In the Raman mode the C—C distance is changed but second neighbor C $\cdots$ C distances remain constant. Just the opposite is the case for the displacements appro-

TABLE II

The Force Constant Ratio for the Diamond Forms of
C, Si, and Ge

|  | C | Si | Ge |
|---|---|---|---|
| $a(c_{11} - c_{12})/4\pi^2 mv_R^2$ | 0.272 | 0.127 | 0.128 |

priate for $(c_{11} - c_{12})$. Table II compares this ratio for diamond [in which $d(C \cdots C) \simeq 2R(C)$] and for the diamond forms of Si and Ge (in which $d > 2R$). The special behavior of carbon is apparent.

It is interesting, too, to fit a Morse potential to the C—C bond energy. We can write this

$$\phi_1(l) = E\{\exp[-2a(l - l_0)] - 2\exp[-a(l - l_0)]\} \qquad (4)$$

We know $\partial^2\phi_1/\partial l^2$ (given above) and also that the heat of atomization $(5.89 \times 10^{-19}$ J/bond) is equal to $-(\phi_1 + 3\phi_2)$ at the equilibrium separation. The parameters in equation (4) can now be determined, giving

$$a = 1.6065 \times 10^{10} \text{ m}^{-1}$$
$$l_0 = 1.299 \times 10^{-10} \text{ m}$$
$$E = 7.544 \times 10^{-19} \text{ J}$$

In Fig. 13 we plot $\phi_1$ and the total energy $\phi_1 + 3\phi_2$ ($\phi_3 = 0$ for the equilibrium angle) as a function of the C—C bond length $l$. Several points deserve attention:

1.   The minimum of $\phi_1$ is at $l_0 = 1.3$ Å—close to that predicted by the potential function of Fitzwalter and Bartell (1976). This is the bond length for a C—C single bond which is *not* being stretched by non-bonded repulsions.

2.   The stretching force constant for this bond (390 Jm$^{-2}$) is not greatly reduced from the value (475 Jm$^{-2}$) for the stretched bond with nonbonded interactions added.

3.   At the minimum of the total energy, nonbonded terms make the major contribution to the stretching force constant.

4.   In this case, the nonbonded repulsion energy is about 10% of the total energy.

5.   The results should not be taken literally. In diamond we should have included more-distant-neighbor interactions. There is no justification either for assuming that $\phi_1$ and $\phi_2$ are valid over such a wide range of distances.

In connection with point (a) above, it might be mentioned that the length of single C—C bonds varies from 1.38 Å in diacetylenes ≡C—C≡ to 1.54 Å in

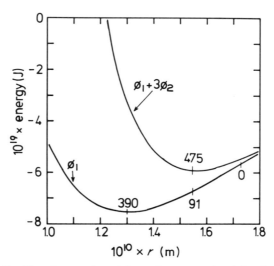

**Fig. 13.** The C—C bond energy $\phi_1$ and the total energy per bond (bonded and nonbonded) $\phi_1 + 3\phi_2$ in diamond. The numbers are curvatures (force constants) at the points indicated (units $Jm^{-2}$).

diamond. Bartell (1962) has already drawn attention to this point and shown that in such molecules the C—C distance is approximately linear with the number of adjacent nonbonded atoms (two and six respectively in the above examples). Extrapolating Bartell's line to zero nonbonded neighbors gives a bond length of 1.31 Å—if nothing more, an amusing coincidence.

We are, of course, aware that there are other explanations for, e.g., the short C—C length in diacetylenes. (However, these must also explain the observation that the —C≡C— bond length is unchanged from that in acetylene.) It is clear though that many of the "explanations" (changes in bond order, conjugation, hyperconjugation, hybridization, back-bonding, ionic character, etc.) are simply different words for saying the same thing. Our thesis is that the conceptually simple notion of nonbonded interactions is equally satisfactory and is susceptible to ready application in a semiquantitative way.

### D.  Force Constants in Silicates

An alternative approach can be taken in dealing with silicates. As Gibbs *et al.* have described in Chapter 9, one may derive reliable force constants for molecules from *ab initio* molecular orbital calculations. Derived Si—O stretch frequencies are in good agreement with experimentally observed ones. Particularly interesting is the observation that one can take properties calculated from molecular fragments over to crystals in which the same configuration occurs.

The angle-bending force constants for Si—O—Si quoted by Gibbs *et al.* can be recast as Si $\cdots$ Si interaction force constants (O'Keeffe *et al.*, 1980), although there must, of course, be contributions from both Si $\cdots$ Si nonbonded pair interactions and from angle-bending forces (to give an equilibrium bond angle of other than 180°). For $(HO)_3Si$—O—$Si(OH)_3$ this force constant is calculated to be 93 $Jm^{-2}$ (i.e., quite "stiff"). Analysis of overlap populations leads clearly to the conclusion that O $\cdots$ O intertetrahedral repulsions are quite unimportant in comparison with the Si $\cdots$ Si interactions. (We have already come to that conclusion from an analysis of the effect of pressure on structure in Sec. IV). Newton *et al.* (1980) took that force constant over to quartz and then calculated the bulk modulus of the crystal on the assumption of rigid $SiO_4$ tetrahedra. The agreement was remarkably good.

The success of this calculation leads us to think that one can develop a "theoretical" approach to assessment of the role of nonbonded interactions in solids that will complement the "experimental" (i.e. based on observation of structure) approach we have taken up until now.

## VI.   WHAT IS THE SIZE OF AN ATOM?

We can, at least partly, answer the question of what the size of an atom is. The answer will depend on the energy scale involved; in order to give a quantitative answer, we might recognize four radii with different uses.

1.   *Van der Waals radius.* This is the largest and corresponds to a balance between attractive and repulsive forces between valence-saturated nonbonded atoms. For light atoms, the energy of interaction is of the order of 1 kJ $mol^{-1}$.

2.   *Nonbonded radii.* (geminal or one-angle). These predict nonbonded distances at which bonding and nonbonded *forces* cancel and correspond to nonbonded energies of the order of 10–20 kJ $mol^{-1}$ (see Fig. 12).

3.   *Slater–Bragg radii.* These predict bonded distances and typically involve interaction energies of several hundred kJ $mol^{-1}$.

4.   *Orbital radii.* These measure the point at which the valence electron pseudopotential goes through zero and thus are roughly related to "core" radius (see Chapters 3–5).

In Fig. 14, these radii are plotted to show the periodic trends which are the same for all four sets (increasing to the left and down the periodic table). Note that in all these radii silicon is larger than oxygen.

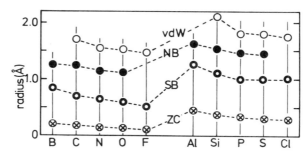

**Fig. 14.** Various radii for atoms in the first two rows of the periodic table: vdW = van der Waals radii (Bondi, 1964), NB = nonbonded radii (Table I), SB = Slater–Bragg radii (Slater, 1965), and ZC = mean of s and p orbital radii of Zunger and Cohen (see Chapter 5).

Bondi (1964) has called attention to an earlier suggestion of Morrison (1955) that the van der Waals radius is proportional to the de Broglie wavelength of the outermost valence electron, and hence proportional to $1/\sqrt{I}$ (where $I$ is the first ionisation potential of the atom). Interestingly, essentially the same correlation was discovered by Pantilides (1975) in the inverse sense that the band gap of inert-gas solids (in essence, $I$) varies as the inverse square of the interatomic distance. The importance of these correlations in the present context is that ionization potentials exhibit a very well known and fundamental dependence on the position of an atom in the periodic table—and this is, of course, just what gives rise to the variation of radius we have described.

We have not included "ionic radii" in our discussion; we feel that their utility is small. (They are better replaced by bond lengths cf Chapter 15). Some tables of ionic radii even require that some ions (for example carbon) have negative radii. It is remarkable that this preposterous notion can be accepted whereas the alternative proposal that carbon and oxygen have comparable size is rejected.

A theme recurring through this book is that the divorce of crystal chemistry (solid-state chemistry) from molecular chemistry was a great mistake and has left both parties the poorer. It is apparent that in this article we have leaned heavily on the ideas of Bartell and others developed long ago for small molecules and which have developed into "molecular mechanics," which is now a sophisticated branch of chemistry (see, e.g., Allinger, 1976). On the other hand, the richness of bonding patterns found in crystals should present fine challenges to molecular theoreticians. To return once again to carbon for a final example, we find it forming six bonds to iron in $Fe_3C$ and eight to Be in $Be_2C$ (the latter is described in one highly respected text book as being "ionic" with cubic close packing of $C^{-4}$!—yet another text explains that $BeF_2$ has the quartz structure because it is "covalent").

## ACKNOWLEDGMENT

We are greatly indebted to Professor L. S. Bartell for calling our attention to much of the early work on molecular mechanics and for helpful correspondence. This work was supported in part by a grant (DMR 78-09197) from the National Science Foundation.

## REFERENCES

Allinger, N. L. (1976). *Adv. Phys. Org. Chem.* **13**, 1.
Bartell, L. S. (1960). *J. Chem. Phys.* **32**, 827.
Bartell, L. S. (1962). *Tetrahedron* **17**, 177.
Bartell, L. S. (1968). *J. Chem. Educ.* **45**, 754.
Bartell, L. S., Doun, S. K., and Goates, S. R. (1979). *J. Chem. Phys.* **70**, 4585.
Bartell, L. S., Fitzwater, S., and Hehre, W. J. (1975). *J. Chem. Phys.* **63**, 2162.
Bondi, A. (1964). *J. Phys. Chem.* **68**, 441.
Bragg, W. L. (1920). *Philos. Mag.* [6] **40**, 169.
Bragg, W. L., and Claringbull, G. F. (1965). "Crystal Structures of Minerals," p. 232. Bell, London.
Chadi, D. J., and Martin, R. M. (1976). *Solid State Commun.* **19**, 643.
Cruickshank, D. W. J. (1961). *J. Chem. Soc.* p. 5486.
d'Amon, H., Denner, W., and Schulz, H. (1979). *Acta Crystallogr., Sect. B* **B35**, 550.
Farrar, J. M., Lee, Y. T., Goldman, V. V., and Klein, M. L. (1973). *Chem. Phys. Lett.* **19**, 359.
Fitzwalter, S., and Bartell, L. S. (1976). *J. Am. Chem. Soc.* **98**, 5107 (see also the appendices to this paper, available as supplementary material).
Glidewell, C. (1975). *Inorg. Chim. Acta* **12**, 219.
Grimm, H., and Dorner, B. (1975). *J. Phys. Chem. Solids* **36**, 407.
Gupta, R. (1979). *J. Phys. Chem. Solids* **40**, 579.
Harrison, W. A. (1980). "Electronic Structure and the Properties of Solids." Freeman, San Francisco, California.
Hazen, R. M. (1976). *Am. Mineral.* **61**, 1280.
Herzberg, G. (1945). "Infrared and Raman Spectra of Polyatomic Molecules." Van Nostrand-Reinhold, Princeton, New Jersey.
Jacob, E. J., Thompson, H. B., and Bartell, L. L. (1967). *J. Chem. Phys.* **47**, 3736.
Jorgensen, J. D. (1978). *J. Appl. Phys.* **49**, 5473.
Keating, P. N. (1966). *Phys. Rev.* **145**, 194.
Kitaigorodsky, A. I. (1973). "Molecular Crystals and Molecules." Academic Press, New York.
Kushwaha, M. S., and Kushwaha, S. S. (1979). *J. Chem. Phys.* **71**, 1440.
Leadbetter, A. J., Smith, T. W., and Wright, A. F. (1973). *Nature (London), Phys. Sci.* **244**, 125.
McDonald, W. S., and Cruickshank, D. W. J. (1967). *Acta Crystallogr.* **22**, 43.
McSkimin, H. J., Andreatch, P., and Glynn, P. (1972). *J. Appl. Phys.* **43**, 985.
Morrison, J. D. (1955). *Rev. Pure Appl. Chem.* **5**, 46.
Musgrave, M. J. P., and Pople, J. A. (1962). *Proc. R. Soc. London, Ser. A* **268**, 474.

Newton, M. D., O'Keeffe, M., and Gibbs, G. V. (1980). *Phys. Chem. Miner.* **6**, 305.

O'Keeffe, M. (1977). *Acta Crystallogr., Sect. A* **A33**, 924.

O'Keeffe, M., and Hyde, B. G. (1976). *Acta Crystallogr., Sect. B* **B32**, 2923.

O'Keeffe, M., and Hyde, B. G. (1978a). *Acta Crystallogr., Sect. B* **B34**, 27.

O'Keeffe, M., and Hyde, B. G. (1978b). *Acta Crystallogr., Sect. B* **B34**, 3519.

O'Keeffe, M., and Hyde, B. G. (1979). *Trans. Am. Crystallogr. Assoc.* **15**, 65.

Pantilides, S. T. (1975). *Phys. Rev. B: Solid State* [3] **11**, 5082.

Parson, J. M., Siska, P. E., and Lee, Y. T. (1972). *J. Chem. Phys.* **56**, 1511.

Pauling, L. (1960). "The Nature of the Chemical Bond," 3rd ed. Cornell Univ. Press, Ithaca, New York.

Shimanouchi, T. (1963). *Pure Appl. Chem.* **7**, 131.

Slater, J. C. (1965). "Quantum Theory of Molecules and Solids," Vol. 2. McGraw-Hill, New York.

Steinfink, H., and Brunton, G. D. (1968). *Acta Crystallogr., Sect. B* **B24**, 807.

Wall, R. S., and Riter, J. R. (1980). *J. Chem. Phys.* **72**, 2886.

Wells, A. F. (1975). "Structural Inorganic Chemistry," 4th ed. Oxford Univ. Press, London and New York.

Williams, D. E. (1967). *J. Chem. Phys.* **47**, 4680.

Wright, A. F., and Leadbetter, A. J. (1975). *Phil. Mag.* **31**, 1391.

# 11

## Molecules within Infinite Solids

### JEREMY K. BURDETT

|      |                                              |     |
|------|----------------------------------------------|-----|
| I.   | Introduction                                 | 255 |
| II.  | Perturbation Theory                          | 256 |
| III. | Geometries of Inorganic Molecules            | 258 |
| IV.  | Simple versus Exact Theories                 | 263 |
| V.   | Molecular Orbitals in Solids                 | 264 |
|      | A. Diamond and Its Defect Structure          | 266 |
|      | B. Distortion of the CsCl Structure          | 269 |
| VI.  | Reactions in the Solid State                 | 274 |
|      | References                                    | 276 |

## I. INTRODUCTION

Molecular orbital ideas have been used very effectively for several years to understand the geometries and electronic structure of molecules, as isolated entities in the gas phase or in low temperature matrices and in crystalline solids where there are no strong bonds between the molecular units. Bond angles and bond lengths, ligand site preferences, and barriers to rotation of coordinated ligands have been some of the molecular properties which can be rationalized both qualitatively and quantitatively using a spectrum of molecular orbital methods (Burdett, 1980; Gimarc, 1979). These range from simple Hückel (Streitweiser, 1961) calculations on organic and inorganic molecules through the widely used semiempirical extended Hückel method (EHMO) (Hoffmann, 1963; Hoffmann and Lipscomb, 1962a,b) and the "NDO" methods (CNDO, INDO etc.) (Segal, 1977) to the full power of modern *ab initio* methods. However, the general rule of thumb, that the more exact the quantum-mechanical calculation the less understandable are the results for the nonspecialist, is unfortunately too often the case (Bartell, 1968). As a result, it is probably true to say that chemists find the simplest models, which are usually drastic simplifications

**255**

Structure and Bonding in Crystals, Vol. I

of the quantum-mechanical "truth", of widest general utility in the laboratory.

The impact of simple molecular orbital theory on the whole structure of experimental chemistry has often been dramatic. In a brilliantly conceived set of papers in the mid-sixties, Woodward and Hoffmann developed a series of symmetry rules which allowed an understanding at a very basic level of the sterochemical course of many types of organic reactions (Hoffmann and Woodward, 1970). This provided the language which brought experimentalists and theorists together, and today, these theoretical concepts are part of the synthetic organic chemists' everyday tools of the trade. These Woodward-Hoffmann rules, as they are known, make extensive use of symmetry and overlap arguments, an approach which we will use in this chapter along with rudimentary ideas of perturbation theory. We now know enough about the electronic structures of a wide range of molecules to begin to apply similar arguments to the infinite "molecules" of the solid state.

## II. PERTURBATION THEORY

In principle we could perform a molecular orbital calculation on every new molecule that is identified experimentally and calculate the orbital energies and wavefunctions. However, it is very profitable to search out the most important factors that affect the orbital interaction energies between atomic orbitals in molecules and treat them in as simple a way as possible. The aim is to be able to transfer our results from one system to another and attempt a global view of the electronic structure problem. Of rather fundamental importance in this respect is the use of perturbation theory (Hoffmann, 1971). If we know the solutions of the Schrödinger wave equation for a particular system and then switch on a perturbation (given by $\mathcal{H}'$), then the approach very readily gives us the new orbital energies and wavefunctions in terms of the old. In this chapter the perturbation we use will most often be a change in molecular geometry as we change the angular disposition of ligands around a central atom or distort the solid or molecule by adjusting the distances between bonded atoms. To second order, the energy of the $i$th level $E_i'$ as a result of the perturbation is given by Eq. 1:

$$E_i' = E_i + \sum_j{}' \frac{\left| \int \psi_i \mathcal{H}' \psi_j \, d\tau \right|^2}{E_i - E_j} \tag{1}$$

where the $E_k$ are the unperturbed energy levels and $\psi_k$ their wavefunctions. The prime over the summation sign means the term for $j = i$ is not included.

The new wavefunctions are given by Eq. 2:

$$\psi_i' = \psi_i + \sum_j{}' \frac{\int \psi_i \mathscr{H}' \psi_j \, d\tau}{E_i - E_j} \psi_j \tag{2}$$

There are two basic rules which stem from these equations which we will use in this chapter. Just consider two orbitals $\psi_{1,2}$ which interact as a result of the perturbation. From Eq. 1,

$$E_1' = E_1 + \frac{\left[\left(\int \psi_1 \mathscr{H}' \psi_2\right) d\tau\right]^2}{E_1 - E_2} \tag{3}$$

and

$$E_2' = E_2 + \frac{\left(\int \psi_1 \mathscr{H}' \psi_2 \, d\tau\right)^2}{E_2 - E_1} = E_2 - \frac{\left(\int \psi_1 \mathscr{H}' \psi_2 \, d\tau\right)^2}{E_1 - E_2} \tag{4}$$

If we assume $E_1 < E_2$ (i.e., that level 1 lies deepest in energy) then, since the numerators are positive, $E_1' < E_1$ and $E_2' > E_2$. This leads to rule 1. When two levels interact, the lower one is stabilized (i.e., drops in energy) and the higher one is destabilized (i.e., rises in energy). As a result of the perturbation the two levels appear to repel each other. The new wavefunctions may be written in a similar way.

$$\psi_1' = \psi_1 + \frac{\int \psi_1 \mathscr{H}' \psi_2 \, d\tau}{E_1 - E_2} \psi_2 \tag{5}$$

$$\psi_2' = \psi_2 - \frac{\int \psi_1 \mathscr{H}' \psi_2 \, d\tau}{E_1 - E_2} \psi_1 \tag{6}$$

Invariably the sign of the integral $\int \psi_1 \mathscr{H}' \psi_2 \, d\tau$ is opposite to that of the overlap integral between the two orbitals. For a positive overlap integral, $\int \psi_1 \mathscr{H}' \psi_2 \, d\tau$ is negative, $E_1 - E_2$ is negative, and so the mixing coefficient of $\psi_2$ into $\psi_1$ to give $\psi_1'$ is positive. Analogously, the mixing coefficient of $\psi_1$ into $\psi_2$ to give $\psi_2'$ is negative. This leads to rule 2. When two orbitals interact, the higher energy orbital mixes into the lower energy orbital in a bonding way and the lower energy orbital mixes into the higher energy orbital in an antibonding way. Armed with these two rules, one describing the new energies and the other describing the new wavefunctions, we are ready to view molecular stereochemistry.

## III.  GEOMETRIES OF INORGANIC MOLECULES

The basic geometries or coordination polyhedra of simple main-group compounds such as $BF_3$, $NF_3$, $CF_4$, and $SF_4$ may readily be predicted using a simple rule of thumb suggested by Sidgwick and Powell in the 1930s and popularized in a more complete form by Nyholm and Gillespie some years later (Gillespie, 1972; Gillespie and Nyholm, 1957). The basic premise of this well known scheme is that the valence electrons around the central atom are arranged in localized pairs containing two electrons of opposite spin, and that these pairs are arranged in space such that the distances between them are maximized. In this valence-shell electron-pair repulsion (VSEPR) method, the mechanism invoked to keep these pairs apart is the *Pauli repulsion force* which operates to prevent electrons of the same spin from occupying the same region of space. An alternative view from valence bond theory is that these electron pairs occupy the lowest energy set of hybrid orbitals which can be constructed from the valence s, p, and d orbitals. The method has often been criticized (Drago, 1973) for its lack of physical reality, but in spite of this is usually very successful in molecular geometry prediction.

The method does, however, fail for certain classes of molecules (Burdett, 1980). For species with a large amount of ionic character, the geometry is often set by the Coulombic repulsion of "ions" rather than by the orientations of electron-pair orbitals. For example, $Li_2O$ with four valence pairs, expected to be nonlinear on the scheme, is actually found as a linear molecule in the gas phase. Clearly the arrangement where the lithium ions are set diagonally around the oxygen ion minimizes the electrostatic repulsion between them. Secondly, the scheme does not appear applicable when looking at systems where there are lone pairs on adjacent atoms. Thus $N_2H_4$ and $P_2H_4$ exhibit the geometry where the lone pairs on the adjacent nitrogen atoms are in the close gauche arrangement rather than the trans form (Wolfe, 1972). Thirdly, the VSEPR scheme often fails for molecules containing a heavy central atom from groups III–VIII of the periodic table. $TeBr_6$ has six valence pairs and has the expected octahedral geometry, but $TeBr_6^{-2}$ with seven pairs is also octahedral, a result not in accord with the scheme. In such cases, the central atom is said to have a stereochemically inert pair. Inert pairs have only been found to date for molecules which contain six and higher coordination. For the species $TeCl_4(thiourea)_2$, two crystalline forms are known, one containing a centrosymmetric Te coordination (Husebye and George, 1969) (with an inert pair) and the other a distorted geometry (Esperås *et al.*, 1975) where it is easy to envisage the presence of a stereochemically active pair.

A powerful way to look at the geometries of molecules to enable us to understand in more detail the energetics associated with molecular geometry changes employs a simple combination of group theory and perturbation theory. Originally described by Bader (1960), the second-order Jahn–Teller approach has been popularized by Bartell (1968) and Pearson (1969). Here we extract the essence of the method but refer the reader elsewhere for a discussion of the mathematical details of a more formal presentation (Burdett, 1980; Pearson, 1969; Bartell, 1968). Let us take the valence-region molecular orbital diagram of a system in a symmetrical geometry such as the trigonal plane for a three-coordinate system or the tetrahedron for a four-coordinate system. In Fig. 1 we label the highest occupied molecular orbital (HOMO) and lowest unoccupied molecular orbital (LUMO). These will invariably be of different symmetry species. The second-order Jahn–Teller recipe allows us to answer the question: "Is there a distortion $q$ of the molecule such that on distortion away from the symmetric geometry the HOMO and LUMO become of the same symmetry species?" Using the language of perturbation theory, described above, can we find a perturbation (via a change in geometry) which mixes HOMO and LUMO together? The result would be a destabilization of the LUMO, a corresponding stabilization of the HOMO, and of course of the energies of the electrons in it. If this is the dominant energetic change on distortion, then the energy of the distorted molecule will be lower than that of the symmetric structure. We can readily obtain the symmetry species of $q$ by examining the symmetry properties of the integral in the

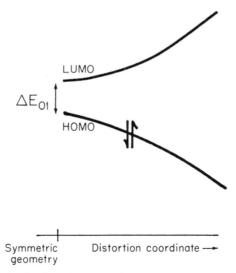

**Fig. 1.**   Behavior of HOMO and LUMO of a symmetric system on distortion such that HOMO and LUMO mix together and "repel" each other.

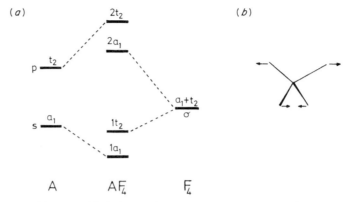

**Fig. 2.** (a) Molecular orbital diagram for a tetrahedral $AF_4$ system. (b) One component of the $t_2$ bending vibration.

numerator of Eq. 7. The Hamiltonian operator $\mathcal{H}$ is totally

$$\Delta E = \frac{\left( \int \psi_{HOMO} \frac{\partial \mathcal{H}}{\partial q} \psi_{LUMO} \, d\tau \right)^2 q^2}{E_{HOMO} - E_{LUMO}} \qquad (7)$$

symmetric to all symmetry operations of the molecule. In order that the integral is nonzero, the direct product $\Gamma_{HOMO} \times \Gamma_q \times \Gamma_{LUMO}$ must also be totally symmetric where we write $\Gamma$ as the symmetry species of the relevant function. This immediately gives us the symmetry of the distortion coordinate as $\Gamma_q = \Gamma_{HOMO} \times \Gamma_{LUMO}$. Two examples will illustrate the application of the method.

Figure 2 shows the MO diagram for a tetrahedral $AF_4$ molecule and one component of the triply degenerate $t_2$ bending vibration. There is also an AF stretching vibration of the same symmetry. For $CF_4$ with four valence electron pairs, the HOMO is of species $t_2$. The LUMO is of species $a_1$ and their direct product leads to $t_2$ as the species of $q$. For $SF_4$, $(1a_1)^2(1t_2)^6(2a_1)^2$, the symmetry species of HOMO and LUMO are reversed, but this leaves the species of $q$ unchanged. $CF_4$ with four valence pairs is a tetrahedral molecule $(T_d)$ but $SF_4$ has a butterfly geometry $(C_{2v})$ readily understandable as a result of a combination of $t_2$ stretching and bending away from the $T_d$ structure (Fig. 3). The difference in behavior of the two molecules, both predicted to distort away from tetrahedral on symmetry grounds, is the much larger HOMO–LUMO gap in the case of tetrahedral $CF_4$ compared to tetrahedral $SF_4$. Clearly the distortion energy decreases as this separation increases, which is why we only usually need consider in analyses such as these the mixing of HOMO and LUMO rather than the mixing of HOMO

**Fig. 3.**   Generation of the $SF_4$ geometry by $t_2$ stretching and bending of the tetrahedral geometry.

and higher energy orbitals in addition. This leads via Eq. 7 to a much smaller distorting force for the four-pair compaired to the five-pair molecule.

Using perturbation theory ideas, we can see how the $2a_1$ orbital of the tetrahedral $SF_4$ structure becomes a lone pair orbital in the distorted geometry (Fig. 4). This is obtained very simply by mixing the HOMO and LUMO of the tetrahedral structure with the phases prescribed by rule 2. Thus the lone pair orbital in $SF_4$ becomes stereochemically active when the number

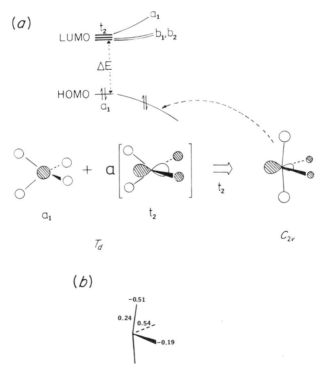

**Fig. 4.**   (a) Mixing of HOMO and LUMO of tetrahedral $SF_4$ to give a lone pair HOMO on distortion to $C_{2v}$. (b) A population analysis of the distorted structure.

of valence electron pairs exceeds the number of ligands, just as in the cases cases of $NH_3$ and $ClF_3$ which can (Bartell, 1968) be treated in an analogous way, and provides a useful link between the molecular orbital description and the VSEPR approach. It is conceivable, however, that even in this case if the HOMO–LUMO gap for a five-valence-pair molecule were large, or the numerator of Eq. 7 were small, the lone pair would be stereochemically inert and not lead to a distorted structure.

The scheme is thus successful in predicting the type of distortion away from the symmetric geometry. Used in this qualitative fashion it does not give us any idea as to the size of the distortion, whether in fact it will even be observable. That task is one that needs to be tackled using quantitative molecular orbital theory. The power of the present approach is that it enables comment on the direction and size of the dominant energy level changes on distortion, and enables us to understand how the orbitals of the distorted structure are related to those of the parent.

Although the admixture of stretching and bending modes allows generation of inequivalent bond lengths in $SF_4$, it is interesting to use a molecular orbital trick to see if in the distorted structure itself different bond lengths are predicted. Figure 4b shows the results of an EHMO calculation where all the SF distances were set equal to each other and identical atomic orbital parameters used for all the ligands so as not to impose any bias. As can be seen, the bond overlap populations (discussed in Chap. 8) for the axial linkages are smaller than those found for the equatorial ones, implying that in the real structure the axial distances should be longer, as indeed found experimentally. Another feature of the results is that the axial ligands carry a higher charge than the equatorial ones. This implies that in substituted systems the more electronegative ligands should occupy the axial sites. This is found for all substituted sulfuranes. In $SF_2(CH_3)_2$, for example, the electronegative fluorine ligands occupy the axial sites and the $CH_3$ the equatorial sites of the trigonal bipyramid of electron pairs. Some other simple ideas as to the origin of the different bond lengths are described elsewhere (Burdett, 1980), but the general result in such species is that bonds trans to a vacancy are invariably shorter than bonds trans to another bond. In solids containing weak secondary bonding, for example $SbCl_3 \cdot C_6H_5NH_2$, a similar effect is seen (Alcock, 1972). There is weak interaction between the N atom and the $SbCl_3$ unit, giving a pseudo-$SF_4$ geometry with a weak axial Sb–N linkage trans to a longer Sb–Cl bond.

A general result of the approach is that occupation of high energy, usually antibonding, orbitals of a system is energetically unfavorable. The system should then distort so that the HOMO is stabilized. In the cases we have examined, this occurs via an angular geometry change and the high energy orbital becomes a lone pair orbital in the new structure. An alternative route, the one found in photodissociation reactions where a high-energy antibonding orbital is temporarily occupied, is loss of a ligand from the coordination sphere. Bond-breaking processes of this type are common when viewing solid-state structures, as we shall see, since the possibilities for angular distortions is somewhat limited.

## IV.  SIMPLE VERSUS EXACT THEORIES

Having rationalized the geometries of these molecules in this way, how does this simple scheme based on HOMO–LUMO coupling on distortion fit in with the results one gets by performing an *ab initio* calculation on the molecule? Such calculations are capable of giving good numerical results for bond angles and bond lengths in molecules. In Chaps. 8 and 9, we can see how well structural parameters of silicates correlate with those from good calculations. However, we find from such studies on molecules that often the dominant energy changes on changing the angular geometry derive from a rather small number of orbitals, invariably the collection of highest energy occupied molecular orbitals. In many cases, the dominant energy change is associated with the HOMO as in our example above. Also, importantly, we find that the directions of the energy changes of these orbitals are well matched by using simple one-electron molecular orbital theory, i.e., by leaving out all the electron–electron repulsions via coulombic and exchange forces included in the *ab initio* calculations. In many cases (for example the transition-metal carbonyls) (Burdett, 1974; Elian and Hoffmann, 1975), a rather good agreement between calculated and observed bond angles is found using the simple one-electron EHMO method. In fact, much of the structural chemistry of both main-group and transition-metal compounds can be viewed in simple terms just by considering the angular dependence of the overlap integrals between ligands and central atom orbitals (Burdett, 1978b). In many cases, the geometry of an isolated system can be understood just by knowing how many valence electrons are present. For example, all low-spin (diamagnetic) four-coordinate $d^8$ transition-metal complexes have square-planar geometries, and the VSEPR scheme just uses as input the number of valence electron pairs.

Prediction of bond lengths, heats of formation, and other quantitative properties such as whether $SF_6$ is more or less stable than $SF_4$ plus two fluorine atoms, lies outside of the area well treated by even quantitative simple molecular orbital theory. While the workings of, for example, the EHMO method are transparent to chemical minds, extraction of the factors influencing the stability of $SF_4$, for example, from high-quality molecular orbital calculations is usually extremely difficult. Indeed, because of correlation effects, the dissociation energy of $F_2$, as calculated using a high-quality (but single-configuration) method, is negative! As evidenced by the material in the other chapters of the book, the coordination-number problem—whether, for example, LiF will exhibit the sphalerite, rocksalt, or CsCl structure—falls into this category. This rather difficult coordination-number is one crystal chemists would like to solve.

As a result we are left with one group of structural parameters which are amenable to understanding on the basis of crude, simple molecular-orbital ideas, and those, which may only be accessible (at present) by considerable effort in "number crunching" by computer. In this volume we see several results of the latter type. The separation of AX systems into bounded regions containing just one structural type (Chapters 4 and 5) is an impressive display of the power of modern quantum mechanics. On the other hand, it is very difficult to get a feel for the chemical or physical factors which differentiate these structures, or to be able to take ideas generated in this study and apply them without more calculation to others. In V,B, we shall use our simple orbital approach to view the red PbO structure which contains 10 electrons per formula unit as a distorted CsCl arrangement. While the approach produces results which may be applied to other systems, it does not answer the problem posed by the existence of other 10-electron systems with different structures such as TlI (two forms, one a distorted rocksalt and the other an undistorted CsCl structure) and InF.

To conclude this section we can write Eq. 8, which describes the *present* state of theoretical treatments of solid-state structure, and perhaps to a somewhat lesser extent molecular structural chemistry: $a$ measures how applicable the qualitative ideas produced by a given theoretical

$$a \cdot b = \text{constant} \tag{8}$$

approach are to unrelated systems, and $b$ measures the accuracy of the calculation, whether it numerically differentiates between two closely related structures perhaps differing only slightly in energy. In the calculations we report, $a$ is high but $b$ is low. The ideas of O'Keeffe and Hyde, although not directly linked to an orbital model, fall into the same category. On the other hand, the elegant pseudopotential calculations described by Cohen and the results of Zunger and of Bloch and Schatteman fall into a regime described by high $b$ but small $a$.

## V.  MOLECULAR ORBITALS IN SOLIDS

Molecular orbital theory has been used for some time in the solid-state context as the tight-binding method. The problem facing us in applying

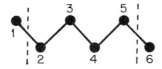

**Fig. 5.** A one-dimensional chain of atoms. The repeat unit of the system is just two atoms.

simple molecular orbital methods to solids is to be able to derive a molecular-orbital picture of such systems so that we may readily apply the tricks learned in understanding molecular geometry to the infinite "molecules" of the extended array. In band structure calculations, use is made of the periodically repeating nature of a fragment in three dimensions by use of the Bloch functions to simplify the theoretical task. In our approach we will use a rather simple way to derive the molecular orbitals of a fragment-within-the-solid.

Figure 5 shows a simple one-dimensional chain of atoms. The repeat unit of the structure is simply the atoms labelled 2 and 3. If we removed a rather larger collection of atoms (say 2 through 5) as representative of the solid-state structure and examined the orbital structure of this unit, a rather poor description of the solid would result. The "loose ends" of the structure at atoms 2 and 5 would give rise to low-lying orbitals which in the real system should interact strongly with the environment. Gibbs and co-workers in Chap. 9 have avoided this problem in tetrahedral structures by tying off these ends with hydrogen atoms to simulate the "crystal field" experienced by the fragment. We will avoid the problem by actually tying together the ends of the molecule in Hilbert space. This is simply done by calculating the interactions between the orbitals on atoms 1 and 2, adding these values to the corresponding interactions between atoms 2 and 5, and performing our calculation as usual on this group of four atoms. Atom 2 feels an interaction of just the right size and spatial direction expected from atom 1, and atom 5 the interaction expected via atom 6. The molecular-orbital diagram which results has the right stoichiometry of the solid and contains a small enough number of orbitals to be readily manageable. The approach is not a new one. It has been used sporadically over many years but has received a detailed analysis by Zunger (1975a,b), to which the enquiring reader may turn.

The relationship between these fragment-within-the-solid orbitals and the band structure of the solid are easiest seen for the $\pi$ orbitals of our simple one-dimensional chain of Fig. 5. Before the ends are tied together, the $\pi$ orbitals of the four-atom unit are those of butadiene with energies, according to simple Hückel theory (Streitweiser, 1961) of $\varepsilon = \alpha \pm \beta(1 \pm \sqrt{5})/2$ where $\alpha, \beta < 0$. With the ends tied together, the $\pi$ orbitals are isomorphous with those of cyclobutadiene with energies of $\varepsilon = \alpha$(twice), $\alpha \pm 2\beta$. The orbitals of a single repeat unit with its ends tied together are at $\varepsilon = \alpha \pm 2\beta$ (Fig. 6).

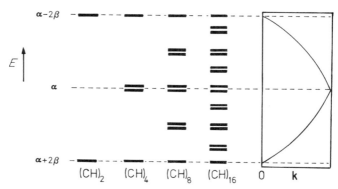

**Fig. 6.** The fragment-within-the-solid $\pi$ orbitals of various $(CH)_n$ fragments. At the right is the $\pi$ band structure of the solid.

The band structure of the chain shows one $\pi$ band with an energy of $\varepsilon = \alpha - 2\beta$ at $\mathbf{k} = 0$ decreasing to $\varepsilon = \alpha$ at the zone edge, and another band with $\varepsilon = \alpha + 2\beta$ at $\mathbf{k} = 0$ which increases to $\varepsilon = \alpha$ at the zone edge. So the molecular orbitals of one repeat unit give the band structure energies at $\mathbf{k} = 0$, the orbitals of two repeat units give in addition the orbitals at the zone edge, and larger units will lead to orbitals which fill in the gaps. Our hypothesis in the next section will be that we can mimic the energy changes of the solid system on distortion by using the orbitals of a number of repeat units of the solid with the cyclic conditions imposed. Clearly, if the band energies change smoothly with $\mathbf{k}$ as in Fig. 6 for the linear chain, then this might be a good approximation. If the band energies do not (as for the case of graphite), then we will have to be very careful in using such approaches. Given this caveat, let us look at three examples which illustrate the application of the method.

### A.  Diamond and Its Defect Structures

This is a subject which has received extensive study via band structure calculations (Levin, 1977). Figure 7 shows a cell of the diamond structure, showing how each carbon atom is tetrahedrally coordinated by four others. With four valence electrons per atom, such an arrangement is readily understandable in terms of traditional bonding ideas using $sp^3$ hybrid orbitals. This cell is not the smallest we could have chosen, but it will be useful to use one of this size in what follows. The repeat unit of such a structure is represented by the collection of solid atoms, and the results of a molecular-orbital calculation on such a fragment before the ends are tied together are shown in Fig. 8. With a total of 32 valence electrons associated with this collection

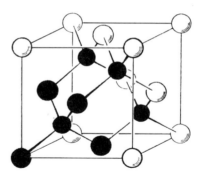

**Fig. 7.** A part of the diamond structure: the repeat unit of this large unit cell is shown as the collection of eight solid atoms.

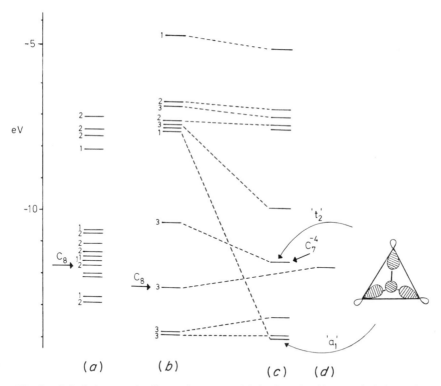

**Fig. 8.** Calculations on the diamond structure: (a) the $C_8$ unit with no ends tied together; (b) the fragment-within-the-solid orbitals for the $C_8$ unit; (c) correlation of these orbitals with the fragment-within-the-solid orbitals for $C_7^{-4}$; and (d) those of an ejected atom. The HOMO in each case is indicated with an arrow and the orbital degeneracy with a small numerical label. The 2s orbital on the ejected atom is not shown. It lies off the bottom of the diagram.

of eight atoms, the HOMO lies in the middle of a closely packed group of orbitals. The presence of these low-lying orbitals is indicative of latent interactions with the environment. Sealing the ends of this fragment by saturating all nonfour-coordinate carbon atoms with electrons fills all these low lying orbitals for $C_8^{-18}$. On tying the ends of the fragment together, this time in three dimensions simplifies the diagram and now the HOMO of $C_8$ is well separated in energy from the LUMO. If we now add eight electrons to this unit, high-energy molecular orbitals are now occupied, and according to our philosophy outlined above, the system will distort so as to create lone pair orbitals of lower energy. One way, which we will not discuss, is to break some of the bonds of this structure to give the arsenic layer structure. Another route is to eject an atom with a complete complement of s and p electrons, i.e. $C^{-4}$. Figure 8 shows how the orbitals of $C_8^{-8}$ correlate with those of $C_7^{-4}$ and the ejected atom. As expected, the high-energy orbitals of the former correlate with lone pair orbitals of the defect structure. These lone pair orbitals are localized around the tetrahedral defect site, and fall into two sets, of local $t_2$ and $a_1$ symmetry as imposed by the site symmetry. It is clear to see that the low-energy orbitals are all filled for the $C_7^{-4}$ configuration and that there is a significant HOMO–LUMO gap.

Two interesting points emerge. The diamond structure was stable for an average contribution of four electrons per atom. This is found in diamond itself and also in derivatives such as ZnS (sphalerite) and $CuFeS_2$ (chalcopyrite). The defect structure is also stable for an average of four electrons per site if when doing the bookkeeping we contribute four electrons from the vacancy. A similar effect occurs if another atom is ejected from the array, stable for $C_6^{-8}$. This is the general result described by the Grimm–Sommerfeld valence rule (Pearson, 1972). If we write the formula of a compound as $0_a 1_b 2_c \cdots 7_h$ where in $N_n$ there are $n$ atoms contributing $N$ electrons ($0 = $ a vacancy written as □ in the structural formula), then

$$\sum n N_n / \sum n = 4 \qquad (9)$$

Thus $\square CdIn_2 Se_4 (0_1 2_1 3_2 6_4)$ is isostructural with our $C_7^{-4}$ example above. We may also perform a population analysis on the $C_7^{-4}$ structure and find (Fig. 9) that the atoms with the largest negative charge are those surrounding the vacancy. These will then be the sites containing the most electronegative atoms. In all defect sphalerite structures we have found, this is always the case. From $\square CdIn_2 Se_4$ to nowackiite (Marumo, 1967) ($\square Cu_6 Zn_3 As_4 S_{12}$; $0_1 1_6 2_3 5_4 6_{12}$), the system contains fewer metal atoms than nonmetal atoms and the vacancy is "coordinated," usually by S or Se. The result also implies that systems which are short of electronegative ligands might adopt some structure other than a tetrahedrally coordinated one.

$$C_7^{-4}$$

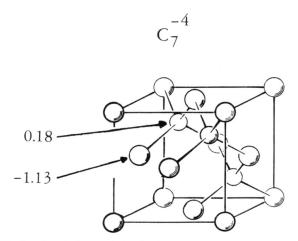

0.18 ——

−1.13 ——

**Fig. 9.**  Population analysis of a $C_7^{-4}$ fragment-within-the-solid.

When there are two defects per cell the situation is similar. With a total of four electrons per site, the two coordinate atoms of this structure carry the highest charges. Interestingly, with two electrons per site the two coordinate atoms carry the smaller charge and the least electronegative atom should occupy these sites. This is the case for the cuprite structure (discussed in V,B), which may be regarded as being derived from the chalcopyrite arrangement by removing all the Fe atoms.

## B.  Distortions of the CsCl Structure

The CsCl structure, shown in Fig. 10, contains eight-coordinate Cs and Cl atoms in symmetry-equivalent positions. We choose, as a repeat unit, the simplest possible, namely the pair of atoms shown in Fig. 10*b*, although similar arguments to the ones we will use apply to larger fragments. Figure 10*c* shows the derivation of the orbital diagram for such a fragment within the solid. It represents the orbitals of the band structure at $\mathbf{k} = 0$. Each atom is in cubal coordination, so the orbital diagram is simple to construct. We show the case where the two atoms have the same atomic orbital parameters. As may be seen, all the low-energy orbitals are occupied for eight electrons per formula unit (e.g., CsCl itself) and the diagram bears a strong resemblance to that of tetrahedral $CF_4$ in Fig. 2*a*. What happens, however, with an extra two electrons per formula unit? Just as these extra electrons led to structural distortion in $SF_4$, we expect an analogous effect here. One way the system

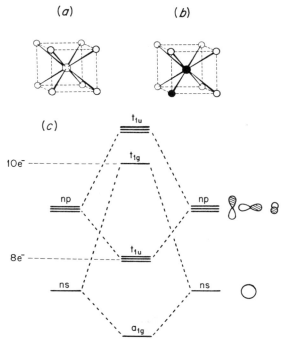

**Fig. 10.** (*a*) The CsCl structure. (*b*) The two atoms of the repeat unit used in our calculations. (*c*) The molecule orbital diagram for this fragment-within-the-solid.

can distort is to move the central atom towards a face of the cube. (Using our EHMO method, this seems to be the lowest energy distortion coordinate, although, as mentioned above, this simple approach often does not give good values of bond lengths). As a result, the HOMO decreases in energy and the system is stabilized (Fig. 11). Perturbation theory gives us a description of the HOMO in the new structure (Fig. 11). Mixing of the $a_{1g}$ and one component of the $t_{1u}$ orbitals of the CsCl structure gives an orbital which is composed of lone pairs on each of the two atoms. A population analysis shows that the originally central atom of the cube carries a positive charge and will thus attract the least electronegative atom. Inclusion of an electronegativity difference into the diagram of Fig. 10c shows that this lone pair is now predominantly located on the least electronegative atom. This is a general effect in molecular orbital theory. The deepest lying orbitals contain the largest amount of atomic orbital contribution from the most electronegative ligands.

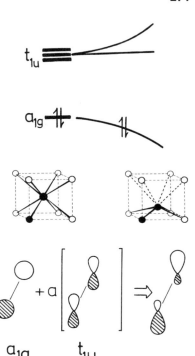

**Fig. 11.**  Energy changes on distortion showing how the nature of the HOMO is understandable from perturbation theory.

For the higher lying orbitals, lying closest in energy to the atomic orbitals of the least electronegative atom, the opposite is true.

The distortion we have described is actually found in the structure of red PbO (Fig. 12), which contains tetragonal pyramidally coordinated lead and tetrahedrally coordinated oxygen atoms. The structure is one often quoted as containing a stereochemically active lone pair on the lead atom. A similar coordination geometry is found (Zhdanov *et al.*, 1956) in the molecule Pb(diethylthiocarbamate)$_2$. Usually, of course, five-valence-pair molecules exhibit the SF$_4$ type of structure rather than the tetragonal pyramid. If the HOMO–LUMO gap is large for the 10-electron CsCl structure, then the system might not distort. The factors influencing this energy separation are not yet completely understood, but one important feature is that an increasing electronegativity difference between the two leads to a smaller gap. On this basis alone, the distorted PbO but undistorted TlI structure is understandable.

One of the features of the CsCl diagram of Fig. 10 is that the central atom is attached to the corner atoms by using t$_{1u}$ and a$_{1g}$ orbitals. It is pertinent to ask: are there any other structural units which may be substituted for the

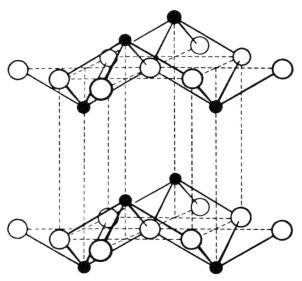

**Fig. 12.**   The structure of red PbO.

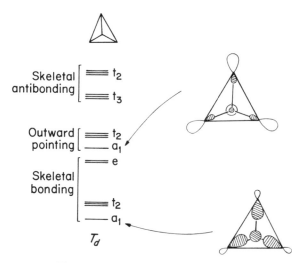

**Fig. 13.**   The orbitals of the tetrahedron.

$Cu_2O$

**Fig. 14.**   The cuprite structure.

central atom and understood in an analogous way? The tetrahedron is one candidate. Figure 13 shows the orbitals of the tetrahedron. To low energy lie six orbitals, involved in bonding within the four atom skeleton. To high energy lie six skeletal antibonding orbitals. In between lie four orbitals pointing outwards from the tetrahedron and which may be used to secure the unit to its surroundings. With a total of 20 electrons, all the skeletal bonding orbitals are filled (six bonds) and the outward-pointing orbitals are lone pairs. This is the structure found for elemental white phosphorous and for the isoelectronic species KGe.

We have not been able to find a system containing an isolated tetrahedron within a cube of atoms, but a related system is that of cuprite (Fig. 14), where the unit cell contains a stuffed tetrahedron within the cube. Other ways of looking at the structure are as two interpenetrating $\beta$-cristobalite-type lattices, or as a defect chalcopyrite structure (leaving out all the iron atoms). Although the presence of the central atom disturbs our simple picture of the tetrahedron, there are still four outward-pointing orbitals of the stuffed unit to interact with the surroundings. Figure 15 shows a simplified molecular orbital diagram of the fragment within the solid. For $Cu_2O$, all the low-energy orbitals are occupied except those involved in Cu–Cu bonding—so there are no strong bonds, at least, between the two interpenetrating $\beta$-cristobalite structures.

Other units which may occupy the cube and which are amenable to a similar treatment include the octahedron which leads to the $CaB_6$ structure. This was treated in a pioneering paper by Longuet-Higgins and de V. Roberts (1954). A more complex example is the square plane, which when inserted into six of eight adjacent cubes in a specific fashion gives the skutterudite ($CoAs_3$) structure.

The use of these methods to relate the molecular-orbital structure of complex systems to much simpler ones appears to be a very useful way of looking at solids. In Chapters 24 and 25, the description of complex structures

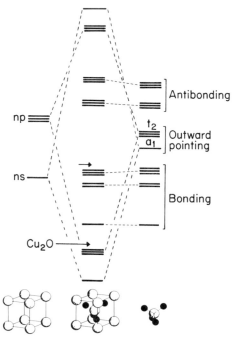

**Fig. 15.** Simplified molecular orbital diagram showing assembly of the cuprite structure from a cube of atoms plus the stuffed tetrahedron.

geometrically in terms of simpler building blocks is an advance in the conceptual ideas of crystal chemistry that we hope can be exploited when describing their electronic structure.

## VI. REACTIONS IN THE SOLID STATE

Structural transformations in the solid state occupy a fascinating branch of solid-state chemistry. They may be induced by pressure, temperature, or sometimes light. A mineralogical example of the latter is the photochemical transformation of realgar to orpiment. In this section we shall look at two chemical reactions in the solid state, one thermally and the other photochemically initiated, although the general principles are applicable to any solid state system.

On passing $S_4N_4$ over silver wire at $200°C$, $S_2N_2$ is produced and may be trapped at liquid nitrogen temperature (Cohen et al., 1976). On warming to $0°C$, the material polymerizes to give the extraordinarily interesting metallic

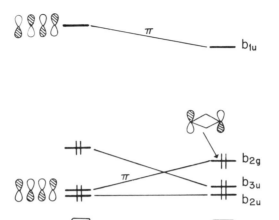

**Fig. 16.**   Orbital correlation diagram for the polymerization of $S_2N_2$.

$(SN)_x$ polymer. It is of interest to correlate the orbitals of the monomeric cyclic $S_2N_2$ as the system is gradually distorted to the polymer geometry where we use the fragment-within-the-solid approach. Figure 16 shows the result. The important feature of this diagram is that all the occupied orbitals of the cyclic monomer smoothly correlate with the occupied orbitals of the polymer. As one S–N bond within the cyclic unit is broken, another is made to the next structural unit of the chain.

A rather different result occurs if the polymerization of diacetylenes $RC\equiv C-C\equiv CR$ is followed. This process occurs in the solid state as a result of UV irradiation of the solid monomer (Cohen, 1975). How the monomer and polymer orbitals correlate is shown in Fig. 17. Here we see that one of the highest energy occupied orbitals of the monomer correlates with a lower energy unoccupied polymer orbital. This means that an unoccupied monomer orbital correlates with an occupied polymer orbital. This sort of orbital consideration was one used by Hoffmann and Woodward (1970) in their discussion of thermally and photochemically allowed reactions of organic molecules. The noncrossing of filled and empty orbitals seen in the $S_2N_2$ case is a classic example of a thermally allowed reaction. The crossing of such orbitals in the diacetylene case is a typical feature of thermally forbidden reactions. By photochemical excitation of an electron from HOMO to LUMO, however, the second process becomes allowed, since the barrier to reaction is now eliminated if one electron drops in energy and the other rises in energy as the reaction moves along the reaction coordinate. The theoretical conclusions are thus in accord with the experimental conditions actually

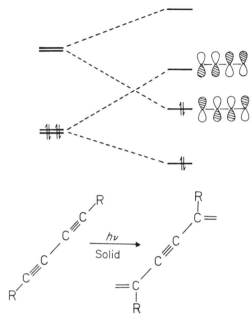

**Fig. 17.** Orbital correlation diagram for the polymerization of diacetylenes.

needed for polymerization. We expect that such a technique using orbital correlation diagrams such as these may be very useful in tracing out the minimum energy pathways for distortions in solids such as those induced by pressure.

## REFERENCES

Alcock, N. W. (1972). *Adv. Inorg. Chem. Radiochem.* **15**, 1.
Bader, R. F. (1960). *Mol. Phys.* **3**, 137.
Bartell, L. S. (1968). *J. Chem. Educ.* **45**, 754.
Burdett, J. K. (1974). *J. Chem. Soc., Faraday Trans. 2* **70**, 1599.
Burdett, J. K. (1978a). *Chem. Soc. Rev.* **7**, 507.
Burdett, J. K. (1978b). *Adv. Inorg. Chem. Radiochem.* **21**, 113.
Burdett, J. K. (1980). "Molecular Shapes." Wiley, New York.
Cohen, M. D. (1975). *Angew. Chem., Int. Ed. Engl.* **14**, 4386.
Cohen, M. J., Garito, A. F., Heeger, A. J., Macdiarmid, A. G., Mikulski, C. M., Saran, M. S.,
    and Kleppinger, J. (1976). *J. Am. Chem. Soc.* **98**, 3844.
Drago, R. S. (1973). *J. Chem. Educ.* **50**, 244.
Elian, M., and Hoffmann, R. (1975). *Inorg. Chem.* **14**, 1058.
Esperås, S., George, J. W., Husebye, S., and Mikalsen, Ø. (1975). *Acta Chem. Scand., Ser. A*
    **A29**, 141.

Gillespie, R. J. (1972). "Molecular Geometry." Van Nostrand-Rheinhold, Princeton, New Jersey,

Gillespie, R. J., and Nyholm, R. S. (1957). *Q. Rev. Chem. Soc.* **11**, 339.

Gimarc, B. M. (1979). "Molecular Structure and Bonding." Academic Press, New York.

Hoffmann, R. (1963). *J. Chem. Phys.* **39**, 1397.

Hoffmann, R. (1971). *Acc. Chem. Res.* **4**, 1.

Hoffmann, R., and Lipscomb, W. N. (1962a). *J. Chem. Phys.* **36** 2179.

Hoffmann, R., and Lipscomb, W. N. (1962b). *J. Chem. Phys.* **37**, 2872.

Hoffmann, R., and Woodward, R. B. (1970). "The Conservation of Orbital Symmetry." Verlag-Chemie, Weinheim.

Husebye, S., and George, J. W. (1969). *Inorg. Chem.* **8**, 312.

Levin, A. A. (1977). "Solid State Quantum Chemistry." McGraw-Hill, New York.

Longuet-Higgins, H. C., and deV. Roberts, M. (1954). *Proc. R. Soc. London, Ser. A* **224**, 336.

Marumo, F. (1967). *Z. Kristallogr., Kristallgeom., Kristallphys., Kristallchem,* **124**, 352.

Pearson, R. G. (1969). *J. Am. Chem. Soc.* **91**, 1252.

Pearson, W. B. (1972). "The Crystal Chemistry and Physics of Metals and Alloys." Wiley, New York.

Segal, G. A. (1977). "Semi-Empirical Methods of Electronic Structure Calculation," Parts A and B. Plenum, New York.

Streitweiser, A. (1961). "Molecular Orbital Theory for Organic Chemists." Wiley, New York.

Wolfe, S. (1972). *Acc. Chem. Res.* **5**, 102.

Zhdanov, G. S., Zvonkova, Z. V., and Rannev, N. V. (1956). *Kristallografiya* **1**, 514.

Zunger, A. (1975a). *J. Chem. Phys.* **62**, 1861.

Zunger, A. (1975b). *J. Chem. Phys.* **63**, 1713,

# 12

## Charge Density Distributions

### ROBERT F. STEWART and MARK A. SPACKMAN

|       |                                                          |     |
|-------|----------------------------------------------------------|-----|
| I.    | Introduction                                             | 279 |
| II.   | Electrostatic Properties from Difffraction Structure Factors | 281 |
| III.  | Pseudoatoms in Diatomic Molecules                        | 283 |
| IV.   | Valence Densities from Pseudoatoms                       | 286 |
| V.    | Electrical Field Gradients                              | 296 |
| VI.   | Conclusions                                             | 297 |
|       | References                                              | 298 |

## I. INTRODUCTION

A crystal is a complex molecule. The number of electrons and nuclei involved in a crystal make the system macroscopic. A quantum chemical treatment may provide some insights, but a statistical mechanical approach is needed to characterize x-ray data even at the most elementary level. A simple crystal structure may give rise to a very complicated charge density distribution. Evidence of the complexity can be found in the Fourier analysis of suitably phased and accurately measured x-ray structure factors. When the nuclear distribution, as deduced from neutron diffraction measurements, is included, the total charge density distribution can be modestly characterized. Under favorable circumstances, integrated Bragg diffraction intensities may be reduced to precise structure factor moduli. These reduced data usually have an accuracy no better than a few per cent. With suitable models, phases may be assigned to the moduli to form a data base of "observed" structure factors. Centric crystals are far more favorable for reliable phase assignments than acentric crystals. The x-ray structure factors are the Fourier components

Structure and Bonding in Crystals, Vol. I

tion. The $F_H$ are phased x-ray structure factors and the $F_N$ are reduced nuclear of the thermally weighted vibrational average of the electron density distribution in the crystallographic unit cell. The neutron structure factors are the Fourier components of the thermally weighted vibrational average nuclear distribution. To convert the latter information to a charge distribution, one may carry out a cumulant analysis (cf. Johnson, 1969) of the neutron diffraction data. Once the cumulant parameters for the several nuclei in the asymmetric unit of the crystallographic cell are determined, the structure factors for the nuclear charge distribution may be formulated by assigning nuclear charges to the nuclear form factor. We imagine that the static charge density of a nucleus is well described by a Dirac delta function. These reduced nuclear structure factors are treated as if the Thomson cross-sections were the same as for a classical electron. When these data are combined with the phased x-ray structure factors, a total charge distribution, within the resolution of an Ewald sphere, may be constructed.

Several strategies are employed to display the charge distributions. Each one, with inherent limitations, can be used to establish a tolerance level for the results. The appropriate level for comparison with theory is not obvious, particularly for a moderately complicated structure. The larger task, however, is to explore the extent to which the experimentally derived (or perhaps contrived) charge distributions provide an understanding of chemical bonding in crystals. We find that the results do not easily yield interpretations of bonding rules by way of valence bond structure, orbital populations, or electronegativities of "atoms." In fact from some displays of charge density distributions, the "atom" appears to be totally absent. The most useful results that can evolve in efforts to determine charge densities from diffraction data of crystals are the bulk physical properties dependent on the distribution. Even this desired goal may not be sufficient to characterize the properties of crystalline materials, since defects can be crucial to the property of interest.

It is probably correct to say that charge density determinations from diffraction measurements have to date provided little new understanding of bonding in crystals. Our own experience, in both an active and passive role, bears witness to few surprises. Many charge density maps have been reported in the crystallographic literature. The maps may appeal to our fantasies, fostered in an early course of beginning chemistry, but their physical content must be critically evaluated.

Clearly, the burden is to use our modest results to resolve disparity between theoretical models, or, better, to complement the data and conclusions of another experiment on the same crystal. It is with these limitations in mind that we will outline several approaches, and a few results, on the mapping of electrostatic properties from diffraction data. The emphasis will focus on charge distributions from pseudoatoms.

TABLE I

**Derived Properties from X-Ray and Neutron Diffraction Data**

| Property | | Dependence of Fourier sum on $|\mathbf{H}| = 2\sin\theta/\lambda$ (Å$^{-1}$) |
|---|---|---|
| $\phi(\mathbf{x})$ | Electrostatic potential | $|\mathbf{H}|^{-2}$ |
| $\sigma_z^d(\mathbf{x})$ | Diamagnetic shielding tensor | $|\mathbf{H}|^{-2}$ |
| $\mathbf{E}(\mathbf{x})$ | Electric field | $|\mathbf{H}|^{-1}$ |
| $\vec{\nabla}\mathbf{E}(\mathbf{x})$ | Electric field gradient | $|\mathbf{H}|^{o}$ |
| $\nabla^2\phi(\mathbf{x})$ | Charge density | $|\mathbf{H}|^{o}$ |
| $\vec{\nabla}(\vec{\nabla}\mathbf{E}(\mathbf{x}))$ | Gradient of electric field gradient | $|\mathbf{H}|^{1}$ |
| $\vec{\nabla}\rho(\mathbf{x})$ | Gradient of charge density | $|\mathbf{H}|^{1}$ |

## II. ELECTROSTATIC PROPERTIES FROM DIFFRACTION STRUCTURE FACTORS

The reduced data (x-ray, reduced nuclear, or both combined) provide a basis for mapping out a variety of electrostatic properties within the crystallographic cell. The strategies discussed here are all based on Fourier summation methods.

Table I is a summary of properties that may be mapped out in the unit cell. Also included is the dependence of the Fourier sum on $|\mathbf{H}|$ (where $|\mathbf{H}| = d^* = 2\sin\theta/\lambda$). For example, the electrostatic potential is constructed by weighting the structure factors by $|\mathbf{H}|^{-2}$, while the charge density and electric field gradients depend on sums of structure factors weighted by $|\mathbf{H}|^{o}$. Explicit formulae for the Fourier sums have been reported elsewhere (Stewart, 1979), except for the diamagnetic shielding tensor. In this case, the shielding tensor in the $z$ direction is

$$\sigma_z^d(\mathbf{x}) = \frac{C}{6\pi V}\sum_{\mathbf{H}} F_{\mathbf{H}}(2 - 3q_z^2)\exp(-2\pi i\mathbf{h}\cdot\mathbf{x})/(\sin\theta/\lambda)^2 \tag{1}$$

where $q_z$ is the direction cosine of $\mathbf{H}$ in the "$z$" direction and $C$ includes the permeability of free space, the square of the charge of an electron, and its rest mass. The box in Table I encloses the properties, namely charge distributions and field gradient results (the latter rather meager), that will be discussed in this paper.

Table II outlines the strategies that are usually employed to implement the evaluation of the several properties given in Table I. The first entails the use of

**TABLE II**

**Coefficients in the Fourier Sums for Properties in Table I**

| | |
|---|---|
| Total charge distribution | $F_{\mathbf{H}} + F_{\mathbf{N}}$ |
| Electron deformation distribution | $F_{\mathbf{H}} - F_{\mathbf{H}}^{c}$ (IAM) |
| Generalized x-ray form factors (projection methods) | $F_{\mathbf{H}} f_{\mathbf{p}}(\mathbf{H}) T_{\mathbf{p}}(\mathbf{H})$ |

$F_{\mathbf{H}} + F_{\mathbf{N}}$ in determining properties that depend on the total charge distribution. The $F_{\mathbf{H}}$ are x-ray structure factors and the $F_{\mathbf{N}}$ are reduced nuclear structure factors. Because of termination effects due to the finite size of $|\mathbf{H}|_{\max}$, Fourier sums over the $F_{\mathbf{H}} + F_{\mathbf{N}}$ have only been used to map out $\phi_{\text{tot}}(\mathbf{x})$, the thermal average electrostatic potential in a unit cell.

The second item in Table II refers to electron deformation distributions. These properties depend on the difference between the thermal mean electron density in the unit cell and a model density constructed by a superposition of isolated atoms, the electron densities of which are convoluted onto its associated nuclear distribution function. The nuclear distributions are inferred from refinements of either x-ray or neutron diffraction data. The isolated atom electron density is usually based upon an Hartree–Fock wavefunction and includes only the spherically symmetrical component of the electron density. (For example, the atomic form factor for $Si(^{3}P)$ does not include the quadrupole moment for this atom.) In this context the results are based on a reference model rather than a reference state. The model of a superposition of thermally delocalized atoms about mean positions is sometimes called the promolecule (Hirshfeld, 1971) or the IAM (Fink and Bonham, 1974). It has the great advantage that it emphasizes the redistribution of charge density due to cohesive factors in the crystal. This includes the nuclear distribution as well. One very practical result of the IAM (independent atom model) is a reduction of series termination ripples in the Fourier synthesis of the $F_{\mathbf{H}} - F_{\mathbf{H}}^{c}$ (IAM) that are used to map out the property of interest. For these conditions the property removed from the mean nuclear position ($>0.5$ Å) is relatively well mapped. The $F_{\mathbf{H}} - F_{\mathbf{H}}^{c}$ (IAM) or $\Delta F$ synthesis methods are quite widely used among charge density investigators. A number of reviews have been written and we recommend the reader consult Coppens (1977).

The third item in Table II deals with rigid pseudoatom models. In this case the $F_{\mathbf{H}}$ are projected onto generalized x-ray scattering factors (Stewart, 1976) $[f_{\mathbf{p}}(\mathbf{H})$ in Table II] which are appropriately modified by the Fourier transform of the nuclear distribution function $[T_{\mathbf{p}}(\mathbf{H})$ in Table II]. For purely harmonic motion of the nuclei, the $T_{\mathbf{p}}(\mathbf{H})$ simply incorporate Debye–Waller factors. The $f_{\mathbf{p}}(\mathbf{H})$ is the Fourier transform of the pseudoatom charge density at nuclear site p in the crystal lattice. In principle it represents the electron charge distribution of the pseudoatom with its associated nucleus at rest. The Fourier sum of the $F_H f_p(H) T_p(H)$ give the weight of the pseudoatom contribution to the property of interest. For electron density distributions, this weight is an electron population parameter. If the pseudoatom is effectively rigid in its motion with the nucleus, the electron population of the pseudoatom density basis functions can be used to estimate the static electron density distribution. The projection methods outlined here are akin to the previously mentioned Fourier methods if the least square error,

$$\varepsilon = \sum_H w_H |F_H - F_H^c(\text{pseudoatom})|^2 \qquad (2)$$

is minimized with respect to the electron population parameters of the pseudoatoms. We emphasize here that $F_H$ (phased) is considered the data base. If the least squares is based on $|F_H|^2$ or $|F_H|$, then the projection methods of least squares no longer have a simple relation to the Fourier methods under discussion. When we refer to results from pseudoatom analysis, we mean a least squares relation as given by (2) and not the usual least squares methods widely used in crystallography. From a statistician's point of view these Fourier techniques suffer from a bias in the observations. In spite of this limitation, a variance study of the Fourier properties can be pursued as outlined by Rees (1977). For appropriate unbiased least squares analysis, a projection onto $|F_H|^2$ is necessary (Wilson, 1979). The explicit relation between minimization of (2) with respect to an electron population parameter and a Fourier synthesis of $F_H f_p(H) T_p(H)$ as indicated in Table II is easily verified. The convolution approximation or assumed rigidity of the pseudoatom charge density with nuclear motion is an essential feature of the model. Other limiting features occur in applications of the model to diffraction data and will be discussed below.

## III.   PSEUDOATOMS IN DIATOMIC MOLECULES

An isolated diatomic molecule is hardly a complex solid and should not be confused with an AB crystal. For some AB molecules, however, relatively accurate wavefunctions have been determined and can serve to educate us on bonding effects. The particular emphasis here is on the difference between an accurate molecular form factor with that comprised of isolated atoms placed (or phased) at the internuclear distance. A Fourier transform of the difference, $F_{mol}(K) - F_{IAM}(K)$, would produce a deformation density map. The task is to find a suitably efficient model to account for the difference or total molecular form factor.

One may represent the total electron density with an infinite multipole expansion about a particular nucleus. Such an expansion is very slowly convergent and does not have local charge density features that can be easily incorporated into a treatment of the nuclear motion. But a second center of infinite multipoles introduces an overcomplete set of density basis functions that leads to indeterminant radial functions. A small finite set of multipoles about the several nuclear centers can be determined, however.

For the simple case of a diatomic molecule it is possible to solve uniquely for the radial form factors of each multipole on the two nuclear centers from an accurate molecular form factor (Stewart *et al.*, 1975). The small finite multipole expansion on each center is called a generalized x-ray scattering factor, and its Fourier transform defines the charge density of the pseudoatom. In applications to first row atom diatomic hydrides and to $CO(^1\Sigma^+)$, $CO(^3\Pi)$, $BF(^1\Sigma^+)$, and $N_2(^1\Sigma_g^+)$, it was found that convergence to the molecular form factor is accurately achieved with rather small multipole expansions (Bentley and Stewart, 1975). With six functions, i.e., $[2|2]$ expansion, the relative root mean square difference is 0.2%. The notation $[J|K]$ refers to the highest order multipole of the $a|b$ pseudoatoms, respectively. Thus a $[2|2]$ expansion includes monopoles, dipoles and quadrupoles on both nuclear centers $a$ and $b$ in the diatomic molecule. The pseudoatoms, or corresponding generalized x-ray scattering factors, have the interesting property that they satisfy all one-center averages of the form $\langle h(r_a)P_j(\cos\theta_a)\rangle$ for all $j \leqslant J$ regardless of the highest multipole order $K$ for the $b$ pseudoatom. The function $h(r_a)$ is arbitrary and $\langle f(\mathbf{r})\rangle$ denotes the average, $\int \rho_{mol}(\mathbf{r})f(\mathbf{r})\,d\tau$. This means that the pseudoatoms in superposition satisfy a very large number of electrostatic properties of the molecule. For example, with a $[0|0]$ expansion, the electron-nuclear potentials $\langle r_b^{-1}\rangle$, the electron charge density and the cusp condition at nuclei $a$ and $b$, the total charge, and the dipole moment are correctly given by the two monopole functions. At a $[2|2]$ expansion, many more properties, such as the electric fields at $a$ and $b$ and electric field gradients about $a$ and $b$, are also correctly represented. On the other hand, the pseudoatoms by themselves do not give unique properties. Atomic charges of the pseudoatoms are not a property unique to a molecule; they are unique only for a given $[J|K]$ expansion. The diffuse radial parts of the pseudoatom on one center are sensitive to the total number of multipoles assigned to the other pseudoatom. A rapid convergence in the mean square fit to the molecular form factor occurs with small expansions, but the pseudoatom properties are strongly dependent on $[J|K]$ (cf. Bentley and Stewart, 1975).

An illustrative example is given here for the phosphorus nitride molecule. We use the near Hartree–Fock wavefunction, reported by McLean and Yoshimine (1967), to compute a molecular form factor from which a $[2|2]$ expansion of generalized x-ray scattering factors is determined. The deformation density, the difference between $\rho_{mol}(\mathbf{r})$ and the densities of the ($^4$S) states for P and N at a separation of 2.818 a.u. (1.491 Å), is shown in Fig. 1. The solid

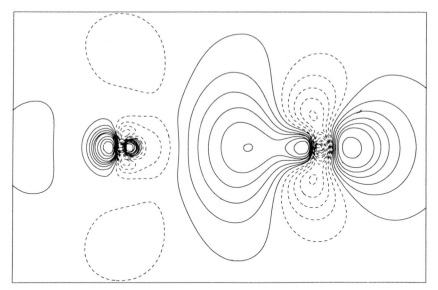

**Fig. 1.** Theoretical deformation density for phosphorous nitride. Solid contours start at +0.01 a.u. and increase in steps of +0.02 a.u.; dashed contours start at −0.01 a.u. and decrease by −0.02 a.u.; 1 a.u. = 6.748 e/Å³.

lines are contours of net electron charge density in units of electrons·bohr⁻³ (1 e/bohr³ = 6.75 e/Å³). Dashed lines are correspondingly deficient electron densities. Upon bond formation, the charge density about nitrogen is built up in the bond and on the back side, with deficiencies in the density above and below the nitrogen nucleus in a region normal to the P-N bond. By contrast, the phosphorous "atom" deforms with a large concentration of charge on the backside of P in the PN molecule and a small migration of charge into the "bonding" region. This feature is also seen in similar deformation maps for SiO and CS (C. Ceccarelli and R. F. Stewart, unpublished work, 1979). We also observe this trend in deformation density maps from diffraction data for $SiO_2$ (low quartz) and $Al_2O_3$ (corundum). A second difference density map is shown in Fig. 2. This map is based on subtraction of [2|2] pseudo-atoms from the molecular density. The maximum contours in the "bonding" regions are the smallest contours (±0.01 e/bohr³) plotted in Fig. 1. The large dashed lines are null points in the map. The small deformation density left about the phosphorous nucleus has octopolar symmetry, while that about nitrogen has octopolar structure in the more remote regions with a hexa-decapole structure close to the nuclues. Inclusion of these angular symmetries in the pseudoatoms would produce a much flatter difference map, but we emphasize that a [2|2] expansion accounts for almost all the deformation density displayed in Fig. 1.

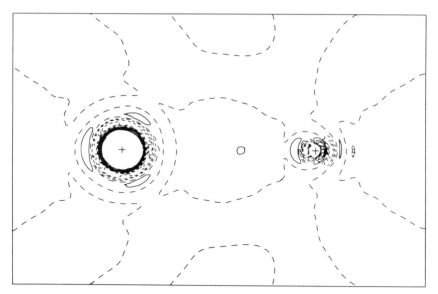

**Fig. 2.** Theoretical difference density of PN with [2|2] pseudoatoms subtracted from the molecular electron density. Large dashed lines are lines of zero density; solid contours are +0.01 a.u., short dash contours are −0.01 a.u.

In our formal study of the least squares expansion of multipoles, we learned that molecular or crystal properties are projected into the pseudoatoms. With a reasonable balance of the density basis functions, a very good mean square fit to the molecular form factor is achieved with a few functions. Properties of the pseudoatoms are not unique, however, and depend on the choice of the several centered multipoles. For actual applications to real diffraction data, this suggests that we look for static charge properties of the crystal system, but we need not dwell excessively on the pseudoatom charge, centroid, or its special charge distribution.

## IV.  VALENCE DENSITIES FROM PSEUDOATOMS

The functional equations that were used to determine generalized x-ray scattering factors in diatomic molecules have been extended to polyatomic systems (Stewart, 1977a). Suitable solutions are tedious and difficult to obtain. Moreover, in applications to real diffraction data, one is forced to use restricted radial functions for the generalized x-ray scattering factors. This loss of flexibility means that only estimates of electrostatic properties are possible. Despite this limitation, we do find that restricted radial functions can provide rather good fits to observed structure factors. These restricted pseudoatoms can then be used to map out a *model*, static electron density distribution, which is derived from measured x-ray diffraction intensities. In the section below we give some valence electron density results for several simple crystals.

The valence electron density can reveal bonding effects, but is rather different from a deformation density map. In order to map out valence electron distributions, one must first *define* the "core" electron density of an atom. In the examples to be given we have taken two slightly different definitions. For the elemental crystals Be, C(graphite), and C(diamond), the core electron distribution is defined as the orbital product of the 1s *canonical* atomic orbital from the Hartree–Fock wavefunction of the ground state atom. This definition was suggested by Stewart (1968). By subtracting out these core distributions from the total density (in our case, based on a pseudoatom model), the residual is the valence electron distribution. The valence densities for the crystals comprised of second-row atoms are based on core densities constructed from *density localized* atomic orbitals (Stewart, 1980).

Our first example is a valence electron distribution for crystalline beryllium (Stewart, 1977b). The data were taken from Brown (1972) and analyzed with generalized x-ray scattering factors that included two monopoles and the spherical surface harmonics $P_5^3(\cos \theta)\sin 3\phi$, $P_6^0(\cos \theta)$, and $P_7^3(\cos \theta)\sin 3\phi$. The space group is $P6_3/mmc$ and the site symmetry for the Be pseudoatom is $\bar{6}m2$. Figure 3 is a plot of the results in the (110) plane in the lattice. Perhaps the most revealing features are displayed in the (110) section. Apparently the "bonding" of the Be pseudoatoms is along the $c$ axis and forms a maximum contribution (0.32 e/Å$^3$) near the tetrahedral "vacancy." A modest concentration of charge (0.28 e/Å$^3$) also occurs in the octahedral vacancy. The small negative region of electron density about the Be nucleus at $z = \frac{1}{4}$ is completely offset by addition of the core electron density distribution. The pseudoatom model predicts a bonding scheme of Be to *its* second nearest neighbor and includes participation by five Be pseudoatoms as the primary contributors to the valence charge density maximum near the tetrahedral vacancy. It is of interest that this model, guided by no theoretical considerations other than respect for the site symmetry of $\bar{6}m2$, is in virtual quantitative agreement with an augmented plane wave calculation reported by Inoue and Yamashita (1973).

Our second example is an elemental crystal also in space group $P6_3/mmc$. The graphite structure of carbon has two atoms in the asymmetric unit; both pseudoatoms have a point symmetry of $\bar{6}m2$. With the diffraction data of Chen *et al.* (1977), the same set of surface harmonics were applied as in the case of crystalline beryllium. We were pleased to learn that the pseudoatom

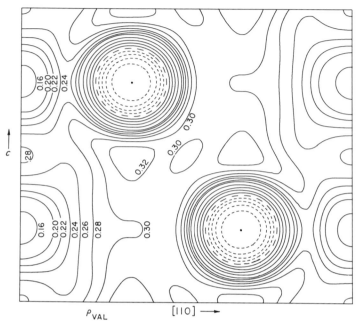

**Fig. 3.**    Valence charge density map for Be in the (110) plane. Border along $c$ extends from $-\frac{1}{2}$ to $\frac{1}{2}$ and for [110] from $(0, 1)$ to $(1, 0)$ for $(x, y)$ lattice coordinates. Units are $e/\text{Å}^3$. (Taken from Stewart, 1977$b$.)

fit to the data was quite different from the Be results. The harmonics of significant population were the core and valence monopoles, a quadrupole term, $P_2(\cos\theta)$, and an octopole, $P_3^3(\cos\theta)\sin 3\phi$. Both carbon pseudoatoms were virtually identical within the estimated standard deviations, and neither atom had a significant electron population of the hexadecapole term, $P_4(\cos\theta)$. Figure 4 is a plot of the pseudoatom model valence density in the (001) plane at $z = \frac{1}{4}$, the mean nuclear position of the carbon atoms. The bonding is between nearest neighbor carbon atoms in this plane and has a maximum charge density of 2.0 $e/\text{Å}^3$ at about one-third of the distance between the atoms. We found no evidence of bonding between the planes; density plots perpendicular to the basal planes appear to be a simple van der Waals packing of infinite sheets of fused benzene rings (cf. Chen *et al.*, 1977). Although the model had the flexibility to allow discrimination between the pseudoatom over the middle of the ring and the one directly over another carbon pseudoatom, the relevant parameter in the least squares fit collapsed to the same radial distribution. We inferred that the diffraction data had no information on this subtle detail on bonding of pseudoatoms in graphite.

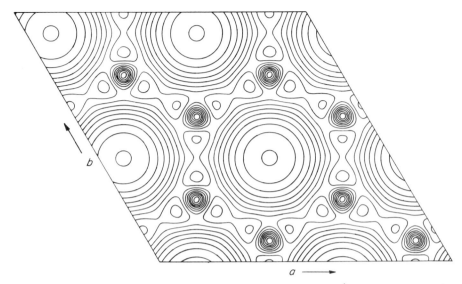

**Fig. 4.**   Valence charge density map for graphite in the plane $z = \frac{1}{4}$. Border on $a$ extends from $-\frac{1}{2}$ to $3/2$ in $x$ and along $b$ from $-\frac{1}{2}$ to $3/2$ in $y$. Contours start at $0.2$ e/Å$^3$ in the middle of the ring and increase in steps of $+0.2$ e/Å$^3$. Maximum contours are $2.0$ e/Å$^3$ in the carbon pseudoatom bonds and decrease to $0.4$ e/Å around the carbon pseudoatom nucleus.

The accuracy of the graphite data ($\sim 3.3\%$) was considerably inferior to the accuracy of the beryllium data ($\sim 0.6\%$).

A less stable form of crystalline carbon is the diamond structure. The powder diffraction data of Göttlicher and Wölfel (1959) has been analyzed with the pseudoatom model (Stewart, 1973). One limitation of powder diffraction data is that the anisotropy of charge distribution information is somewhat limited. The diamond crystal of carbon is space group $Fd3m$ with the site symmetry of the carbon atom $\overline{4}3m$. The first nonvanishing spherical surface harmonics are an octopole, $P_3^2(\cos 2\theta)\sin 2\phi$, and a sum of two hexadecapoles, $\frac{1}{4}[P_4^4(\cos 2\theta)\cos 4\phi + P_4(\cos 2\theta)]$. The pseudoatom models contained these two functions plus a valence monopole term. The relative root mean square fit to the data (.0084) was comparable to the Be results (.0081). A plot of these results in the $(\overline{1}10)$ plane is shown in Fig. 5. The bonding is between nearest neighbor carbon pseudoatoms with a maximum charge density of $2.0$ e/Å$^3$ at about one-third of the distance as in graphite. The charge density at the mid-point between nearest neighbor carbons is $1.7$ e/Å$^3$, which is less than the corresponding result of $1.9$ e/Å$^3$ found in graphite. The pseudoatom model, in a modest way, does reveal more charge density in a "single plus one-half double bond" picture, expected for graphite.

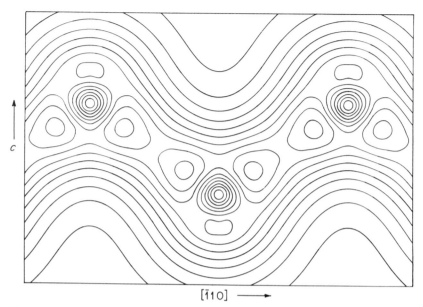

**Fig. 5.** Valence charge density map for diamond in the ($\bar{1}$10) plane. Border on $c$ varies from $-3/8$ to $3/8$ in $z$; border on [$\bar{1}$10] moves from $(-1/16, -1/16)$ to $13/16, 13/16)$ in the $(x, y)$ fractional coordinates. Lowest contour starts at 0.2 e/Å$^3$ and increases in units of 0.2 e/Å$^3$. Maximum contours are 2.0 e/Å$^3$ in the C pseudoatom bonds and decrease to 0.4 e/Å$^3$. Maximum contours are 2.0 e/Å$^3$ in the C pseudoatom bonds and decrease to 0.4 e/Å$^3$ around the C pseudoatom nucleus.

We now turn to examples that consist of second-row atoms. Elemental Si crystallizes in space group $Fd3m$, and the pseudoatom has a site symmetry of $\bar{4}3m$. One may use the same harmonics as for diamond. We have used the compiled data set for Si reported by Yang and Coppens (1974) as a basis for a pseudoatom analysis. The compiled set include Pendellösung fringe measurements by Aldred and Hart (1973) and Hattori *et al.* (1965), as well as the weak 222 reflection (Roberto and Batterman, 1970) and 442 reflection (Trucano and Batterman, 1972) measurements. These structure factor data are probably the most accurate ($\sim 0.1\%$) yet obtained. With a pseudoatom model similar to diamond, we were able to fit the data with a relative root mean square error of 0.24%. A pseudoatom valence density map is displayed in Fig. 6. The section is in the ($\bar{1}$10) plane as for diamond in Fig. 5. Note that the contour level is at one-quarter the intervals used for C(diamond). The bonding is between nearest neighbors with a maximum of 0.55 e/Å$^3$ at approximately one-third the distance to the nearest Si pseudoatom. This is considerably less than for the two carbon structures discussed above. The midpoint

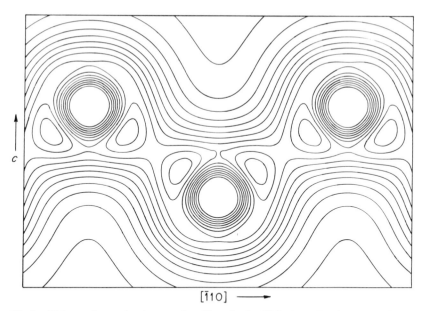

**Fig. 6.**   Valence charge density map for silicon in the (Ī10) plane. Borders cover the same space as described in Fig. 5. Smallest contours start at 0.05 e/Å³ and increase in steps of 0.05 e/Å³. Maximum contours are 0.55 e/Å³ in the Si pseudoatom bonds and decrease to 0.1 e/Å³ about the Si pseudoatom nucleus.

region between the nearest Si pseudoatoms has a charge density of 0.47 e/Å³. This is about one-fourth the density found in the similar valence electron distribution of diamond. As a general rule, the second-row atoms display rather less valence electron density than first row atoms.

Simple oxides of Si and Al have also been analyzed with pseudoatoms. For these systems, however, the electron populations of *localized* K, L, and M monopole shells (Stewart, 1980) were allowed independent variations. This added flexibility can lead to rather more structure in the "valence" density near the nucleus. In contrast to the elemental crystals discussed above, the pseudoatom monopoles contain total electron charges that differ from their nuclear charge. The higher multipoles always integrate to zero, so that these shape the charge distribution of a pseudoatom, but do not play a direct role in "charge transfer."

The structure factors for low-quartz as reported by LePage and Donnay (1976) have been analyzed with generalized x-ray scattering factors. The space group is $P3_221$; the Si pseudoatom has a site symmetry of 2 and O is in a general position of the cell. It was found that octopoles on the Si psueodatom were heavily populated, while quadrupole and hexadecapole

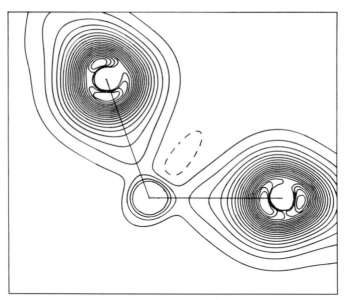

**Fig. 7.** Valence charge density map for low quartz in the plane of Si and two O pseudoatoms. The Si—O distance is 1.609 Å; the two O pseudoatoms are related by a two-fold axis bisecting the O—Si—O bond angle. Dashed contour is zero. Positive contours increase by 0.4 e/Å³. Maximum contour about the Si pseudoatom is 1.6 e/Å³; maxima about O are 8.0 e/Å³.

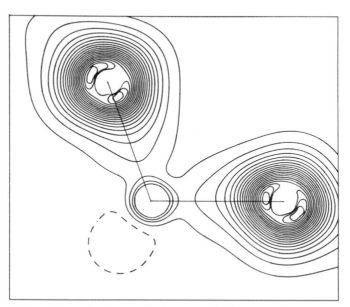

**Fig. 8.** Valence charge density map for low quartz in the plane of Si and two O′ pseudo-atoms. The Si—O′ distance is 1.611 Å; the two O′ pseudoatoms are related by a two-fold axis bisecting the O′—Si—O′ bond angle. Dashed contour is zero. Positive contours increase by 0.4 e/Å³. Maximum contours about Si and O′ are the same as in Fig. 7.

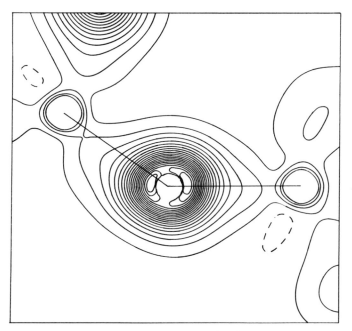

**Fig. 9.**   Valence charge density map for low quartz in the plane of O and two Si pseudoatoms. The Si—O distance is 1.611 Å, and the O—Si′ distance is 1.609 Å. Dashed contour is zero. Positive contours increase in units of 0.4 e/Å³; maxima about Si, O, and Si′ pseudoatoms are the same as in Fig. 7.

functions played the dominant role in shaping the oxygen pseudoatom. The present relative root mean square fit to these data is 0.75%. Several sections of the valence density map in the cell have been constructed. Figure 7 contains the plane of the Si and two equivalent O atoms with the "short" bond of 1.609 Å. The bond is strongly directed along the Si–O nearest neighbors. The maximum about Si is 1.6 e/Å³, but the valence charge about O reaches a maximum of ~8 e/Å³. A very similar valence density is shown in Fig. 8. This plane is defined by O—Si—O atoms with the long bond length of 1.611 Å. The pseudoatom model builds up a charge density about Si with a near tetrahedral shape. In Fig. 8 there is a slight hint of the valence density contours describing a bent bond. This directional feature is more pronounced in the plot given in Fig. 9. The plane contains the Si—O—Si pseudoatoms with the long bond (1.611 Å) to the left and the shorter bond (1.609 Å) to the right. The line of least descent from Si down to 0.8 e/Å³ and then least ascent up towards the O pseudoatom to 8 e/Å³ is clearly an arc that deviates from the straight line that connects the 1.611 Å 0–Si bond. The least descent

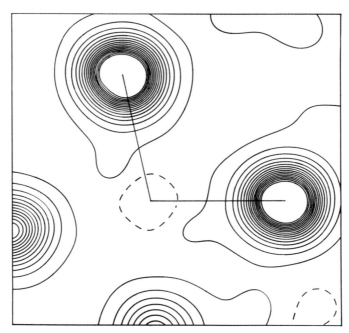

**Fig. 10.** Valence charge density map for corundum in the plane of Al and two O pseudo-atoms. The Al—O distances are equal and make the short bond length of 1.854 Å. Dashed contour about the Al pseudoatom is zero. Positive contours are in units of 0.4 e/Å³ and increase to a maximum 6.4 e/Å³ about the oxygen pseudoatoms.

from O to Si on the short bond side appears to follow a straight line. The valence maps in Figs. 7, 8, and 9 do not reveal direct bonding between O pseudoatoms or between Si pseudoatoms.

The corundum structure, $Al_2O_3$, has also been analyzed with generalized x-ray scattering factors. The diffraction data were collected by Y. LePage (Personal communication, 1978). The space group is $R\bar{3}c$ and the site symmetries for Al and O are 3 and 2 respectively. Similar angular components of the multipoles used for $SiO_2$ were also utilized in mean square fits to the $Al_2O_3$ data. The relative root mean square error was 0.76%, so that the quality of fit is comparable to low quartz. The valence densities, however, differ markedly from quartz. One section is shown in Fig. 10. This plane contains the Al and two oxygens with equal short bonds of 1.854 Å. The valence charge density about Al is virtually null, while that about oxygen is nearly spherically symmetrical and reaches a maximum of 6.5 e/Å³. The valence charge appears to be slightly polarized by the Al, but not in

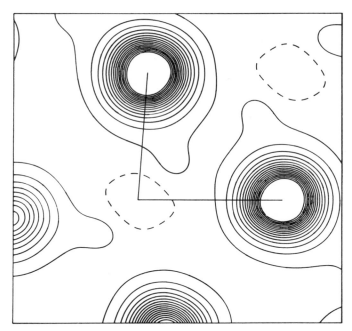

**Fig. 11.** Valence charge density map for corundum in the plane of two Al and two O pseudoatoms. The O—Al vertical distance is 1.854 Å; the Al—O horizontal distance is 1.972 Å. The midpoint between the two oxygens (and two aluminums) is at $(\frac{1}{2}, \frac{1}{2}, \frac{1}{2})$ in the cell with point symmetry of $\bar{1}$. Dashed contours about the aluminums are zero. Positive contours are the same as for Fig. 10.

the direction of the Al–O bond. A second section of the valence electron density is displayed in Fig. 11. This plane contains the O—Al—O moiety where the near vertical O–Al bond is 1.854 Å and the horizontal Al–O bond is 1.972 Å. The gross features resemble the results in Fig. 10. There is some hint in Fig. 11 that oxygens may form weak bonds. But the dominant picture, as illustrated in Figs. 10 and 11, is one of Al cations immersed in a sea of oxygen anions.

Throughout this discussion, we have avoided tabulation of the many parameters of the pseudoatom multipoles, but have instead emphasized the static *model* crystal electron distributions constructed from superposition of the sundry terms. In our studies of diatomic molecules (*vide supra*), we learned that pseudoatom properties in isolation can be misleading. An example in crystal structure analysis occurs here in comparing quartz with corundum. The valence density in Si appears to be highly covalent; that in quartz is a far more covalent picture (Figs. 7, 8, and 9) than the "ionic" bonding features that appeared in $Al_2O_3$ (Figs. 10 and 11). The pseudoatom charges found in $SiO_2$ are $q_{Si} = 1.027$ ($\pm 0.094$) and $q_O = -.514$ ($\pm 0.048$), while in corundum we found $q_{Al} = 0.41$ ($\pm 0.23$) and $q_O = -.274$ ($\pm 0.065$). The caveat, here, is that one must plot out the pseudoatom results to gain information about the crystal system; the pseudoatom charges alone are not sufficient information to resolve questions of ionicity or covalency.

The monopole functions centered on Al, Si and O consist of three shells which were constructed from Hartree–Fock atomic orbitals. The third monopole term was allowed to vary in its "size" or radial extent. Each shell was given a separate electron population parameter. The net charges, cited above, refer to the total charge content in the monopole functions which extend to infinity in the radial coordinate. At a radius of 4.0 Å from the Al nuclear site, the net charge is $+0.42$, and from the Si site the net charge is $+1.03$. At these distances the pseudoatom density functions shield the nucleus to near saturation so that when viewed from 4.0 Å or larger the pseudoatom appears to be a point charge of $+0.41$ (Al), 1.03 (Si), and so forth. The distribution functions we have shown in Figs. 7–11, on the other hand, are at much higher resolution and one mentally tends to integrate the distributions in volume elements much smaller than a 4 Å sphere. For example, in a sphere of 0.8 Å about the Al site of corundum, the net charge is $+3.0$, while a sphere of $+0.58$ Å about Si in low quartz contains a net charge of $+4$. In a sphere of 1.61 Å the $q_{Si}$ is $+1.72$; in spheres 1.85 Å and 1.97 Å about Al, the net charges are $+1.02$ and $+0.91$, respectively. We see that the pseudoatom net charges represent an integrated property in volume elements that are large compared to spheres of radii that are comparable to bond lengths. Because of the continuous variation of the electron distribution in molecules or crystals, the counting of electron charge is hopelessly dependent on the nature of the distribution function over which the integration is taken. Clearly, the charge of an "atom" in a molecule or crystal is not an observable; it is arbitrary.

## V.  ELECTRICAL FIELD GRADIENTS

There is some prospect that pseudoatom results may be used to characterize electrical field gradients in crystals. If this model successfully deconvolutes the electron distribution from the nuclear motion, then one may be optimistic in using pseudoatom results to complement information from NQR experiments. Probably the optimal systems for such studies are Al silicates, for which mean square amplitudes of vibration are both small and nearly isotropic. Although absolute values of electric field gradients from x-ray data may prove difficult to determine, the orientations of the field gradients are most amenable to a reliable determination. Detailed results for silicates have yet to be reported.

Electric field gradient estimates from a pseudoatom model have been reported for the organic molecular crystals of 1,1'-azobiscarbamide and melamine (Cromer *et al.*, 1976). A somewhat surprising result occured for the former compound. The amide N(2) in 1,1'-azobiscarbamide apparently has a principal field gradient component, $\lambda_{zz}$, *perpendicular* to the amide plane. The lone pair electron density associated with N(2) is built up by a superposition of dipoles and octopoles, which make no contribution to the field gradient at N(2). Definitive results on the static charge densities in these crystals, however, are hampered by the large amplitudes of motion of the nuclei. The extent of deconvolution is not known, so that thermal motion can be confused with components of the static electrical field gradient. For molecular crystals, a reduced temperature study is clearly desirable and necessary before quantitative determinations of electric field gradients from x-ray data can be seriously undertaken.

The estimate of the electric field gradient about Fe in iron pyrite has also been reported (Stevens *et al.*, 1980). The valence contribution was based on a pseudoatom analysis of the x-ray diffraction data. Stevens *et al.* found that the pseudoatom model predicted a splitting of 0.8 ($\pm 0.5$) mm/s compared with the observed splitting of 0.635 mm/s in the Mössbauer spectrum. Although the value based on x-ray data lacks the accuracy of the Mössbauer measurement, it gives the proper sign and correct order of magnitude to provide some insight into the origin of the observed quadrupole splitting.

## VI. CONCLUSIONS

Electron distributions may be modestly determined from x-ray diffraction data that are accurate to within a few percent. Best results are achieved with well ordered crystal structures that have relatively small mean square amplitudes of atomic vibrations. Bonding details are more easily seen in first-row atom materials but second-row atoms are also amenable to study. Transition-metal structures (the first row) have not been covered in this paper. In this report we have emphasized that the pseudoatom model is one means to extract static density information from x-ray diffraction data. The complexity of the structure is not an inherent limitation, but to date the model has been applied to simple crystal structures.

The reliability of the pseudoatom model can only be established by correlating results from several structures of similar chemical bonding or functionality. The model may also be used to estimate physical properties which can be used to complement or confirm results from a separate experiment. It may well turn out that electron distributions determined from x-ray data in complex solids will give new insights into chemical bonding. For simple crystal structures, however, most results have confirmed our previous ideas on chemical bonding.

## ACKNOWLEDGMENT

This research has been supported by a National Science Foundation Grant CHE-77-09649.

## REFERENCES

Aldred, P. J. E., and Hart, M. (1973). *Proc. R. Soc. London, Ser. A* **332**, 223–238.
Bentley, J. J., and Stewart, R. F. (1975). *J. Chem. Phys.* **63**, 3794–3803.
Brown, P. J. (1972). *Philos. Mag.* **26**, 1377–1394.
Chen, R., Trucano, P., and Stewart, R. F. (1977). *Acta Crystallogr., Sect. A* **A33**, 823–828.
Coppens, P. (1977). *Angew. Chem., Int. Ed. Engl.* **16**, 32–40.
Cromer, D. T., Larson, A. C., and Stewart, R. F. (1976). *J. Chem. Phys.* **65**, 336–349.
Fink, M., and Bonham, R. (1974). "High Energy Electron Scattering." Van Nostrand-Reinhold, Princeton, New Jersey.
Göttlicher, S., and Wölfel, W. (1959). *Z. Electrochem.* **63**, 891–901.
Hattori, H., Kuriyama, H., Katagawa, T., and Kato, N. (1965). *J. Phys. Soc. Jpn.* **20**, 988–996.
Hirshfeld, F. (1971). *Acta Crystallogr., Sect. B* **B27**, 769–781.
Inoue, S. T., and Yamashita, J. (1973). *J. Phys. Soc. Jpn.* **35**, 677–683.
Johnson, C. K. (1969). *Acta Crystallogr., Sect. A* **A25**, 187–194.
LePage, Y., and Donnay, G. (1976). *Acta Crystallogr., Sect. B* **B32**, 2456–2459.
McLean, A. D., and Yoshimine, M. (1967). *IBM J. Res. Dev.* **12**, Suppl. 82.
Rees, B. (1977). *Isr. J. Chem.* **16**, 180–186.
Roberto, J. B., and Batterman, B. W. (1970). *Phys. Rev. B: Solid State* [3] **2**, 3220–3226.
Stevens, E. D., DeLucia, M. L., and Coppens, P. (1980). *Inorg. Chem.* **19**, 813–820.
Stewart, R. F. (1968). *J. Chem. Phys.* **48**, 4882–4889.
Stewart, R. F. (1973). *J. Chem. Phys.* **58**, 4430–4438.
Stewart, R. F. (1976). *Acta Crystallogr., Sect. A* **A32**, 565–574.
Stewart, R. F. (1977a). *Isr. J. Chem.* **16**, 124–131.
Stewart, R. F. (1977b). *Acta Crystallogr., Sect. A* **A33**, 33–38.
Stewart, R. F. (1979). *Chem. Phys. Lett.* **65**, 335–342.
Stewart, R. F. (1980). *NATO Adv. Study Inst. Ser. Ser. B* **48**, 427–431.
Stewart, R. F., Bentley, J. J., and Goodman, B. (1975). *J. Chem. Phys.* **63**, 3786–3793.
Trucano, P., and Batterman, B. W. (1972). *Phys. Rev. B: Solid State* [3] **6**, 3659–3666.
Wilson, A. J. C. (1979). *Acta Crystallogr., Sect. A* **A34**, 474–475.
Yang, Y. W., and Coppens, P. (1974). *Solid State Commun.* **15**, 1555–1559.

# 13

## Some Aspects of the Ionic Model of Crystals

### M. O'KEEFFE

| | | |
|---|---|---|
| I. | Introduction | 299 |
| II. | The Energy | 300 |
| III. | Bond Lengths | 301 |
| IV. | Ionic Radii | 303 |
| V. | Structural Predictions | 304 |
| VI. | Close Packing (Eutaxy) | 307 |
| VII. | van der Waals Energy and Structure | 308 |
| VIII. | The Bulk Modulus and Elastic Constants | 308 |
| IX. | The Volume (Density) | 310 |
| X. | Polarizability and Polarization | 311 |
| | A. Basic Terms | 311 |
| | B. Numerical Values | 313 |
| XI. | Madelung Potentials and Energy Levels | 316 |
| | A. Potentials | 316 |
| | B. Electronic Energy Levels | 319 |
| XII. | Some Conclusions | 319 |
| | References | 320 |

## I. INTRODUCTION

The ionic model of cohesion in solids is remarkably successful when applied to calculating the cohesive energy (lattice energy) of simple crystals such as NaCl. It also provides a convenient foundation for discussion of structural features (coordination numbers, interatomic distances) in simple solids. Because it is easy to understand and apparently easy to apply, it is used very frequently—often in situations where it is not likely to be the most appropriate model. Often, too, it is used incorrectly with the result that either erroneous predictions are attributed to the model or unwarranted conclusions about the nature of the bonding in crystals are drawn.

This introductory chapter reviews the salient features of the model and attempts to define the range of applicability of the theory (both to types of

**299**

Structure and Bonding in Crystals, Vol. I

problem and types of crystal). It seems worth repeating the basic principles, as quite often one sees them misstated in texts, but only the barest outline is necessary; more details and references to early work can be found in the classical reviews by Born and Huang (1954), by Pauling (1960), and by Tosi (1964). Some of the material included, however, is new.

## II. THE ENERGY

The essence of the model, as applied to high symmetry crystals (such as those with the NaCl structure), is to express the energy of interaction of two ions A and B as

$$E = z_A z_B e^2/(4\pi\varepsilon_0 r_{AB}) + E_{rep} \qquad (1)$$

where $z_A e$, $z_B e$ are the charges on ions separated by $r_{AB}$ and $E_{rep}$ is a short-range repulsive term. In general, it is common to add the effects of polarization of the ions in low symmetry environments, and a van der Waals (London) attraction term. However, as detailed below, it is my opinion that when these terms are sufficiently large to require specific evaluation, the theory can no longer be applied in a quantitative way.

For simplicity following Pauling (1928, 1960), the repulsive energy is written

$$E_{rep} = \beta_{AB}\beta_0 e^2 (r_A + r_B)^{n-1}/r_{AB}^n \qquad (2)$$

Here $\beta_0$ is a characteristic proportionality constant; $\beta_{AB}$ is taken as unity for cation–anion repulsions and equal to 0.75 (1.25) for anion–anion (cation–cation) repulsions; $n$ is a parameter that depends (only) on the nature of the ion pair; and $r_A$ and $r_B$ are parameters (basic radii) characteristic of the ions, that measure their size. Clearly $\beta_0$ and $(r_A + r_B)$ are not independent parameters. This expression for $E_{rep}$ can be criticized on the grounds that the values of $\beta_{AB}$ are somewhat arbitrary and more particularly that it will not in general be correct to use the same $n$ for repulsions between both anions and cations. Other authors prefer a different functional form for $E_{rep}(r_{AB})$, such as the Born-Mayer exponential form (see Tosi, 1964). Recently a less empirical approach to the repulsion energy has been developed (Gordon and Kim, 1972). This is the so-called modified electron gas treatment in which the electron density is assumed to be the superposition of free-ion electron densities. The ion–ion repulsions are then evaluated using an electron gas treatment. This method works quite well for the energy, elastic constants, and phase transitions of the alkali halides (Cohen and Gordon, 1975) but, not surprisingly (cf. below), does not provide very useful results for oxides (Tossell, 1980). For simplicity and for its heuristic value, Eq. (2) is used in this chapter.

The lattice energy (the energy necessary to separate the static crystal to noninteracting ions) is for a binary crystal,

$$U = A/r + B/r^n \tag{3}$$

$$A = \alpha(r)(ze)^2/4\pi\varepsilon_0 \tag{4}$$

and $\alpha(r)$ is the familiar Madelung constant referred to the A–B distance $r$, and $z$ is now the highest common factor of the ionic valences ($z = 2$ *for* MgO, ThO$_2$ etc.).

If repulsions from only nearest-neighbour unlike and like (e.g. anion–anion) ion pairs are considered,

$$B = (r_A + r_B)^{n-1}\beta_0 \left[ m + \frac{1.25m_+}{\gamma_+^n}\left(\frac{2\rho}{1+\rho}\right)^{n-1} + \frac{0.75m_-}{\gamma_-^n}\left(\frac{2}{1+\rho}\right)^{n-1} \right] \tag{5}$$

where $\rho = r_A/r_B$; $m$ is the number of cation–anion nearest neighbors (the coordination number); $m_+$ ($m_-$) is the number of cation–cation (anion–anion) nearest neighbor interactions at distance $\gamma_+ r$ ($\gamma_- r$). (Note that, e.g., $m_+$ in AB is the number of cation–cation interactions per A ion and is equal to half the number of A atoms nearest to given A.)

## III. BOND LENGTHS

The condition that at (zero pressure) equilibrium $\partial U/\partial r = 0$ gives from Eq. (3), for the equilibrium interatomic distance

$$r_0 = (nB/A)^{1/(n-1)} \tag{6}$$

$$r_0 = (r_A + r_B)\frac{n\beta_0^{1/(n-1)}}{A}$$

$$\times \left[ m + \frac{1.25m_+}{\gamma_-^n}\left(\frac{2\rho}{1+\rho}\right)^{n-1} + \frac{0.75m_-}{r_-^n}\left(\frac{2}{1+\rho}\right)^{n-1} \right]^{1/(n-1)} \tag{7}$$

For $0.5 < \rho < 1.5$, the factor in square brackets is only a weak function of $\rho$. Then, *for a given structure type*, one can choose $\beta_0$ such that $r_0 = r_A + r_B$. Pauling (1928, 1960) in this way obtained basic radii for alkali and halogen ions that accounted very well for interatomic distances in all the alkali halides, the departure from $r_0 = r_A + r_B$ when $\rho \neq 1$ being well accounted for by the full expression in Eq. (7).

To the extent that the terms in $\rho$ in Eq. (7) can be neglected, the theory gives for two different crystal structures $i$ and $j$ an expression first given by Zachariasen (1931):

$$r_{0_i}/r_{0_j} = (m_i\alpha_j/m_j\alpha_i)^{1/(n-1)} \tag{8}$$

The dominant term here is the ratio of coordination numbers, so it has long been appreciated that ionic radii (or better, their sum—the bond length) depend significantly on coordination number (increasing bond length with increasing coordination number).

It is important to note that Eq. (8) was derived for binary crystals (comparing two possible structures for, e.g., MgO) and is not readily generalized to ternary (etc.) crystals. Equally important, essentially the same prediction concerning the variation of interatomic distance with coordination number arises from semiempirical bond length–bond strength correlations that are independent of an ionic description of the bonding (Brown and Shannon, 1973) and which can be generalized to ternary crystals.

Interatomic distances in insulating crystals are usually (arbitrarily, but see below), divided into a sum of cation and anion contributions ("radii") and it is found empirically (Shannon and Prewitt, 1969; see also Shannon, Chap. 16) that most of the bond length variation can be ascribed to a change in cation radius with coordination number.

A definitive set of such radii for use in oxides and fluorides is that of Shannon and Prewitt (1969) as revised by Shannon (1976). They have been enormously useful in oxide crystal chemistry but, as the following example shows, they should not be used uncritically in mixed-anion compounds. We take the example of bond lengths in the lanthanide oxyfluorides, which illustrate nicely the factors determining bond lengths in crystals (O'Keeffe, 1979; see also Zachariasen and Penneman, 1980).

In the ordered forms of LaOF (with structures derived from that of fluorite), each lanthanum ion is surrounded by four oxygen and four fluorine ions, and the anions are correspondingly surrounded by four cations. As Table I illustrates, if we take uncritically Shannon's radii for these coordi-

TABLE I

**Observed and Calculated Bond Lengths in LaOF**

|  |  | La—O | La—F |
|---|---|---|---|
| Observed |  | 2.42 | 2.60 |
| Calculated: | ionic radii[a] | 2.54 | 2.47 |
|  | ionic radii[b] | 2.41 | 2.67 |
|  | bond strength[c] | 2.43 | 2.58 |
|  | bond strength[d] | 2.41 | — |

[a] Shannon's (1976) radii, La coordination number 8.

[b] Shannon's (1976) radii, La coordination number 6 for 0 and 12 for F (see text).

[c] Zachariasen (1978).

[d] Brown and Wu (1976).

nation numbers, the predicted La—O and La—F distances do not agree with experiment; indeed, the order of the bond lengths is reversed.

An alternative analysis of the bond lengths starts from Pauling's (1929) concept of (electrostatic) bond strength and the celebrated "rule" that the sum of the bond strengths at the anion must be equal to the anion valence. If the bonds from La were assigned a bond strength equal to the charge of the cation divided by the coordination number (viz, $\frac{3}{8}$) as Pauling originally proposed, then the sums at fluorine and at oxygen would both be $\frac{3}{2}$. However, with the generalized concept of bond strength (see particularly Pauling, 1947; Brown and Shannon, 1973; Zachariasen, 1978), we can make the sums at the anions (and cation) correct by assigning a bond strength $\frac{1}{4}$ to the La–F bond and $\frac{1}{2}$ to the La–O bond. The calculated (Table I) bond lengths using empirical bond length–bond strength equations (Brown and Wu, 1976; Zachariasen, 1978) are now in close agreement with the observed distances.

Shannon's radii could still have been used in this example if we recognise that a La–O bond of strength $\frac{1}{2}$ corresponds to La six-coordinated by oxygen, and likewise a La–F bond of strength $\frac{1}{4}$ corresponds to La surrounded by 12 fluorine ions. Using radii appropriate to these coordination numbers, one now finds reasonable agreement with experiment (Table I).

It should be noted that it is very difficult to distinguish between fluorine and oxygen by x-ray diffraction, and early workers on the oxyfluorides had oxygen and fluorine incorrectly assigned on the basis of a misuse of ionic radii (see O'Keeffe, 1979). Note also that when oxygen and fluorine (or hydroxyl) have different coordinations, Pauling's valence sum rule almost invariably distinguishes between them (cf. Donnay and Allmann, 1970).

## IV.  IONIC RADII

The arguments given above suggest that bond lengths are best discussed in terms of bond strengths, and tables of ionic radii considered simply as tables of bond lengths for certain rational bond strengths. However, the temptation to assign a definite size to an ion is irresistible. Early divisions of bonds were made in ratios proportional to properties of *free* ions—a mistaken procedure as the electrostatic (Madelung) potential acts to expand cations and contract anions (Tosi, 1964; O'Keeffe, 1977, 1979). These and other fallacious arguments led to scales of ionic radii with anions that were almost certainly too large and cations too small (O'Keeffe, 1977).

It is very doubtful that an unambiguous and useful procedure will ever be devised to divide a bond length into cation and anion parts. Although the electron density (both calculated and experimental) has a minimum at a point between the cation and anion, it is a shallow one. Furthermore, if the minimum of electron density is taken as a dividing point, one finds in theoretical studies, at least, that the charges of the ionic regions so defined are significantly less than the formal charges (see, e.g., Jennison and Kunz, 1976). Attempts have been made to divide experimental electron density maps into regions of integral charge and in this way obtain experimental ionic radii. The results, subject to large experimental uncertainties, tend to support the idea that early sets of radii had anions that were too large and cations that were too small (see Tosi, 1964).

Tosi and Fumi (1964) made an exhaustive analysis of "basic radii" [cf. Eq. (2)] within the Born–Mayer framework and derived a set of radii for alkali halides in good accord with the "experimental" x-ray results. This has provided the basis for the "crystal radii" of Shannon and Prewitt (in which, for example, tetrahedral oxygen and fluorine have radii of 1.24 Å and 1.17 Å, respectively), which are certainly to be preferred to older sets (with, e.g., the radius of oxygen at 1.4 Å).

In Chap. 16, Volume 2, Shannon provides further discussion of ionic radii. There he shows that one cannot use cation radii appropriate for oxides in sulfides. I (O'Keeffe, 1979) have shown that the same is true in nitrides. Thus, if cation ionic radii appropriate for oxides are used to calculate an ionic radius for nitrogen in the rock-salt structure nitrides, it is found that the apparent radius of nitrogen has a range of almost 0.2 Å! In fact, it seems quite clear that one should not ascribe a constant radius to an ion, particularly to anions such as $O^{2-}$ and $N^{3-}$ that do not exist as free ions. Before leaving this topic, I wish to point out that very great difficulty has arisen from the use of the term "radius", which implies that the size of an ion is the same in all directions, i.e., that the same "radius" is appropriate for bonded and nonbonded interactions; as Hyde and I discuss in Chap. 10, this is clearly absurd.

## V.  STRUCTURAL PREDICTIONS

From Eqs. (3) and (6), the equilibrium (zero pressure) lattice energy of a binary crystal is

$$E = (n - 1)n^{n/(1-n)}(A^n/B)^{1/(n-1)} \tag{9a}$$

so that when one substitutes for $B$ from Eq. (5), one has for the relative energies of two different structures $i, j$ for a given pair of ions (i.e., for constant $r_A, r_B, n$):

$$\frac{E_{0_i}}{E_{0_j}} = \left(\frac{\alpha_i}{\alpha_j}\right)^{n/(n-1)}\left(\frac{M_j}{M_i}\right)^{1/(n-1)} \tag{9b}$$

TABLE II

Parameters for the NaCl and ZnS Structures

|      | $\alpha$ | $m$ | $\gamma_+ = \gamma_-$ | $m_+ = m_-$ |
|------|----------|-----|-----------------------|-------------|
| NaCl | 1.74757  | 6   | $\sqrt{2}$            | 6           |
| ZnS  | 1.63806  | 4   | $\sqrt{8/3}$          | 6           |

with

$$M_i = m_i + \frac{1.25m_{+i}}{\gamma_{+i}^n}\left(\frac{2\rho}{1+\rho}\right)^{n-1} + \frac{0.75m_{-i}}{\gamma_{-i}^n}\left(\frac{2}{1+\rho}\right)^{n-1} \qquad (10)$$

For a given pair of structures, Eq. (9b) contains just two parameters characteristic of the ions, namely, $\rho$ and $n$. A comparison of the ZnS (B3) and the NaCl (B1) structure (for which the parameters are given in Table II) gives the result of Fig. 1, which shows the regions of $n$ and $\rho$ for which the NaCl and the ZnS structures are more stable. For smaller $\rho$ and $n$, the ZnS structure is the most stable. Thus for smaller cations and a "softer" interaction, one has tetrahedral rather than octahedral coordination—very roughly in accord with experience. The often-quoted value of $\rho = 0.414$ as the dividing

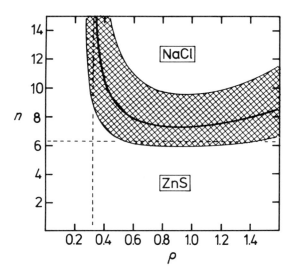

Fig. 1.   The field of stability of the NaCl and ZnS structures for various values of the radius ratio ($\rho$) and the repulsion coefficient $n$ [i.e., the heavy line is the values of $\rho$ and $n$ that satisfy Eq. (9)]. The shaded area represents the range of uncertainty when a 2% error in the solution is allowed. The vertical line is the "hard sphere" critical radius ratio, and the horizontal line is the critical value of $n$ if only nearest neighbor (cation–anion) repulsions are considered.

line between tetrahedral and octahedral coordination can be seen to have no special significance; in fact, for hard spheres ($n \to \infty$), the critical value of $\rho$ is 0.325 (see, e.g., Born and Huang, 1954). If, alternatively, all repulsions other than nearest-neighbor (cation–anion) were neglected, one would find the ZnS structure more stable for $n < 6.3$ for all values of $\rho$.

The heavy line in Fig. 1 represents the values of $n$ and $\rho$ for which the right-hand side of Eq. (9b) is equal to unity, i.e., $E_{0_i} = E_{0_j}$. It is instructive to look also at the values of $n$ and $\rho$ for which the same quantity is 0.98 or 1.02 to simulate an uncertainty in the predictions of the theory due to a 2% error in the energy ratio. Figure 1 shows that a wide range of $\rho$ and $n$ fall in this uncertain band, which in fact covers essentially all the range appropriate to, e.g., the alkali halide crystals (for which typically $n = 9$; see, e.g., Fig 2). Indeed the ionic model predicts that LiF should have the zinc blende (actually more likely the wurtzite) structure (Narayan and Ramaseshan, 1979). For oxides such as ZnO and MgO, the difference in energy between the structures with tetrahedral and octahedral coordination is about 45 kJ/mol (Navrotsky and Phillips, 1975)—i.e., about 0.1% of the lattice energy!

Another interesting example of the difficulty of making structural predictions is that of $Cu_2O$ (O'Keeffe, 1963). The cuprite structure can be sep-

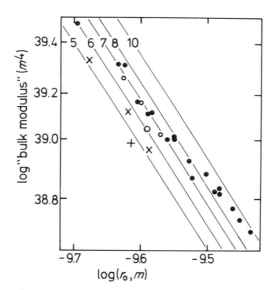

**Fig. 2.**    The logarithm of the "bulk modulus" [actually $K\alpha(r)(ze)^2/36\pi\varepsilon_0\gamma$] plotted against the logarithm of interatomic distance $r_0$. The straight lines are for the constant values of $n$ (shown in the top left-hand corner) according to Eq. 13. Filled circles are alkali halides with the NaCl structure; open circles, alkaline earth fluorides with the fluorite structure; crosses, alkaline earth oxides; large open circle, $PbF_2$; +, $ThO_2$.

arated into two substructures with the $C9$ ("idealized high cristobalite") structure. The ionic model with spherical ions predicts a decrease in electrostatic energy upon this separation [$\alpha(r)$ increases from 4.443 to 4.453] at the same time the number of nearest neighbor Cu $\cdots$ Cu and O $\cdots$ O interactions is halved. The ionic model would require that some, at least, of these nonbonded interactions be attractive for the cuprite structure to be the more stable one. The question of the stability of the cuprite structure is also taken up in Sec. X,B.

## VI.   CLOSE PACKING (EUTAXY)

It is often stated in textbooks that a feature of ionic crystal structures is that one often has a close packing of large anions in contact with each other with small cations in either the octahedral or the tetrahedral interstices according to "radius ratio" rules. It takes only a very cursory examination of the facts to find out that in reality the situation is rather different (O'Keeffe, 1977).

Because of the frequent occurrence of octahedral or tetrahedral coordination of cations by anions, one often finds that the anion arrangement is topologically that of close packing. An example is provided by the structure of rutile ($TiO_2$). The anion packing is far removed from close packing—nevertheless the arrangement of $TiO_6$ octahedra is topologically the same as in a structure derived by filling half the octahedral voids in hexagonal close-packed oxygen, and for some purposes (related to the use of topological arguments) this is a very useful way to describe the structure (Bursill et al., 1972). In this topological sense, even the very open arrays of the cristobalite structure (O'Keeffe and Hyde, 1976) and the $ReO_3$ structure (O'Keeffe and Hyde, 1977) are "close-packed" as they can be deformed into structures with close-packed anions without breaking any cation–anion "bonds."

The geometrical arrangement of close packing can arise as a solution to the problem of packing equal spheres into the smallest possible volume. It can arise in quite different context also. I have shown (O'Keeffe, 1977) that a number of simple ionic crystal structures (including some in which close-packing occurs) can be derived as a constrained *maximum* volume structure. The constraint is that of fixed cation–anion distance. The rutile and bixbyite structures (the latter having an anion arrangement very different from close packing), for example, are rather accurately derived in this way. Not surprisingly, these structures are also close to those of maximum Madelung constant. I have proposed the term *eutaxy* (adjective *eutactic*) to describe arrangements in which the atom centers are arrayed in approximately the same way as the centers of spheres in close packing.

Not all simple ionic crystal structures are maximum volume ones. Simple counter-examples are those of the cristobalite ($SiO_2$) and the $PdF_3$ structures. These are not structures of maximum Madelung constant either, but presumably could be at least formally included within the ionic description by allowing polarization energy to stabilise nonlinear cation–anion–cation configuration.

## VII.  VAN DER WAALS ENERGY AND STRUCTURE

In more sophisticated treatments of the lattice energy, attractive $r^{-6}$ and $r^{-8}$ terms are included to account for the van der Waals (or London) attractive forces. This usually only makes a small difference to the calculated energy, as to a large extent the van der Waals term can be subsumed into the $B/r^n$ repulsive term. However, this masks the fact that in heavy-atom crystals the van der Waals term can be rather large—even comparable to the repulsion energy (so that when neglected one has a large $n$ in the $B/r^n$ term). Although theoretical evaluations have been made for the coefficients of the van der Waals terms they are subject to much the same criticisms (made below) of calculations of polarizabilities, and must be corrected for multibody effects which may not be trivial (Dalgarno and Davison, 1966; Margenau and Kestner, 1969).

It is very likely the van der Waals energy that accounts for the occurrence of certain crystal structures. The CsCl structure, for example, is only found in heavy-atom nonmetallic crystals at normal pressures (i.e. CsCl, CsBr, CsI and CsAu). On the other hand, it is an extremely common alloy structure. Note that it is not a "maximum volume" structure in the sense of the previous section, but considered as a packing of all the ions, the arrangement is a superstructure of the dense bcc arrangement. (See also the space-filling diagrams of Parthé, 1961, in this respect.)

The $PbCl_2$ structure type plays the analogous role for compounds of stoichiometry $AB_2$. Only $PbF_2$ of the fluorides and no oxides adopt this structure, which again is restricted to heavy atoms in nonmetallic crystals (e.g., $BaCl_2$ etc.), but it is a rather common alloy type (usually referred to then as the $Ni_2Si$ type).

## VIII.  THE BULK MODULUS AND ELASTIC CONSTANTS

It is straightforward, but tedious, to derive expressions for the elastic constants of ionic crystals. Explicit expressions are given by Anderson (1970). The theory is reasonably successful but is normally restricted to a pair potential of the type of Eqs. (1) and (2) (i.e., "central" forces) so that (in a structure in which every atom is at a center of symmetry) there will be equalities between certain elastic constants (Cauchy relations; see Born and Huang, 1954). For a cubic crystal there is just one equation: $c_{12} = c_{44}$. In practice, for crystals like LiF or MgO $c_{44}/c_{12} \approx 1.5$ and for the silver halides $c_{44}/c_{12} \approx 0.2$, so that it is clear that in general the central force model will not predict elastic constants.

The next step is to allow for polarization of the ions, but as discussed in Sec. X it is doubtful that this can be done in a quantitative way. It is soon found that there are a number of parameters that have to be determined empirically. In fact, when ionic models are constructed for the full lattice dynamics they tend to be rather baroque constructions with a number of parameters that are of very dubious significance and which are certainly not transferable from the immediate problem at hand. As discussed below for cuprous oxide (Sec. X,B) in the $Ag^+$ salts, there should be a large quadrupole effect in the elastic constants (see also Mott, 1962, in this respect).

The bulk modulus $K_T = (c_{11} + 2c_{12})/3 = V(\partial P/\partial V)_T$ is at zero pressure and temperature,

$$K_0 = V(\partial^2 E/\partial V^2) \tag{11}$$

For some simple cubic structures one can write

$$V = \gamma r^3 \tag{12}$$

with $\gamma$ a constant depending on the structure—thus if $V$ and $E$ refer to an ion pair, $\gamma = 2$ for the NaCl structure. One then has from Eqs. (3) and (4),

$$K_0 = \frac{(n-1)\alpha(r)(ze)^2}{36\pi\varepsilon_0\gamma r^4} \tag{13}$$

This expression is often used to determine $n$ (or the relevant parameter if another functional form is used for the repulsion energy).

An elementary point which is sometimes overlooked is that Eq. (12) only holds with constant $\gamma$ for the very limited class of cubic structures in which all atoms are in fixed positions. A good example of the opposite extreme, with $V$ almost independent of $r$, is that of quartz (O'Keeffe et al., 1980).

In Fig. 2, compressibility data (all from Landolt-Börnstein, 1966, except for SrO from Johnston et al., 1970) for a number of cubic crystals are plotted according to Eq. (13). As $n$ tends to increase with $r_0$, one has approximately that $K_0 \propto r^{-3}$ for the alkali halide series (Harrison, 1980). A related observation on "compressibilities" of bond lengths has been made by Hazen and Finger (see Chap. 19). The data for $Cu_2O$ (Hallberg and Hanson, 1970) fall off the range of the graph but would fall very close to the line for $n = 7$.

Anderson (1970) has shown that the ionic model (using the simple $1/r^n$ repulsive potential) also provides a good account of the high pressure behavior of the volume. A simple, but useful, result is at zero pressure

$$\left(\frac{dK}{dP}\right)_0 = \frac{n+7}{3} \tag{14}$$

and, for many purposes, it is adequate to assume that $K(P)$ is linear up to $P/K_0 \sim 0.2$.


## IX.  THE VOLUME (DENSITY)

If again $V$ is the volume per ion pair (or per "molecule" $A_x B_y$), one may readily derive from Eqs. (6) and (8) a simple expression for the relative volumes of two structures (subscripts $i, j$)

$$\frac{V_i}{V_j} = \left[\frac{\alpha_i(V^{\frac{1}{3}}) \, E_j(r_0)}{\alpha_j(V^{\frac{1}{3}}) \, E_i(r_0)}\right]^3 \tag{15}$$

where $\alpha(V^{\frac{1}{3}})$ is the Madelung constant referred now to the cubic root of the volume: $\alpha(V^{\frac{1}{3}}) = \alpha(r)V^{\frac{1}{3}}/r$. Thus for materials that are polymorphic at low temperatures and pressures [so that $E_i(r_0) \approx E_j(r_0)$], $[\alpha(V^{\frac{1}{3}})]^3$ is proportional to the crystal volume. Quite good estimates of relative densities can in practice be obtained in this way. Values of $[\alpha(V^{\frac{1}{3}})]^3$ may be calculated very simply from published Madelung constants (e.g., Johnson and Templeton, 1961). One finds that $\alpha(V^{\frac{1}{3}})$ invariably decreases with increasing coordination number for a given stoichiometry (cf. Table III) so that, as is well known, higher density structures usually involve higher coordination number.

An interesting application of the numbers in Table III is to oxides $M_2O_3$ with either the corundum $(Al_2O_3)$ structure or the bixbyite structure of $Y_2O_3$ etc. In both of these structures the cation coordination number is six and the coordination polyhedra are rather irregular. The bixbyite structure is adopted in oxides of cations with larger radius than those which form the corundum structure—one would normally expect then that the bixbyite structure would be the high pressure structure. However, the values of $\alpha(V^{\frac{1}{3}})$ suggest that the corundum structure should be denser by approximately 4%. This is confirmed quite accurately in the case of $In_2O_3$ (Shannon, 1966). In framework structures that can reduce volume by partial collapse of the framework (e.g., the cristobalite framework, Chap. 10), the coordination number of inserted ("stuffing") ions may well decrease with increasing pressure. This is very likely the case in perovskites $ABX_3$ for example, especially when $z_B > z_A$ (O'Keeffe et al., 1979).

TABLE III

Madelung Constants in Some Simple Structures

| Structure | Coordination[a] | $\alpha(r_0)$ | $[\alpha(V^{\frac{1}{3}})]^3$ |
|---|---|---|---|
| B3 (ZnS) | 4 | 1.638 | 13.53 |
| B1 (NaCl) | 6 | 1.748 | 10.68 |
| B2 (CsCl) | 8 | 1.763 | 8.43 |
| C8 ($\alpha$-quartz) | 4 | 4.403 | 779.9 |
| C3 ($Cu_2O$) | 4 | 4.443 | 539.9 |
| C4 ($TiO_2$) | 6 | 4.770 | 460.3 |
| C1 ($CaF_2$) | 8 | 5.039 | 394.0 |
| $D0_9$ ($ReO_3$) | 6 | 8.962 | 5755 |
| $D0_{12}$ ($PdF_3$) | 6 | 8.860 | 4135 |
| ($YF_3$) | 8 | 8.899 | 3328 |
| $D0_5$ ($LaF_3$) | 9 | 9.111 | 3190 |
| $D5_1$ ($Y_2O_3$) | 6 | 25.10 | 99900 |
| $D5_2$ ($Al_2O_3$) | 6 | 24.24 | 95910 |
| $D5$ ($La_2O_3$) | 8 | 24.18 | 87290 |

[a] Coordination number of the minority species.

For phase transitions that require very high pressures, the factor $E_j(r_0)/E_i(r_0)$ in Eq. (14) is significantly less than unity and the volume decrease is less than for lower pressure phase transitions. A related observation, that volume change in the NaCl → CsCl transition depends on radius ratio, has been made by Jamieson (1977) [but recall that Eq. (15) refers to zero-pressure volumes only].

## X.  POLARIZABILITY AND POLARIZATION

### A.  Basic Terms

When the ionic model is applied in low symmetry situations it is essential to include the effects of polarization of the ions by the electric field. This is often done in a rather unrealistic way, so in this section I call attention to some important features that are often overlooked.

We need to start by defining terms. The component of the dipole moment in the $i$ direction ($\mu_i$) of an assembly of charges $q_n$ at positions $r_n$ is

$$\mu_i = \sum_n q_n r_{ni} \tag{16}$$

The *dipole polarizability* of an entity (atom, molecule, ion) is usually defined in terms of the dipole moment induced by a field $F_i = -\partial\phi/\partial r_i$,

$$\mu_i = \alpha_{ij}F_j \tag{17}$$

In Eq. (17) the repeated ("dummy") suffix j implies summation, so that for, e.g., $i = x$, the $x$ component of the dipole moment is

$$\mu_x = \alpha_{xx}F_x + \alpha_{xy}F_y + \alpha_{xz}F_z \tag{18}$$

The polarizability is a second-rank tensor and will in general have three principal values; however, for a free atom or ion or one in a crystal on a site of cubic symmetry, there is just one independent value $\alpha_D$ (i.e., $\alpha_{ij} = \alpha_D\delta_{ij}$). As applied to charged species, the dipole moment defined in the usual way [Eq. (16)] depends on the origin of coordinates (indeed one can always choose an origin so that $\mu = 0$ for an ion!). One can avoid the difficulty by using the polarization energy, given in the isotropic case by

$$\Delta E = -\alpha_D F^2/2 \tag{19}$$

to define $\alpha_D$ and then define the dipole moment (and thus the origin of coordinates) by $\mu = \alpha_D F = 2\Delta E/F$.

Ionic polarizabilities are usually obtained from the familiar Clausius-Mossotti equation

$$3\varepsilon_0 V(n^2 - 1)/(n^2 + 2) = \alpha \tag{20}$$

where $V$ is the volume per formula unit and $\alpha$ is assumed to be a sum of ionic contributions. We discuss below some difficulties (similar to those that arise with ionic radii) attendant on partitioning $\alpha$ into components. Here we emphasize that the electric fields involved in measuring $n$ are typically of the order of $1$ $Vm^{-1}$, while on the atomic scale they may be ten orders of magnitude larger. It has long been known (Coulson *et al.*, 1952) that Eq. (17) is then no longer adequate, and should be replaced in general by (Buckingham and Orr, 1967):

$$\mu_i = \alpha_{ij}F_j + \tfrac{1}{2}\beta_{ijk}F_jF_k + \tfrac{1}{6}\gamma_{ijkl}F_jF_kF_l \tag{21}$$

where $\beta$ and $\gamma$ (third- and fourth-rank tensors) are known respectively as the *first* and *second hyperpolarizabilities*. The components of $\beta$ and $\gamma$ are particularly simple for spherical symmetry. We have then that all $\beta_{ijk} = 0$ and $\gamma_{xxxx} = \gamma_{yyyy} = \gamma_{zzzz} = 3\gamma_{xxyy} = 3\gamma_{yyzz} = 3\gamma_{zzxx} = \gamma$; $\gamma$ is the quantity often described without qualification as "the hyperpolarizability" (e.g., Langhoff *et al.*, 1966). In the absence of a center of symmetry there will be at least one nonzero component of $\beta$. Although this topic is not pursued here, it might be mentioned that this term is the one responsible for second-harmonic generation in noncentrosymmetric crystals, and is important in a variety of effects (Kerr effect, etc.) in molecular gases and liquids. For the symmetrical case

Eq. (19) must be replaced by

$$\Delta E = -\tfrac{1}{2}\alpha_D F^2 - (1/24)\gamma F^4 \tag{22}$$

In principle at least, higher-order multipoles can be of significance—they certainly are for molecular crystals (Buckingham, 1959). Following Buckingham, one can define the $ij$ component of the quadrupole moment $Q$ by

$$Q_{ij} = \tfrac{1}{2} \sum_n q_n(3r_{ni}r_{nj} - r_n^2\delta_{ij}) \tag{23}$$

With suitably oriented axes $Q_{ij} = 0$ for $i \neq j$ and further $Q_{xx} + Q_{yy} + Q_{zz} = 0$. It is usually assumed that a quadrupole moment is induced in an ion (atom or molecule) by a field gradient—but see Buckingham (1959) for the general expression, i.e.,

$$Q_{ij} = \alpha_{ijkl}F'_{kl} \tag{24}$$

where $\alpha$ is now the field gradient *quadrupole polarizability* and $F'_{kl}$ is $-\partial^2\phi/\partial r_k \partial r_l$.

Again for spherical symmetry, there is very considerable simplication: $\alpha_{xxxx} = \alpha_{xxyy} = 2\alpha_{xyxy} = \cdots = \alpha_Q$; where $\alpha_Q$ is the quantity normally quoted as the quadrupole polarizability (e.g., of an ion). In an axially symmetric (about $z$) system, $F'_{zz} = -2F'_{xx}$ and $F_{ij} = 0$ for $i \neq j$; the polarization energy is then

$$\Delta E_{\text{quad}} = -\tfrac{1}{4}\alpha_Q(F'_{zz})^2 \tag{25}$$

## B.   Numerical Values

Without apologies, a brief list of conversion factors is given in Table IV. (One otherwise excellent review uses three systems of units with incorrect conversion factors!) Polarizabilities $\alpha_D$, $\alpha_Q$, etc. are often given in units of $\text{Å}^m$ (where $m = 2n + 1$ for a $2^n$-pole polarizability)—the quantity reported is

TABLE IV

Units and Conversion Factors[a]

| To S.I. | From e.s.u. | From a.u. | From (other) |
|---|---|---|---|
| $\mu$ (Cm) | 3.336 ($-12$) | 8.478 ($-30$) | 3.336 ($-30$) (D) |
| $Q$ (Cm$^2$) | 3.336 ($-14$) | 4.447 ($-40$) | |
| $\alpha_D$ (Cm$^2$J$^{-1}$) | 3.336 ($-7$) | 1.029 ($-22$) | 1.113 ($-40$) (Å$^3$) |
| $\alpha_Q$ (Cm$^4$J$^{-1}$) | 3.336 ($-11$) | 2.882 ($-43$) | 1.113 ($-60$) (Å$^5$) |
| $\beta$ (C$^3$m$^3$J$^{-2}$) | 3.711 ($-21$) | 3.206 ($-53$) | |
| $\gamma$ (C$^4$m$^4$J$^{-3}$) | 1.238 ($-25$) | 6.236 ($-65$) | |

[a] To convert to S.I., multiply by the factor shown (power of ten in brackets).

really $\alpha/4\pi\varepsilon_0$. (All equations in this chapter give correct results when S.I. units are used and $1/(4\pi\varepsilon_0)$ is set equal to $8.987 \times 10^9 C^2 J^{-1} m^{-1}$.)

"Experimental" numerical values of dipole polarizabilities of ions are obtained from Eq. (20) with $\alpha = x\alpha_+ + y\alpha_-$ for a binary crystal $A_x B_y$. For the alkali halide series it is often assumed that $\alpha_{Li}$ is known and a set of individual ionic polarizabilities derived—much as sets of ionic radii are. A very large literature on this topic has developed, but it is subject to two major criticisms:

(a) It does not work. The evidence is now very strong that, e.g., for the alkali halides one must allow the anion polarizabilities to vary from crystal to crystal (see, e.g., Mahan, 1980a). This is a well-documented effect and closely parallels the variation of apparent ion radius (O'Keeffe, 1979).

(b) It is not correct in principle. Pantiledes (1975) has put forward evidence that a major contribution to the optical susceptibility of an alkali halide crystal comes from transitions between valence states that are essentially anion atomic (ionic) states to cation states. Actually this argument is much more compelling for narrow band gap crystals (e.g. oxides such as ZnO and $Cu_2O$ that often are discussed within the framework of the ionic model). In the alkali halides the cation states are very close to the vacuum level anyway (see below).

One can even separate the static polarizability of crystals [obtained by replacing $n^2$ in Eq. (20) by the static dielectric constant] into ionic components (Roberts, 1951), which proves useful in discussions of ferroelectricity, but it is hard to find any theoretical justification for this procedure.

Theoretical values of dipole, quadrupole, and octupole polarizabilities for a number of ions *in crystals* have recently been calculated by Mahan (1980a), who also gives a useful list of semiempirical dipole polarizabilities for halides and chalcogenides that are compatible with optical data.

Most of our knowledge of hyperpolarizabilities ($\gamma$) of ions comes from the classical work of Langhoff *et al.* (1966). Ha (1976) has recently compiled polarizabilities and hyperpolarizabilities for atoms, ions, and some small molecules.

Typical values for an ion such as $O^{2-}$ are $\alpha_D \approx 2 \text{ Å}^3 = 2.2 \times 10^{-40}$ $Cm^2J^{-1}$; $\alpha_Q \approx 2 \text{ Å}^5 = 2.2 \times 10^{-60} \text{ } Cm^4J^{-1}$; and $\gamma = 6.2 \times 10^{-21} \text{ } C^4m^4J^{-1}$. Let us consider as an example a "point" ion of charge $-|e|$ with such properties at a distance 2 Å along the $z$ axis from another point charge of $+|e|$. The potential at the polarizable ion is $+7.2$ V and the field is $F_z = -3.6 \times 10^{10}$ $Vm^{-1}$. The energies of interaction are

| | |
|---|---|
| monopole | $E = 1.15 \times 10^{-18}$ J (7.20 eV) |
| dipole polarizability | $1.43 \times 10^{-19}$ J (0.89 eV) |
| dipole hyperpolarizability | $4.34 \times 10^{-20}$ J (0.27 eV) |
| quadrupole | $7.13 \times 10^{-20}$ J (0.45 eV) |

These numbers are rather informative. They tell us that in unsymmetrical situations, on an atomic scale, polarization terms may be a sizable fraction of the total electrostatic energy. They tell us, too, that if a realistic estimate of this energy is to be made, not only must the dipole polarizability be known accurately, but also higher-order and nonlinear terms must be considered. Indeed, an expansion of the energy in multipole terms must be shown to converge.

My appreciation of the situation at present is both that our knowledge of the appropriate parameters is rudimentary and that the theory is not entirely appropriate, so that when significant polarization effects are anticipated, no attempt should be made to attach quantitative significance to calculations. This warning has been given before in the context of estimating quantities such as hydration energies (Buckingham, 1959). It should apply equally to low-symmetry configurations (e.g., postulated defect structures) in solids.

On a more positive note, inclusion of polarizability does have at least the heuristic value of allowing one to rationalize the occurrence of less symmetrical configurations in the ionic model. A simple example is provided by the "molecule" $A_2^+ B^{2-}$ (see Fig. 3) with a polarizable B ion. It is easy to show that if $\alpha$ is the B ion polarizability and $l$ the A—B distance, the molecule will be bent when

$$\alpha/4\pi\varepsilon_0 > l^3/8 \tag{26}$$

and the bond angle $\theta$ will then be

$$\theta = 2 \sin^{-1}[l(\pi\varepsilon_0/2\alpha)^{\frac{1}{3}}] \tag{27}$$

One could, at least in principle, rationalize in this way the occurrence of structures such as that of cristobalite (with bent Si–O–Si bonds) that are not structures of a maximum in Madelung constant. The difficulty alluded to above, arriving at accurate numerical values, remains, however.

A final example is one in which quadrupole effects are important. F. Jellinek (private communication) has suggested that quadrupole polarization might be important in the crystal chemistry of $Cu^+$. Indeed, $\alpha_Q$ is large for $Cu^+$ ($\approx 1.4 \times 10^{-60}$ $Cm^4 J^{-1}$; Sternheimer, 1957) and the major contribution

**Fig. 3**   A triatomic molecule with a polarizable atom, defining the terms in Eqs. (26) and (27).

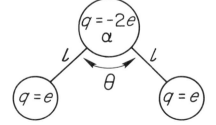

comes from d → s transitions. For $Cu_2O$, R. Bersohn (quoted by Das and Hahn, 1958) using a point ion model has calculated a field gradient at a $Cu^+$ ion site of $7.9 \times 10^{20}$ $Vm^{-2}$ (note that this is considerably larger even than in the numerical example given above). This value agrees well with that calculated from the $^{63}Cu$ nuclear quadrupole coupling constant. The quadrupole polarization energy [cf. Eq. (25)] calculated from these numbers is 130 kJ/mol—an amount so large (comparable to the heat of formation) that it should not be taken literally. I think that, rather, the indication is that the discussion of bonding in $Cu_2O$ and similar compounds should be recast in the familiar terms of d–s hybridization.

Mahan (1980b) has shown that in tetrahedral solids (e.g., CuCl) there is an octopole polarization energy (mainly due to the anion polarization) of comparable energy ($\sim 100$ kJ/mol).

## XI. MADELUNG POTENTIALS AND ENERGY LEVELS

### A. Potentials

One usually focusses on the total energy of an ionic crystal, but sometimes it is very useful to consider separately the contributions to the electrostatic energy of each kind of ion in the structure (Hoppe, 1970). Thus if ion of type $i$ has charge $z_ie$, the electrostatic energy can be written

$$E = \tfrac{1}{2}\sum z_ie\phi_i \tag{28}$$

Here $\phi_i$ is the electrostatic (Madelung) potential at the ion site due to all the other ions in the crystal.

The $\phi_i$ can be expressed as

$$\phi_i = -z_ie\alpha_i/4\pi\varepsilon_0 r \tag{29}$$

where $\alpha_i$ is a number (between 1 and 2). The Madelung constant is

$$\alpha = \tfrac{1}{2}\sum z_i^2\alpha_i \tag{30}$$

The $\alpha_i$ for binary crystals (in which all ions of a given type are crystallographically equivalent) depend to a good approximation only on the coordination $m_j$ of the counterion $j$. Thus in the fluorite ($CaF_2$), zinc blende (ZnS), and cuprite ($Cu_2O$) structures in which the *anions* are in tetrahedral coordination, the potential at the *cation* site is given by Eq. (29) with $\alpha_+ = 1.6381$. The $\alpha_+$ are in fact identical in these three structures, as can readily be verified by expressing $\phi$ in terms of Hund's basic potentials (see, e.g., Zucker, 1975, for details).

An error often found in the literature (including recent work) is the assumption that $\alpha_+ = \alpha_-$. This is not true in general (see, for example, $CaF_2$). Values of $\alpha_+$ and $\alpha_-$ are not generally available in the literature, so for convenience a short list for binary structures is included in Table V. [Note, however, that the $\alpha_i$ can be obtained readily from the "PMF" of Hoppe (1970)—the values for $PdF_3$ in Table V come from this source.]

The data of Table V for $m_j \leqslant 6$ obey rather accurately the expression

$$\alpha_i = Cm_j^M \tag{31}$$

with $C = 1.20$ and $M = 1/4.6$.

One may also write

$$m_j = z_j/s \tag{32}$$

where $s$ is the Pauling bond strength. Further, the bond length is related to the bond strength by (see Chapter 14)

$$r = r_0/s^N \tag{33}$$

in particular for six-coordination of the counterion $j$

$$r_6 = \frac{r_0}{(z_j/6)^N} \tag{34}$$

TABLE V

Madelung Potential Constants $\alpha_i$ for given $m_j$
in Binary Crystal Structures

|  |  | Structure | $\alpha_i$ |
|---|---|---|---|
| $m_j = 8$ | $\alpha_+ = \alpha_-$ | CsCl (B2) | 1.7627 |
|  | $\alpha_-$ | $CaF_2$ | 1.7627 |
| $m_j = 6$ | $\alpha_+ = \alpha_-$ | NaCl (B1) | 1.7476 |
|  | $\alpha_-$ | $Al_2O_3$ | 1.767 |
|  | $\alpha_-$ | $TiO_2$ (rutile) | 1.747 |
|  | $\alpha_-$ | $PdF_3$ | 1.745 |
| $m_j = 4$ | $\alpha_+ = \alpha_-$ | ZnS | 1.6381 |
|  | $\alpha_+$ | $CaF_2$ | 1.6381 |
|  | $\alpha_+$ | $Cu_2O$ | 1.6381 |
|  | $\alpha_+$ | $Al_2O_3$ | 1.632 |
|  | $\alpha_-$ | $\alpha$-quartz | 1.621 |
| $m_j = 3$ | $\alpha_+$ | rutile | 1.514 |
| $m_j = 2$ | $\alpha_-$ | $Cu_2O$ | 1.4022 |
|  | $\alpha_+$ | $\alpha$-quartz | 1.352 |
|  | $\alpha_+$ | $PdF_3$ | 1.382 |
|  | $\alpha_+ = \alpha_-$ | chain[a] | 1.3863 |

[a] For an infinite one-dimensional array of equally-spaced positive and negative charges $\alpha_+ = \alpha_- = 2 \ln 2$.

When Eqs. (31)–(34) are substituted back into Eq. (29), one finds

$$\phi_i = \frac{-z_i e C6^N}{4\pi\varepsilon_0 r_6}\, m_j^{M-N} \tag{35}$$

It is found experimentally that $N \simeq M$—for example, for bonds between second-row cations and oxygen, $N = 1/4.3$ (Chapter 14). One has then that $m_j^{M-N} \simeq 1$. To this approximation, Eq. (35) becomes

$$\phi_i \simeq -\frac{1.77 z_i e}{4\pi\varepsilon_0 r_6} \tag{36}$$

This is quite a remarkable result. It says that for a given ion (e.g., oxygen with $z_i = 2$), the potential at that ion site depends only on the size of the counterion (as measured by the bond length to oxygen appropriate for six-coordination). In particular, there is no dependence on structure, or even on the charge of the counter ion.

This result was found (O'Keeffe, 1979) empirically for the case of oxygen (see Fig. 4). An obvious generalisation from Eq. (27) is that the electrostatic energy is almost independent of structure, a conclusion that we have hinted at above. It is in fact an old result in this form, and has considerable use in rough thermochemical calculations (Kapustinskii, 1956).

The basic observation leading to Eq. (36) is that the Madelung potential constants, $\alpha_i$, for stable structures depend on coordination number in approx-

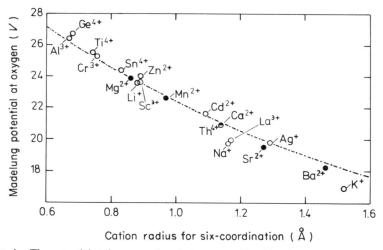

**Fig. 4.** The potential at the oxygen ion site in some oxides. The abscissa is the cation radius for six-coordination.

imately the same way that bond lengths do. There must surely be an underlying reason for this, but it remains obscure.

### B.  Electronic Energy Levels

In ionic crystals, with fairly localized electrons, a good approximation to
the relative positions of the energy levels in a crystal can be made by starting
with free ion levels that are then modified by the crystal Madelung potential
(Seitz, 1940). This has proved a useful way to discuss oxides such as NiO and
MnO (Morin, 1958; van Houten, 1960; O'Keeffe and Valigi, 1970). There is,
however, a major difficulty in making corrections for lattice polarization,
and one must use different levels for optical and thermal excitations (see
especially O'Keeffe and Valigi, 1970). An interesting point in the present
connection is that as the Madelung potential is not very structure sensitive,
the energy levels will not be either.

The band gap in the alkali halides corresponds to transition of an electron
in an anion p state to a *neighboring* cation s state. One must allow therefore
for the electron-hole interaction energy of $e^2/4\pi\varepsilon_0 r$. When this is done one
gets good agreement with optically determined band gaps (Mott and Gurney,
1940). What is quite remarkable is the observation (Harrison, 1980) that one
gets at least as good agreement by using atomic energy levels—thus in KCl
the band gap is the difference between the energy of the Cl 3p and the K 4s
atomic states. Harrison's explanation of this is that the shift in atomic energy
levels on ionization of an atom is largely compensated by the Madelung
potential (e.g., that the potential due to the extra electron in $Cl^-$ is largely
compensated by the positive crystal potential in KCl). Actually, although the
band gap is given by a difference in atomic levels, the individual levels are
shifted somewhat so that in the alkali halides the cation s state is close to the
vacuum level (so that the band gap is close to the photothreshold). A comprehensive review of experimental and theoretical studies of the electronic
structure of the alkali halides is given by Poole, Jenkin, Liesegang, and
Leckey (1975) and Poole, Liesegang, Leckey, and Jenkin (1975).

### XII.  SOME CONCLUSIONS

In this chapter I have tried to point up some of the uses and limitations of
the ionic model of crystals. It works quite well for calculating energies, but
because energy is not very structure-dependent the ionic model fails to make
useful structural predictions. The failure is in part a consequence of the
omission of angular-dependent forces, which also results in failure to predict

accurately quantities such as shear elastic constants. Harrison (Chapter 6) outlines one approach (the "chemical grip") to remedying this omission.

It is common to state that this or that compound is ionic or covalent and indeed even to assign a quantitative degree of "ionicity" or "covalency." Although this is often fruitful (cf. Chapter 1), it is well to bear in mind that the ionic and covalent descriptions are really not very different and certainly not mutually exclusive (Shull, 1962). As Harrison (1980) points out, an ionic cohesive energy calculation with integral charges will give close to the correct result, even if the final electron distribution is closer to that of free atoms as there is really only a small energy difference between that distribution and that corresponding to free ions. This point has been discussed fully by Slater (1965), particularly with respect to ionic radii and their significance.

One is forced to conclude, then, that when we say that a crystal is "ionic," what is implied is that some simple properties of the crystal are well described by the ionic model. We should not consequently draw conclusions about "quantities" such as ionic radii and ionicity, which, although sometimes useful concepts, are in principle not determinable. We should not be surprised either if alternative models work equally well for the same material; indeed, they often do.

## ACKNOWLEDGMENT

My own work described herein is supported largely by the National Science Foundation (DMR 78-09197). I am grateful to G. D. Mahan for a prepublication copy of his work on polarizabilities and for some helpful correspondence.

## REFERENCES

Anderson, O. L. (1970). *J. Geophys. Res.* **75**, 2719.
Born, M., and Huang, K. (1954). "Dynamical Theory of Crystal Lattices." Oxford Univ. Press (Clarendon), London and New York.
Brown, I. D. (1977). *Acta Crystallogr., Sect. B* **B33**, 1305.
Brown, I. D., and Shannon, R. D. (1973). *Acta Crystallogr., Sect. A* **A29**, 266.
Brown, I. D., and Wu, K. K. (1976). *Acta Crystallogr., Sect. B* **B32**, 1957.
Buckingham, A. D. (1959). *Q. Rev., Chem. Soc.* **13**, 183.
Buckingham, A. D., and Orr, B. J. (1967). *Q. Rev., Chem. Soc.* **21**, 195.
Bursill, L. E., Hyde, B. G., and O'Keeffe, M. (1972). *NBS Spec. Publ. (U.S.)* **364**.
Cohen, A. J., and Gordon, R. G. (1975). *Phys. Rev. B: Solid State* [3] **12**, 3228.
Coulson, C. A., Maccoll, A., and Sutton, L. E. (1952). *Trans. Faraday Soc.* **48**, 106.
Dalgarno, A., and Davison, W. D. (1966). *Adv. At. Mol. Phys.* **2**, 1.
Das, T. P., and Hahn, E. L. (1958). "Nuclear Quadrupole Resonance Spectroscopy," Solid State Phys. Suppl. 1. Academic Press, New York.
Donnay, G, and Allmann, R. (1970). *Am. Mineral.* **55**, 1003.

Gordon, R. G., and Kim, Y. S. (1972). *J. Chem. Phys.* **56**, 3722.

Ha, T.-H. (1976). *In* "Molecular Electro-Optics" (C. T. O'Konski, ed.), p. 471. Dekker, New York.

Hallberg, J., and Hanson, R. C. (1970). *Phys. Status Solidi* **42**, 305.

Harrison, W. A. (1980). "Electronic Structure and the Properties of Solids." Freeman, San Francisco, California.

Hoppe, R. (1970). *Adv. Fluorine Chem.* **6**, 387.

Jamieson, J. C. (1977). *In* "High Pressure Research" (M. H. Manghnani and S.-I. Akimoto, eds.), p. 209. Academic Press, New York.

Jennison, D. R., and Kunz, A. B. (1976). *Phys. Rev. B: Solid State* [3] **13**, 597.

Johnson, Q. C., and Templeton, D. H. (1961). *J. Chem. Phys.* **34**, 2004.

Johnston, D. L., Thrasher, P. H., and Kearny, R. J. (1970). *J. Appl. Phys.* **41**, 427.

Kapustinskii, A. F. (1956). *Q. Rev., Chem. Soc.* **10**, 283.

Landolt-Börnstein (1966). "Zahlenwerte und Funktionen Kristall- und Festkörperphysik, Vol. 1.

Langhoff, P. W., Lyons, J. D., and Hurst, R. P. (1966). *Phys. Rev.* [2] **148**, 18.

Mahan, G. D. (1980a). *Solid State Ionics* **1**, 29.

Mahan, G. D. (1980b). *Chem. Phys. Lett.* **76**, 183.

Margenau, H., and Kestner, N. R. (1969). "Theory of Intermolecular Forces." Pergamon, Oxford.

Morin, F. J. (1958). *Bell Syst. Tech. J.* **37**, 1048.

Mott, N. F. (1962). *Rep. Prog. Phys.* **25**, 218.

Mott, N. F., and Gurney, R. W. (1940). "Electronic Processes in Ionic Crystals." Oxford Univ. Press (Clarendon), London and New York.

Narayan, P., and Ramaseshan, S. (1979). *Phys. Rev. Lett.* **42**, 992.

Navrotsky, A. N., and Phillips, J. C. (1975). *Phys. Rev. B: Solid State* [3] **11**, 1583.

O'Keeffe, M. (1963). *J. Chem. Phys.* **38**, 3035.

O'Keeffe, M. (1977). *Acta Crystallogr., Sect. A* **A33**, 924.

O'Keeffe, M. (1979). *Acta Crystallogr., Sect. A* **A35**, 776.

O'Keeffe, M., and Hyde, B. G. (1976). *Acta Crystallogr., Sect. B* **B32**, 2923.

O'Keeffe, M., and Hyde, B. G. (1977). *Acta Crystallogr., Sect. B* **B33**, 3802.

O'Keeffe, M., and Valigi, M. (1970). *J. Phys. Chem. Solids* **31**, 947.

O'Keeffe, M., Hyde, B. G., and Bovin, J.-O. (1979). *Phys. Chem. Miner.* **4**, 299.

O'Keeffe, M., Newton, M. D., and Gibbs, G. V. (1980). *Phys. Chem. Miner.* **6**, 305.

Pantiledes, S. T. (1975). *Phys. Rev. Lett.* **35**, 250.

Parthé, E. (1961). *Z. Kristallogr., Kristallgeom., Kristallphys., Kristallchem.* **115**, 52.

Pauling, L. (1928). *J. Am. Chem. Soc.* **50**, 1036.

Pauling, L. (1929). *J. Am. Chem. Soc.* **51**, 1010.

Pauling, L. (1947). *J. Am. Chem. Soc.* **69**, 542.

Pauling, L. (1960). "Nature of the Chemical Bond," 3rd ed. Cornell Univ. Press, Ithaca, New York.

Poole, R. T., Jenkin, J. G., Liesegang, J., and Leckey, R. C. G. (1975). *Phys. Rev. B: Solid State* [3] **B11**, 5179.

Poole, R. T., Liesegang, J., Leckey, R. C. G., and Jenkin, J. G. (1975). *Phys. Rev. B: Solid State* [3] **B11**, 5190.

Roberts, S. (1951). *Phys. Rev.* [2] **81**, 865.

Seitz, F. (1940). "The Modern Theory of Solids." McGraw-Hill, New York.

Shannon, R. D. (1966). *Solid State Commun.* **4**, 629.

Shannon, R. D. (1976). *Acta Crystallogr., Sect. A* **A34**, 751.

Shannon, R. D., and Prewitt, C. T. (1969). *Acta Crystallogr., Sect. B* **B25**, 925.

Shull, H. (1962). *J. Appl. Phys.* **33**, 290.

Slater, J. C. (1965). "Quantum Theory of Molecules and Solids," Vol. 2. McGraw-Hill, New York.

Sternheimer, R. M. (1957). *Phys. Rev.* **107**, 1565.

Tosi, M. P. (1964). *Solid State Phys.* **16**, 1.

Tosi, M. P., and Fumi, F. G. (1964). *J. Phys. Chem. Solids* **25**, 45.

Tossell, J. A. (1980). *Am. Mineral.* **65**, 163.

van Houten, S. (1960). *J. Phys. Chem. Solids* **17**, 4.

Zachariasen, W. H. (1931). *Z. Kristallogr., Kristallgeom., Kristallphys., Kristallchem.* **80**, 137.

Zachariasen, W. H. (1978). *J. Less-Common Met.* **62**, 1.

Zachariasen, W. H., and Penneman, R. A. (1980). *J. Less-Common Met.* **69**, 369.

Zucker, I. J. (1975). *J. Phys. A: Math. Gen.* **8**, 1734.

# Index

Note references to a given *substance* are given under the main heading *Compounds and elements,* references to a given *structure* under *Structure types.* Bold face indicates the volume.

## A

Alloys
  based on f.c.c. lattice, **2,** 118–129
  complex cubic structures, **2,** 244–256, 292–294
  rare earth, **2,** 259–296
  tetrahedrally close packed, **2,** 235–244
Aqueous solution, **2,** 225, 231

## B

Band structure, **2,** 90–91, 283–289
  silicon, **1,** 99
Blandin–Friedel–Saada Theory, **2,** 283–289
Bond angle
  in cristobalites, **1,** 229–236
  O—P—O, **2,** 47
  Si—O—Al, **1,** 211, 213–216
  Si—O—Si, **1,** 175–192, 202–210, 213–216
Bond charge, **1,** 67, 79, 102
Bond energy, **1,** 161, 167
Bond length
  cation–cation distance, **1,** 237; **2,** 48

C—C, **1,** 249
  and coordination number, **2,** 59, 63–65, 89
  coordination polyhedra edges, **2,** 46, 48
  correlation with bond angle, **1,** 213
  correlation with bond strength, **2,** 4
  in ionic crystals, **1,** 301–303
  Madelung potential and, **1,** 318
  in molecules and solids, **1,** 216–221
    table of, **1,** 220
  silicon–oxygen, **1,** 190, 201–205, 210, 213–216
  in sulfides, **2,** 53–69
  table of, **2,** 43, 56–58
  variation of, **2,** 36, 39, 42, 44
  variation with temperature and pressure, **2,** 110–112
Bond strength, **1,** 211, *see also* Bond valence
Bond valence
  basic concepts, **2,** 2
  correlations with bond length, **2,** 4
  correlations with force constants, **2,** 6
  table of parameters, **2,** 19–29
  valence balancing, **2,** 45

323

Bulk modulus
  ionic crystals, **1**, 306
  quartz, **1**, 207
Burnside's lemma, **2**, 137

## C

Charge density, **1**, 279–298, *see also* Valence
    charge density
  GaAs, **1**, 34
  Ge, **1**, 34
  Si, **1**, 32, 42
  Si surface, **1**, 41
  ZnSe, **1**, 34
Chemical grip, **1**, 144
Chevrel phases, **2**, 299
Compounds and elements
  AB compounds, **1**, 117–131, 145, *see also*
      Octet compounds
  $Al_2O_3$, **1**, 294
  aqueous, *see* water species
  Be, **1**, 22, 288
  BeO, **1**, 66, 243
  C, **1**, 247–250, 267, 289, 290
  chalcogenides, **2**, 297–348
  $(CH)_n$, **1**, 266
  $CO_2$, **1**, 150
  $Cu_4Cd_3$, **2**, 249–256
  CuCl, **1**, 67
  $Cu_2O$, **1**, 273, 306, 316
  diatomic molecules, $M_2$, **1**, 169
  disiloxane, *see* $(SiH_3)_2O$
  GaAs, **1**, 30, 32, 100
  Ge, **1**, 30, 32
  $H_2O$, *see* water species
  $H_6SiAlO_7$, **1**, 210, 214, 216
  $H_6Si_2O_7$, **1**, 184–192, 202–210, 213, 251
  KCl, **1**, 138
  $KMoO_3$, **1**, 151
  $KTaO_3$, **1**, 151
  LaOF, **1**, 302
  $Li_2SiO_3$, **1**, 244
  metals, **1**, 155–174
  MgO, **2**, 99
  $MgSiO_3$, **2**, 47, 81–84, 104–106
  $Mg_2SiO_4$, **2**, 47, 86, 104–106
  Mo, **1**, 10
  $Na_2SiO_3$, **1**, 244
  nitrides, **1**, 304; **2**, 38
  octet compounds, **1**, 16, 64
  orthosilicic acid, *see* $Si(OH)_4$

PbO, **1**, 271
phosphates, **2**, 197–232
PN, **1**, 284
pnictides, **2**, 297–348
polycompounds, **2**, 297–348
pyrosilicic acid, *see* $H_6Si_2O_7$
rare-earth alloys, **2**, 259–296
$Rh_7Mg_{44}$, **2**, 248, 292
$SF_4$, **1**, 260
Si, **1**, 10, 32, 40–46, 291
SiC, **1**, 66
$(SiH_3)_2O$, **1**, 184–192, 213, 216, 237
silicates, **2**, 197–232
silicic acid, *see* $Si(OH)_4$
$SiO_2$, **1**, 151, 229–237, 291; **2**, 100–104
$Si(OH)_4$, **1**, 200–202, 216
$SiP_2O_7$, **1**, 241
$(SN)_x$, **1**, 274
$SrTiO_3$, **1**, 151, 153
sulfides, **2**, 53–70, 297–348
tetrelides, **2**, 297–348
$V_4O_7$, **2**, 5
W, **1**, 101
water species, **2**, 11–16
ZnSe, **1**, 30, 32
Compressibility, *see* Bulk modulus
  bond, **2**, 111–112
Covalency, **1**, 15, 142, 177, *see also* Ionicity
Crystal structure, *see* Structure types

## D

Deformation density, **1**, 282
  for PN, **1**, 284
Density matrix, **1**, 179
Dielectric constant, **1**, 16, 30
Dielectric theory, **1**, 16, 30, 75
Disorder, **2**, 87, 89
Distortion theorem, **2**, 6–7

## E

Earth, temperature and pressure in, **2**, 97
Elastic constants
  diamond, **1**, 247
  ionic crystals, **1**, 306
  instability in rutile-structure difluorides
      and, **2**, 101–103
  NaCl–CsCl transition and, **2**, 78
  silicon, **1**, 41, 45

tetrahedral compounds, **1,** 153
tetrahedral structures, **1,** 64
Electric field gradient, **1,** 296
Electronegativity, **1,** 57, 107
   table of, **1,** 58
Engel–Brewer theory of metals, **1,** 76, 170–173
Eutaxy, **1,** 307

## F

Force constant
   bond valence theory and, **2,** 6
   C—C, **1,** 246
   diamond, **1,** 247
   distance least squares and, **2,** 32–33
   nonbonded interactions, **1,** 246
   O—Si—O bend, **1,** 201
   phase transitions and, 89–90, 78
   related to ionicity, **1,** 17
   silicates, **1,** 250
   Si—O stretch, **1,** 201, 207
   Si—O—Si bend, **1,** 201, 206

## G

Glasses, *see* Melts and glasses
Graph theoretic enumeration of structures, **2,** 133–163

## H

Heat of atomization of elements (table), **1,** 157, 158
Hybridization, **1,** 146, 152, 180–184, 187–191
Hydrated compounds, **2,** 11–14

## I

Independent atom model, **1,** 282
Ionic crystals, **1,** 138
Ionicity, **1,** 15, 17, 31, 34, 177, 186

## J

Jahn–Teller effect, **2,** 40

## L

Lattice energy, **1,** 301; **2,** 77
Lattice vibrations, silicon, **1,** 45

Lewis acid, **2,** 3, 8–10
Ligand field effects, **2,** 67, 78, 82
Lone pair, **2,** 16–17, 40

## M

Madelung constant, table **1,** 311
Madelung potential, **1,** 316–319
   table of constants, **1,** 317
Melts and glasses, **2,** 211–222, 223, 224, 226, 230
Miedema model of intermetallic compounds, **1,** 21, 59, 129; **2,** 251
Modules (structural slabs, building blocks), **2,** 88, 109–118, 160–171, 235, 268–270
Molecular orbitals, **1,** 175–192, 195–222, 252–277
   CsCl structure, **1,** 269
   $Cu_2O$, **1,** 273
   diamond structure, **1,** 267
   highest occupied molecular orbital (HOMO), **1,** 259
   localized molecular orbitals, **1,** 180
   lowest unoccupied molecular orbital (LUMO), **1,** 259
   PbO, **1,** 271
   perturbation theory and, **1,** 256
   $SF_4$, **1,** 261
   solids, **1,** 264–276
   tetrahedral molecule, **1,** 260
Mooser–Pearson rule, **2,** 298, 329

## N

Nowotny phases, **2,** 338–339

## O

Overlap populations, **1,** 179, 187, 199, 203

## P

Pauling's rules, **1,** 211, 303; **2,** 41
Phase transitions
   in $ABO_3$ compounds, **2,** 82–85
   band structure and, **2,** 90
   elastic constants and, **2,** 78, 101–103
   enthalpy of, **2,** 76, 79, 81, 86
   entropy of, **2,** 78, 85–86, 88–91
   four-coordination to six-coordination, **2,** 72–74, 90

free energy of, **2**, 73, 74, 75, 76, 80, 81, 97
ilmenite to corundum, **2**, 90
ilmenite to perovskite, **2**, 89, 104
olivine to spinel or modified spinel, **2**, 85–87, 105
pressure-temperature slopes, **2**, 83, 85, 88–91
pyroxene to pyroxenoid, **2**, 79–81
rocksalt to cesium chloride, **2**, 76–79, 89, 99
rocksalt to nickel arsenide, **2**, 74–76
in rutile difluorides, **2**, 100–104
rutile to fluorite, **2**, 90
semiconductor to metal, **2**, 91
vibrational force constants and, **2**, 78, 89–90
volume of, **2**, 72, 77, 86, 87, 96, 99, 100, 104
zincite to rocksalt, **2**, 90
Polarizability, **1**, 311–316
Polya's theorem, **2**, 139–142
Polycompounds
definition, **2**, 297–298
derivation from metal structures, **2**, 340–345
Mooser–Pearson rule in, **2**, 289, 329
semiconductor to metal transitions in, **2**, 345–347
table listing, **2**, 302–324
Polytypes, **2**, 153–157; **2**, 168–194
planar defects in, **2**, 190–191
stacking formula, **2**, 173, 180–182
stacking variants in rare-earth alloys, **2**, 277–281
Population analysis, **1**, 178, 185
Promotion energy, **1**, 159–167, 183
Pseudoatoms, **1**, 282, *see also* Pseudopotentials
in diatomic molecules, **1**, 283
Pseudopotentials, **1**, 8, 15, 24–48, 50, 73–153
density functional pseudopotentials, **1**, 29, 35, 37, 83–102
empirical pseudopotential method, **1**, 28
hard core, **1**, 35, 52, 79, 89
soft core, **1**, 36, 78

### R

Radii
Bragg–Slater, **1**, 228, 251
ionic, **1**, 139, 229, 303; **2**, 36, 67–70
halides, **2**, 54–55
nitrides, **2**, 38
oxides, **2**, 66
rare-earth alloys, **2**, 270–272
sulfides, **2**, 53–69
tables of, **2**, 61–62, 66
nonbonded, **1**, 208, 237–244, 251
table of, **1**, 242
orbital, **1**, 37, 49–71, 76, 80, 251; **2**, 73
table of, **1**, 62, 95
tetrahedral, **1**, 61, 111
van der Waals, **1**, 245, 252

### S

Semiconductor to metal transitions, **2**, 90–91, 345–347
Solid electrolyte, **2**, 89
Stacking variant, *see* Polytypes
Structure, computer simulation of crystal, **2**, 31
examples, **2**, 33–34
Structure maps, **1**, 37, 69, 117–131
Structure prediction
AB compounds, **1**, 117–131
acid-base compounds, **2**, 15–16
alloy superstructures, **2**, 117–132
brookite, **1**, 9
cristobalites, **1**, 229–237
CsCl-derived structures, **1**, 269
dielectric theory, **1**, 18, 30
Engel–Brewer model, **1**, 171–174
high-pressure structures
AB compounds, **2**, 75–78, 99
$ABO_3$ compounds, **2**, 79–84
$SiO_2$, **2**, 100–104
ionic model and, **1**, 304
interatomic distances and, **2**, 35–39
metals, **1**, 171–174
octet compounds, **1**, 38, 69, 117–131
phosphates, **2**, 218–221
silicates, **1**, 199–202, 239–243; **2**, 218–231
silicon, **1**, 41, 45
silicon surface, **1**, 41–45
wurtzites, **1**, 68, 243
Structure types
arsenic, **2**, 151–152, 159
borides (FeB and CrB), **1**, 21; **2**, 283
bracelet and pinwheel, **2**, 145–147
cesium chloride, **2**, 76–79, 99, 269
cristobalites, **1**, 229–236

cuprous oxide, **1,** 273, 306, 316
diamond, **1,** 267, 289
fluorite, **2,** 101, 103, 150, 338
garnet, **2,** 81
graphite, **1,** 287
graph-theoretic enumeration, **2,** 133–163
ilmenite, **2,** 82
lead chloride, **2,** 103
lead oxide, **2,** 271
marcasite, **2,** 150–151, 330
mica, **2,** 169–171, 175–177, 179, 182–187
nickel arsenide, **2,** 74–76, 103, 188, 332
olivine, **2,** 85–88
oxyfluorides, **1,** 302
perovskites, **1,** 153; **2,** 82–84, 89
phosphates, **2,** 197–232
phosphorus (black), **2,** 152
polytypes, **2,** 153–157, 168–194
pyrite, **2,** 150–151, 159, 327
pyroxene, **2,** 79–81, 84, 204–211, 224
pyroxenoid, **2,** 79–81, 204–211, 224
quartz, **1,** 291; **2,** 90, 188
rocksalt, **2,** 74–79, 99, 150–153, 159, 187
rutile, **2,** 90, 101, 103, 148, 158, 326
silica polymorphs, **2,** 90, 188
silicates, **1,** 240; **2,** 37, 197–232
silicon, **1,** 290
sphalerite, **2,** 72–74, 159–160, 188
spinel, **2,** 85–88, 60, 65
spinelloid, **2,** 85–88, 114–115
superstructures, **2,** 117–129
tetrahedral structures, **1,** 243
wallpaper, **2,** 161
wurtzite, **2,** 72–74, 159–160, 188
    axial ratio $c/a$ of, **1,** 68, 243

Surface
    GaAs, **1,** 101
    silicon, **1,** 41
Superstructures
    cluster variation method, **2,** 118, 122–125
    coherent phase diagram, **2,** 117, 118
    configuration polyhedron, **2,** 120–122
    correlation function, **2,** 119–120
    face centered cubic lattice, ground states,
        **2,** 119–131
    many body interactions, **2,** 127–129
    pair interactions, **2,** 125–127

**T**

Term values, table of free-atom, **1,** 139
Tetrahedral anions, **1,** 152
Thermal expansion, of bonds, **2,** 110–111
Twinning, unit cell, **2,** 277–283

**V**

Valence, *see* Bond valence, Pauling's rules
Valence charge density, **1,** 286, *see also*
    Charge density
    for $Al_2O_3$, **1,** 294
    for Be, **1,** 287
    for diamond, **1,** 289
    for graphite, **1,** 287
    for quartz, **1,** 291
    for silicon, **1,** 290

**W**

Wurtzite structure, axial ratio, $c/a$, **1,** 68,
    243